Davidson
2002

Monastic Spaces
and their Meanings

MEDIEVAL CHURCH STUDIES

1

Monastic Spaces and their Meanings

Thirteenth-Century English Cistercian Monasteries

by

Megan Cassidy-Welch

BREPOLS

A catalogue-in-publication record for this book is available from
the British Library.

© **2001 Brepols Publishers n.v., Turnhout, Belgium**

All rights reserved. No part of this publication may be reproduced,
stored in a retrieval system or transmitted, in any form or by any means,
electronic, mechanical, photocopying, recording, or otherwise,
without prior permission of the publisher.

D/2001/0095/73
ISBN 2-503-51089-2

Printed in the EU on acid-free paper.

Contents

List of Illustrations vii

Acknowledgements xiii

Abbreviations xv

Introduction 1

1. Boundaries and Memories: The Space of Transition 23

2. Metaphorical and Material Space: The Cistercian Cloister 47

3. The Cistercian Church 73

4. Community, Discipline, and the Body: The Cistercian Chapter House 105

5. Blood, Body, and Cosmos: The Infirmary 133

6. Status, Space, and Representation: The Cistercian Lay Brother 167

7. Apostasy and Contravention of Monastic Space 195

8. Sites of Death and Spaces of Memory 217

Epilogue 243

Bibliography 255

Index 283

List of Illustrations

Introduction
Fig. i: Map of northern England showing the eight Cistercian houses 13

Chapter One
Fig. 1.1: The precinct of Fountains abbey 24
Courtesy, Dr Glyn Coppack, Ancient Monuments Inspector, English Heritage

Fig. 1.2: The precinct of Byland abbey 26
Courtesy, English Heritage

Fig. 1.3: The precinct of Kirkstall abbey 27
Courtesy, West Yorkshire Archaeology Service

Fig. 1.4: Byland abbey, two views of the remains of the outer gatehouse 29
Photos, author

Chapter Two
Fig. 2.1: Plan of St Gall, showing the cloister south of the church 50
Courtesy, English Heritage

Fig. 2.2: Ground floor plan of Fountains abbey, showing a standard Cistercian claustral design 51
Courtesy, English Heritage

Fig. 2.3: Reconstruction of part of the cloister arcade at Rievaulx abbey 52
Photo, author

Fig. 2.4: The main cloister at Rievaulx abbey and the infirmary cloister at Rievaulx abbey 53
Photos, author

Fig. 2.5: Plan of Rievaulx abbey, showing the cloister and the ranges 54
Courtesy, English Heritage

Fig. 2.6: Plan of Byland abbey, showing the lay brothers' lane west of the cloister 55
Courtesy, English Heritage

Fig. 2.7: Jervaulx abbey, entrance to the chapter house; Kirkstall abbey, entrance to the chapter house; Fountains abbey, entrance to the chapter house 56
Photos, author

Fig. 2.8: Rievaulx abbey, south range, entrance to the refectory 57
Photo, author

Fig. 2.9: Plan of Rievaulx abbey, showing the lavers (basins) used for the *mandatum* against the refectory wall 62
Courtesy, English Heritage

Chapter Three

Fig. 3.1: Villard de Honnecourt, plan of a paradigmatic Cistercian church 75
From E. Stiegman, 'Analogues of the Cistercian Abbey Church', in A. MacLeish ed., The Medieval Monastery, Medieval Studies in Minnesota 2 (St Cloud, MN: North Star Press of St Cloud, 1988), pp. 17-33 at p. 22.

Fig. 3.2: Byland abbey, plan of church 77
Courtesy, English Heritage

Byland abbey, view to east end of church
Photo, author

Fig. 3.3: Byland abbey, altar, detail of tiled steps; Byland abbey, cloister tiles, detail from west cloister arcade 78
Photos, author

Fig. 3.4: Byland abbey, west façade of church, remains of rose window 79
Photo, author

Fig. 3.5: Jervaulx abbey, plan of church 82
From S. Davies, Jervaulx Abbey, *London [n.d.]*

Byland abbey, east end of church
Courtesy, English Heritage

List of Illustrations ix

Fig. 3.6: Rievaulx abbey, church, thirteenth-century chapels in the west end of the nave; Rievaulx abbey, church, detail of chapel and altar in the west end of the nave 83
   Photos, author

Fig. 3.7: Kirkstall abbey, twelfth-century presbytery; Rievaulx abbey, thirteenth-century east end of the church 84
   Photo, author

Fig. 3.8: Fountains abbey, plan of the church prior to the building of the Chapel of Nine Altars 85
   Courtesy, Dr Glyn Coppack, Ancient Monuments Inspector, English Heritage

Fig. 3.9: Fountains abbey, plan of church showing the Chapel of Nine Altars 86
   Courtesy, Dr Glyn Coppack, Ancient Monuments Inspector, English Heritage

Fig. 3.10: Fountains abbey, Chapel of Nine Altars, exterior; Fountains abbey, detail of interior of the Chapel of Nine Altars, exterior 87
   Photos, author

Chapter Four

Fig. 4.1: Rievaulx abbey chapter house, showing hemicyclic apse 109
   Photo, author

Fig: 4.2: Rievaulx abbey, tiered seating in the chapter house 110
   Photo, author

Fig. 4.3: Rievaulx abbey, shrine to abbot William in the chapter house 110
   Photo, author

Fig. 4.4: Fountains abbey, abbatial tombs in the chapter house 114
   Plan from R. Gilyard-Beer, 'The Graves of the Abbots of Fountains', *Yorkshire Archaeological Journal, 59 (1987), 45-50* at 50.
   Courtesy, The Yorkshire Archaeological Society

Fig. 4.5 Byland abbey, tomb in the chapter house 115
   Photo, author

Chapter Five

Fig. 5.1: plan of Kirkstall abbey showing the infirmary to the south-east of the cloister 135
   Courtesy, West Yorkshire Archaeology Service

Fig. 5.2: Rievaulx abbey, the infirmary cloister; Rievaulx abbey, detail of the infirmary cloister arcade 136
   Photos, author

Fig. 5.3: Byland abbey, plan showing the site of the infirmary 138
Courtesy, English Heritage

Fig. 5.4: Roche abbey, plan of the precinct showing the choir monks and the lay brothers' infirmaries 139
Courtesy, English Heritage

Fig. 5.5: Fountains abbey, the infirmary 140
Photo, author

Fig. 5.6: 'Bloodletting man', London BL MS Harley 3719, fols 158v-159r 152
Courtesy, British Library, London

Fig. 5.7: 'Disease man', London BL MS Arundel 251, fol. 37 153
Courtesy, British Library, London

Fig. 5.8: 'Wound man', London Wellcome MS 290, fol. 53v 154
Courtesy, The Wellcome Library, London

Fig. 5.9: Man positioned in relation to the cosmos, Wien, Österreichische Nationalbibliothek, MS 12,600, fol. 29r 155
Courtesy, Bildarchiv der ÖNB, Wien

Fig. 5.10: The cosmos, London BL MS Sloane 795, fol. 20 156
Courtesy, British Library, London

Chapter Six

Fig. 6.1: Fountains abbey, two photographs showing the west range, interior 172
Photos, author

Fig. 6.2: Byland abbey, the west range, interior 174
Photo, author

Fig. 6.3: Byland abbey, the lay brothers' lane 174
Photo, author

Chapter Eight

Fig. 8.1: Rievaulx abbey, tombs in the Galilee porch 234
Photos, author

Epilogue

Fig. 9.1: Western facade of Fountains abbey 244
Photo, author

Fig. 9.2: The gardens of Studley Royal 246
Photos, author

List of Illustrations xi

Fig. 9.3: West range stairs at Jervaulx abbey and entrance to west range 247
> *Photos, author*

Fig. 9.4: The cloister at Jervaulx abbey; stacks of masonry at Jervaulx abbey 248
> *Photos, author*

Fig. 9.5: Kirkstall abbey, east end of church and tower 250
> *Photo, author*

Fig. 9.6: Kirkstall abbey, barred doorway 251
> *Photo, author*

Acknowledgements

This book started life as a PhD thesis in the History Department at the University of Melbourne in 1994. The writing of the original thesis was aided immeasurably by the vibrant intellectual culture of the History Department, under the headships of Professor Patricia Grimshaw and Professor Peter McPhee. I am also grateful for the initial financial assistance of an Australian Postgraduate Research Award through the University of Melbourne together with contributions of the Faculty of Arts and the School of Graduate Studies at the University of Melbourne.

I would also like to thank a number of other institutions and archives for their generosity in opening research facilities to me during the last six years. At the University of Melbourne, I thank the staff of the Baillieu Library, the Architecture and Planning Library; the Ormond College Theological Library; and the Brownless Medical Library. I would also like to thank the British Library, London and particularly the staff of the Western Manuscripts Room; the Public Record Office, Kew; the Wellcome Institute for the History of Medicine; Dr. David McKitterick of the Wren Library, Trinity College, Cambridge; Dr. Frances Willmoth of Jesus College Library, Cambridge; the Masters and Fellows of Corpus Christi College; the Cambridge University Library; the Duke Humfries Library and the Bodleian Library in Oxford; Mr. Keith Parker at the Institute for Advanced Architectural Studies, York University; and the Bibliothèque Nationale, Paris, particularly the staff of the Salle des Manuscrits; and the Library of Congress, Washington D.C.

Various individuals have also provided invaluable encouragement and advice throughout this project. There are too many to list exhaustively, but I would especially like to thank Professor David Bell, Father Michael Casey and the Cistercian community at Tarrawarra abbey, Dr. James Girsch, Professor E. Rozanne Elder, the participants of the 1997 and 1999 Cistercian Studies Conferences, Western Michigan University, Kalamazoo, my former colleagues in the School of History and Classics at the University of Tasmania, especially Professor Michael Bennett and Emeritus Professor Rod Thomson. Portions of chapter one are reproduced with the kind permission of *Cistercian Studies Quarterly*. Particular thanks are due to Professor Patrick Geary and Mr. Peter Draper, both of whom encouraged the publication of this book and generously offered me the benefit of their advice and wisdom. Thanks also are due to Associate Professor Charles Zika and Associate Professor Anne Gilmour-Bryson, who acted as the supervisors of the thesis from which this book arises. At Brepols, I thank Simon Forde for his encouragement and advice, Elizabeth Wall for her diligence and eagle eye, and the anonymous reader for a number of helpful suggestions.

To my family—my parents and my two sisters—I dedicate this book. We discovered the beauty of northern England together many years ago, and throughout this project, our memories of that period have been relived. I thank my father for taking me to Fountains and all those years ago, and for advising me on the architectural aspects of this book; I thank my mother for her care and her compassion; I thank Charlotte for her humour and her help with various aspects of this project; and I thank Brigid for her tireless enthusiasm for this thesis and for her invaluable help with the illustrations. I owe them all my deepest gratitude.

Abbreviations

ASOC	*Analecta Sacri Ordinis Cisterciensis*
CChR	*Calendar of Charter Rolls*
CCR	*Calendar of Close Rolls*
CFS	Cistercian Fathers Series
COCR	*Collectanea Cisterciensis Ordinis Reformatorum*
CPR	*Calendar of Patent Rolls*
CSS	*Cistercian Studies Series*
Mem. F.	*Memorials of the Abbey of St. Mary of Fountains*, ed. by J. R. Walbran, Publications of the Surtees Society vols 42, 67 and 130 (London: published for the Society by Andrews & Co., 1863-1918)
PL	*Patrologiae Cursus Completus, Series Latina*, ed. by J.-P. Migne, 221 vols (Paris, 1844-64)
Statuta	*Statuta Capitulorum Generalium Ordinis Cisterciensis ab anno 1116 ad annum 1786*, ed. by J. M. Canivez, 8 vols (Louvain: Bureaux de la Revue d'Histoire Ecclesiastique, 1933-41)

Introduction

Prologue

My first encounter with the monasteries of northern England was as a child, when my architect father took the family on tours of interesting sites and buildings around north Yorkshire where we lived for a time. I remember the enchantment of the formal gardens around Fountains abbey, and I remember the strangeness of the neat and orderly ruins at Rievaulx, rising from the banks of the river Rye. Like Emily Brontë's Mr Lockwood, I lingered 'under that benign sky; watched the moths fluttering among the heath and the harebells; listened to the soft wind breathing through the grass; and wondered how anyone could ever imagine unquiet slumbers for the sleepers in that quiet earth'.[1] This book began from my memories of these spaces, and became about the many layers of meaning embedded deep in the lie of the land.[2]

Physical space and imagined or abstract space may be distinctive and separate, although the distinction is not inflexible. Space may refer to the physical dimensions of sites such as buildings or streets or fields, all of which are defined, fixed and demarcated geographically or topographically. Space can also be understood as referring to more elusive or intangible geographies. Heaven, purgatory, and hell may all be described as spaces, yet they are also spaces which intangible or invisible in a strict material sense. In this book, I particularly concentrate on the creation and meaning of such spaces in the Cistercian monasteries of Yorkshire during the thirteenth century. In these monasteries the relationship between the regulation of material space and the

---

[1] Emily Brontë, *Wuthering Heights*, ed. by D. Daiches (Harmondsworth: Penguin Books, repr. 1985), p. 367.

[2] For a similar view of the layers of meaning in the landscape, see Simon Schama, *Landscape and Memory* (London: Harper Collins, 1995).

liberation associated with abstract spaces may be uncovered. An interdisciplinary approach to the study of these sites reveals that the relationship between freedom and enclosure was defined and negotiated by these medieval Cistercians in spatial terms. Space was at once practised and perceived in many different ways.

I begin this introduction by looking at some of the ways in which historians and other scholars have approached a critical study of space, concentrating particularly on those who have sought to look at the production of space and the meanings given to material and imagined space by its inhabitants, users, and creators. Some explanation for my focus on thirteenth-century Cistercian monasteries also needs to be given in the context of this historiography and Cistercian historiography in general. A brief résumé of the major sources used in this book will be provided at the end of this preamble.

Space

The distinction between material and imagined spaces has been the concern of a number of scholars who have sought to define and interpret the ways in which such spaces are produced. Henri Lefebvre's work on the social production of space was one of the first attempts to distinguish between material space and abstract space, and provides a useful reminder that understandings of space and its representations are always historically contingent.[3] Lefebvre begins his elaboration of the meanings of abstract/imagined spaces by defining such spaces as 'absolute'. Absolute space, according to Lefebvre, derives from the moment when a particular place is defined as being transcendent or sacred, 'magical and cosmic'.[4] An example of absolute space may be, for instance, a Christian cemetery, which occupies a fixed topographical site which can be measured and described. Yet a cemetery also describes relationships between the living and the dead and systems of belief and ritual, while the subterranean focus of a cemetery reminds the living that underneath the topography that they presently occupy lurks the space of death. A church building provides a similar example, where the fixed location of the site signifies only part of the transcendent function of that space.

Lefebvre devotes a small part of his discussion to the Middle Ages, and again, he initially concentrates on religious understandings of space, arguing:

> What a narrow, indeed mistaken view it is which pictures the Church as an entity having its main 'seat' in Rome and maintaining its presence by means of clerics in individual 'churches' or villages and towns, in convents, monasteries, basilicas and so forth. The fact is that the 'world'—that imaginary/real space of shadows—was inhabited, haunted by the Church.[5]

---

[3] H. Lefebvre, *The Production of Space*, trans. by D. Nicholson-Smith (Oxford: Basil Blackwell, 1991), esp. pp. 229–91.

[4] Lefebvre, *Production of Space*, p. 234.

[5] Lefebvre, *Production of Space*, pp. 254–55.

Thus, Lefebvre sees the medieval 'world' as the space of Christendom, which was characterized by links between the 'magico-mystical', the imaginary and the real. In this context, space is institutionally, spiritually, and psychologically meaningful, while the anchoring of such meanings in buildings or places alone, tells only part of the story. Medieval space, in Lefebvre's paradigm, was absolute space; it was of the earth and it transcended the earth.

There are, however, several difficulties with Lefebvre's subsequent views on the medieval period and changing perceptions of space during that time.[6] Lefebvre identifies the emergence of a 'new' form of space during the twelfth century—secular space—which was identifiable in the growth of urban centres.[7] He then establishes a dichotomous relationship between the absolute space of Christendom and the commercially driven spaces of towns, arguing that 'the Papacy sought to defend itself against these developments', which essentially separated spaces of ideas from social spaces.[8] This argument is problematic on several grounds. Lefebvre seems to accept unquestioningly the existence of a particular and identifiable medieval understanding of pre-secular space, without any real contextual or even empirical foundation, while the secular/spiritual dialectic he claims for the distinction between urban and Christian space is overstated.[9] Bernard of Clairvaux, Abelard and Abbot Suger are the only examples given by Lefebvre of resistance to new ideas of space, and even their so-called opposition to changing world views is not explained, nor are the reasons for the papacy's alleged 'resistance'. Just who was receptive to these various ideas of space is also ignored by Lefebvre, who is extremely attentive elsewhere to the class hierarchies present in the production of urban and residential spaces.[10]

Lefebvre's historical narrative of the Middle Ages is driven by an overall Marxist interpretation of the medieval period as an age of accumulation, in which the marketplace, money, and the exchange of commodities became central to the transformation of urban space. These developments, in Lefebvre's view, were considered 'subversive', as they foreshadowed and even ushered in secularization, while urban space itself was 'a tool of terrifying power' which commandeered and enveloped the countryside. The argument behind *The Production of Space* is thus that the history of space is the history of the shift from 'absolute' space to space as a political tool and the instrument of capitalism.

Whether or not the reader is convinced by Lefebvre's Marxist analysis of concepts and changing meanings of the production of space, *The Production of*

---

[6] Lefebvre's chronologies could also be questioned—he accepts without reservation such descriptions as 'medieval', 'antiquity', and 'Renaissance'.

[7] Lefebvre, *Production of Space*, pp. 256–57.

[8] Lefebvre, *Production of Space*, p. 260.

[9] Later on, Lefebvre qualifies the starkness of this dialectic, acknowledging that 'alongside religious space, and even within it, there was room for other spaces—for the space of exchange, for the space of power', Lefebvre, *Production of Space*, p. 266.

[10] Lefebvre, *Production of Space*, pp. 315–17.

Space remains an enduring and significant heuristic tool with which to consider space and its meanings critically. Lefebvre's work is important in that it attempts to understand space in terms of history. Spaces do not signify the same concerns year after year, or century after century, and Lefebvre recognizes these contingencies. Despite the many lacunae in his analysis of medieval and Christian understandings of space, the conceptual premise on which his work is based remains sound: that space is a meaningful and insightful concept through which to consider the histories of societies and cultures.

More recently, Michel de Certeau has also approached ideas of space in an historical way.[11] Like Lefebvre, de Certeau believes that a distinction ought to be made between fixed topographies and abstract or complex spaces. De Certeau argues that places and spaces are essentially different things. A place implies 'an indication of stability', a location, or what de Certeau describes as an 'instantaneous configuration of positions'—what may also be termed a site.[12] A space, on the other hand, 'exists when one takes into consideration vectors of direction, velocities and time variables'. Thus space is composed of 'intersections of mobile elements' and is produced or created, in de Certeau's view, by movement from one place to another.

The Practice of Everyday Life is predominantly concerned with urban culture, and the first example de Certeau gives of the creation of space by movement from one place to another is the example of pedestrians walking the streets. Walking, according to de Certeau, has a number of functions in the production of space. First, the act of walking the streets is a process of 'appropriation of the topographical system on the part of the pedestrian'.[13] Second, walking is a 'spatial acting out of place', and third, the act of walking 'implies relations among differentiated positions'.[14] The walker himself may be constrained by certain topographical features (such as walls), but at the same time invents spaces as he moves about places (by creating detours, for instance, which redefine the fixed street pattern). The production of such space depends on the body of the walker moving from one site to another, occupying the space he has created fleetingly and then linking it to another 'liberated space that can be occupied'.[15]

De Certeau argues that distinctions between space and place are important in revealing how boundaries are fabricated, how frontiers and bridges are constructed, and what it means when transgression of place occurs. These concerns and de Certeau's theories of place/space are also extended in *The Practice of Everyday Life* to investigate narratives, or 'spatial stories', which use topographical metaphors to provide (for instance) closure in introducing an impassable frontier, or boundaries in 'setting the scene'. On a more general

---

[11] Michel de Certeau, *The Practice of Everyday Life*, trans. by S. Rendall (Berkeley and Los Angeles: University of California Press, 1984).

[12] De Certeau, *Practice of Everyday Life*, p. 117.

[13] De Certeau, *Practice of Everyday Life*, p. 97.

[14] De Certeau, *Practice of Everyday Life*, p. 98.

[15] De Certeau, *Practice of Everyday Life*, p. 105.

level, de Certeau's understanding that space is a practised place is significant in acknowledging uses and functions of topographies, whether urban or rural. Places are made meaningful by the people who move through them and transform them into spaces. The critical element in the production of space, in de Certeau's view, is the body of the individual. Whereas Henri Lefebvre argues for the historical existence of absolute space, de Certeau argues for the continual creation of space in the practices and movements of everyday life.

Underpinning de Certeau's arguments for the production of space is the idea that spaces are filled with meanings which people impute to them. This has been the more specific concern of other historians, who have identified particular sets of meanings for physical sites, and the definition of physical space in terms of its significations and abstract references. Three important examples may be found in the work of Pierre Nora, Pierre Bourdieu, and Michel Foucault.[16] In his groundbreaking work on sites of memory, Nora claims that historians and the academic discipline of history are responsible for the 'conquest and eradication' of 'milieux de mémoire', or environments of memory.[17] Real, collective memory, according to Nora:

> [I]s life, borne by living societies founded in its name. It remains in permanent evolution, open to the dialectic of remembering and forgetting, unconscious of its successive deformations, vulnerable to manipulation and appropriation, susceptible to being long dormant and periodically revived [...] Memory is a perpetually actual phenomenon, a bond tying us to the eternal present [...] Memory [...] only accommodates those facts which suit it; it nourishes recollections that may be out of focus or telescopic [.][18]

The organizing agenda of history has destroyed this spontaneity, in Nora's view, and replaced the fluidity of social remembrance with fixed sites (*lieux*) of memory. These sites of memory may be '[m]useums, archives, cemeteries, festivals, anniversaries, treaties, depositions, monuments, sanctuaries, fraternal orders'.[19] Without exception, these sites enshrine particular memories and ways of remembering; they are deliberate and artificial creations, made of paper or stone, and designated by historians or those attempting to represent history as the spatial and proper embodiment of collective remembrance.

In this critique of history and historians, space itself has three characteristics. Sites of memory are material, symbolic, and functional. Material space may be the site of a monument or archive; symbolic actions give meaning to such

---

[16] For the purposes of this introduction, I will concentrate on a particular example of spatial concerns from each of these scholars: Pierre Nora, 'Between Memory and History: les lieux de mémoire', *Representations*, Special Issue 26 (1989), 7–25; Pierre Bourdieu, 'The Kabyle House or the World Reversed', in P. Bourdieu, *Algeria 1960*, trans. by R. Nice (Cambridge and Paris: Cambridge University Press, 1979), pp. 133–53; Michel Foucault, *Discipline and Punish: The Birth of the Prison*, trans. by A. Sheridan (Harmondsworth: Penguin Books, 1977).

[17] Nora, 'Les Lieux de mémoire', p. 8.

[18] Nora, 'Les Lieux de mémoire', p. 8.

[19] Nora, 'Les Lieux de mémoire', p. 12.

physical locations—the minute of commemorative silence on Remembrance Day, for example, while functional space uses a specific site for the performance of a ritual, such as a veterans' reunion.[20] Generally, in Nora's view, sites of memory provide fixed frames of reference for the practice of historical commemoration, although they operate as 'purely, exclusively self-referential signs'.[21] Nora's thesis has been criticized for its utopian view of pre-modern collective memory,[22] while the subversive qualities he attributes to historians may also be challenged.[23] In terms of the symbolic and referential qualities of particular sites and space, Nora's thesis nevertheless provides a useful reminder that the construction of material sites and places may also be replete with designated symbolic meanings, while the use of those places may consolidate and strengthen these.

The ordering of space and the meanings given to that order have also been the concern of anthropologists and sociologists such as Pierre Bourdieu, whose work on the anthropology of the Algerian Kabyle house looks closely at the nature of physical space within a specific domestic environment. Bourdieu focuses on the structure of the internal space of the house, which is demarcated initially in terms of light and darkness. The part of the house filled with light is the upper section of the two-part house, and the part inhabited by people. The darker part of the house is the stable, while a dividing wall partially conceals one area from the other.[24] Bourdieu analyses the demarcation and organization of the light and dark areas of the house in terms of the symbolic ordering of the world—the 'natural' world is expressed in the use of the stable area for not only the keeping of animals, but also for the women's and children's sleeping quarters, a space, according to Bourdieu, where childbirth, sleep, and sex take place. By contrast, the lighter and larger domestic sphere is associated with protection (in the presence of a loom),[25] with honour (in being the space in which guests are received), and with masculinity (as this is where tools and weapons are housed). The dividing wall is especially significant, according to Bourdieu, as it supports the entire house structurally, while symbolically providing the line of demarcation for various activities and practices.

---

[20] Nora, 'Les Lieux de mémoire', p. 19.

[21] Nora, 'Les Lieux de mémoire', p. 23.

[22] Peter Burke, 'History as Social Memory', in *Memory, History, Culture and the Mind*, ed. by Thomas Butler (Oxford and New York: Basil Blackwell, 1989), p. 97; and Peter Burke, 'French Historians and their Cultural Identities' in *History and Ethnicity*, in E. Tonkin, M. McDonald, and M. Chapman (London and New York: Routledge, 1989). For a different view of the domination of the written over memory, see Jacques Le Goff, *History and Memory*, trans. by S. Rendall and E. Claman (New York: Columbia University Press, 1992).

[23] See for instance, the comments on the importance of the social and cultural role of the historian by Patrick Geary, *Phantoms of Remembrance: Memory and Oblivion at the End of the First Millennium* (Princeton: Princeton University Press, 1994), pp. 11–12.

[24] Bourdieu, 'The Kabyle House', p. 134 for a plan of the house.

[25] This, as Bourdieu explains, has a particular protective symbolism in Algerian culture.

The specific rituals that take place within and without the Kabyle house are also explored by Bourdieu, who sees the physical site of the house as mirroring wider cosmological and cultural principles. Space is important not only in the internal organization of the house, but also in its exterior positioning, the direction it faces and the situation of its entrances and exits.[26] Bourdieu's anthropological methodology allows him to take into account both the significance of the construction of material space, but also the wider *practice* of space, which underpins social mores, ideas of masculinity, and the opposition of the natural and man-made worlds. Bourdieu's theoretical framework is successful in allowing a very detailed analysis of a particular space and the meanings contained in its regulation, a 'thick description' of the regulation and symbolism imbued in this domestic space.[27] Rituals and the performance of space are thus fundamental issues to be raised in any consideration of an inhabited, material site.

Finally, the concept of space as an institutional tool of discipline, correction, power, and knowledge has been another significant and influential theoretical framework to emerge over the last twenty-five years. This is due initially to the work of Michel Foucault, whose history of the modern prison, in particular, considers the regulation of space to be of primary importance in the playing out of disciplinary practices of surveillance, correction, and rehabilitation.[28] Foucault emphasizes not so much the production of space by the people who inhabit it, but rather the ways in which space can be manipulated and fabricated in very specific ways by institutions. The ways that space is used by institutions such as the prison, the army, the school, and the clinic, according to Foucault, are strategies designed to underpin the technology of power which Foucault describes as 'discipline'. This sort of 'power' is productive, in Foucault's view—it produces docile subjects, 'domains of objects and rituals of truth'. In short, power 'produces reality'.[29]

The primary example given in *Discipline and Punish* of the manipulation of space in order to produce a conformist subject is Bentham's panoptic model prison.[30] In this building, two principles of the dungeon—to deprive the prisoner of light and to hide him from view—are reversed. The Panopticon is fully lit while the shadowy figure of each inmate is clearly seen by the occupant of the central tower. Only the principle of enclosure survives from the old dungeon model. The way that the building is constructed, with the central tower surveying the annular building surrounding it, makes the operation of Foucault's

---

[26] Bourdieu, 'The Kabyle House', p. 152.

[27] For the idea of 'thick description', see C. Geertz, 'Thick Description: Toward an Interpretive Theory of Culture', in his *The Interpretation of Cultures: Selected Essays* (New York: Basic Books, 1973), pp. 3–30.

[28] Foucault, *Discipline and Punish*, passim. See also Michel Foucault, 'Espace, savoir et pouvoir', in Michel Foucault, *Dits et écrits 1954–1988*, 4 vols (Paris: Editions Gallimard, 1994), 4, pp. 270–85.

[29] Foucault, *Discipline and Punish*, p. 194.

[30] Foucault, *Discipline and Punish*, pp. 195–228.

discipline entirely visible. The inmate cannot see the warder in the tower, so is never able to know when he is under surveillance. The success of this trick of light and space, in Foucault's view, is that the inmate will ultimately survey himself and modify his own behaviour. Foucault's description of the Panopticon serves as a way of illustrating his own conviction that the building itself is 'a mechanism of power reduced to its ideal form [...] a figure of political technology'.[31] In other words, the Panopticon expresses and replicates the way in which society operates.

Foucault's own political agenda should not eclipse the very important general point that he makes in relation to uses of space. That is, there may be many discursive regimes at work in the production of physical space, and many implications for the meanings thus attributed to those sites. The Panopticon should not only be understood in terms of the microcosmic operation of discipline, as Foucault argues; rather, it should also be considered as an example of how the articulation of space can successfully maintain and express institutional interests. A comparable instance is the monastery, which uses principles of enclosure to express institutional requirements of segregation from the non-monastic world.[32] We may choose to read this in political terms, but it is also productive to interpret formation and definition of physical space within the *specific* institutional (and historical) contexts at issue.

I have given this brief résumé of theoretical approaches to the history and philosophy of space in order to demonstrate that there are many ways in which to proceed in exploring spatial concepts. The major themes I have stressed in this historiographical sample have been the production of space, and the ways in which meanings may be attributed to certain spaces by virtue of abstract reference, performative practice, and institutional direction. These are themes to which I return throughout this book. It is also important to establish at this juncture that although I recognize the distinction between material and abstract space made by theorists such as de Certeau and Lefebvre, I do not see this distinction as fixed or unnegotiable. One important purpose of my ensuing exploration of medieval monastic space is to explore how this distinction is negotiated through historical time and circumstance.

Theories of space such as those outlined above have rarely been applied to the study of medieval monastic sites, although there are some notable exceptions. The disciplines of archaeology and architecture have traditionally undertaken work on the structural fabric of physical monastic sites, and it is within these disciplines that the few examples of a wider understanding of monastic space may be glimpsed. Archaeological approaches, for instance, tend to focus on the identification of material remains; explanation of these remains in their particular context, however, often fails to recognize wider changes in monastic *mentalités*.[33] Work by archaeologists such as Kathleen Biddick and Roberta

---

[31] Foucault, *Discipline and Punish*, p. 205.

[32] This is one of many strategies, which will be discussed more fully in the body of this book.

[33] For a critique of this trend in archaeology with respect to the monasteries and convents of northern England, see Harold Mytum, 'Functionalist and Non-Functionalist

Introduction 9

Gilchrist has more recently sought to elicit contextual and wide-ranging meanings from particular sites. Roberta Gilchrist's *Gender and Material Culture* is one example of the ways in which archaeological readings of monastic sites may be extended to uncover the gendered concerns of their establishment and operation.[34] In particular, Gilchrist uses the physical evidence of north-facing cloisters in medieval English women's houses as a starting point for iconographic and spatial explanations of how women's religious houses ought to be studied on their own terms, rather than within the framework of histories of men's houses.[35] In Gilchrist's analysis, space is not only articulated by the material remains uncovered by archaeological investigation, but is also articulated by the meanings we may find within those remains.[36]

Likewise, Kathleen Biddick has explored the possibilities of more nuanced readings of space in her work on rural space and the culture of the medieval 'peasant'.[37] Biddick argues that the distinction between space and place described by Michel de Certeau in *The Practice of Everyday Life* is a useful way of challenging the conventional spatial frameworks within which archaeologists work. Biddick claims that in order to deconstruct descriptive social categories such as 'the medieval peasant', it is important to move beyond the boundaries that created these categories. By removing the archaeologist's traditional dependence on location or place (represented in the confining practice of excavating toft and croft or grid), and by excavating 'in-between' zones instead, archaeology may be liberated from replicating the material culture of particular places, and consider the broader meanings that may arise from considerations of space. This approach to rural space is vastly different to the geographical and economic focus of scholars such as R. A. Donkin, whose prolific work on the Cistercians of northern England has dominated scholarship on rural landscape

---

Approaches in Monastic Archaeology', in *The Archaeology of Rural Monasteries*, ed. by R. Gilchrist and H. Mytum (Oxford: British Archaeological Reports British Series 203, 1989), pp. 339–57.

[34] Roberta Gilchrist, *Gender and Material Culture. The Archaeology of Religious Women* (London and New York: Routledge, 1994).

[35] Gilchrist, *Gender and Material Culture*, p. 24: 'The archaeology of nunneries has remained unwritten because monasteries for women have been judged against standards which are male'.

[36] See also Roberta Gilchrist, 'Community and Self: Perception and Use of Space in Medieval Monasteries', *Scottish Archaeological Review*, 6 (1988), 55–64.

[37] Kathleen Biddick, 'Decolonizing the English Past: Readings in Medieval Archaeology', *Journal of British Studies*, 32 (1993), 1–23; see also *Archaeological Approaches to Medieval Europe,* Studies in Medieval Culture, vol. 18, ed. by Kathleen Biddick (Kalamazoo: Medieval Institute Publications, 1984); Kathleen Biddick, 'People and Things: Power in Early English Development', *Comparative Studies in Society and History*, 32:1 (1990), 3–23; Kathleen Biddick, 'Malthus in a Straitjacket? Analyzing Agrarian Change in Medieval England', *Journal of Interdisciplinary History*, 20:4 (1990), 623–35.

and the Cistercian order.[38] Biddick's claims are a potentially useful addition to the empirical focus of Donkin's rural world.

Architectural historians have been quicker to address ideas and meanings of space in their work on structural remains of particular buildings within monastic sites. Some examples may be found in the vast historiography on the medieval Gothic church, which forms part of the third chapter of this book. Iconographic approaches to the Gothic church have attempted to uncover symbolic meanings not only for the interior space of the church, but also for church furnishing, decoration, sculpture, and so forth.[39] Scholars such as Erwin Panofsky have concentrated on even less tangible elements of the church, such as light, while others have concentrated on the importance of liturgy in architecture.[40] More recently, there have been attempts to deconstruct altogether terms like 'Gothic', which are said to reflect more the interests of the academy which produced them than any objective and integrated style.[41] Architectural historians have also attempted more synthesized readings of other particular monastic buildings. Peter Fergusson, for instance, has linked the building of the refectory in the twelfth-century Cistercian houses at Byland and Rievaulx to a desire to replicate biblical spaces within the physical fabric of the monastery.[42] Another example is Fergusson and Harrison's work on the Rievaulx abbey chapter house which

---

[38] See *inter alia* R. A. Donkin, *The Cistercians: Studies in the Geography of England and Wales* (Toronto: Pontifical Institute of Mediaeval Studies, 1978); R. A. Donkin, 'The Site Changes of Medieval Cistercian Monasteries', *Geography*, 44 (1959), 251–58; R. A. Donkin, 'Settlement and Depopulation on Cistercian Estates during the Twelfth Century, Especially in Yorkshire', *Bulletin of the Institute of Historical Research*, 33 (1960), 141–65; R. A. Donkin, 'The Cistercian Grange in England in the Twelfth and Thirteenth Centuries, with Special Reference to Yorkshire', *Studia Monastica*, 6 (1964), 95–144; R. A. Donkin, 'The Cistercian Order and the Settlement of Northern England', *Geographical Review*, 54 (1969), 403–16.

[39] For an introductory article, see Richard Krautheimer, 'Introduction to an Iconography of Medieval Architecture', *Journal of the Warburg and Courtauld Institutes*, 5 (1942), 1–33. For more recent developments in this field, see Michael Camille, 'Mouths and Meanings: Towards an Anti-Iconography of Medieval Art', in *Iconography at the Crossroads: Papers from the Colloquium Sponsored by the Index of Christian Art*, ed. by B. Cassidy (Princeton: Index of Christian Art, Dept. of Art and Archaeology, Princeton University, 1993).

[40] Peter Draper, 'Architecture and Liturgy', in *Age of Chivalry: Art in Plantagenet England 1200–1400*, ed. by J. Alexander and P. Binski (London: Royal Academy of Art in association with Weidenfeld and Nicholson, 1987), pp. 83–91.

[41] See for example, Madeline H. Caviness, 'Artistic Integration in Gothic Buildings: A Post-Modern Construct', in *Artistic Integration in Gothic Buildings*, ed. by V. C. Raguin, K. Bush, and P. Draper (Toronto and Buffalo: University of Toronto Press, 1995), pp. 249–61; Brigitte Bedos-Rezak, 'Towards a Cultural Biography of the Gothic Cathedral; Reflections on History and Art History', in Raguin et al., *Artistic Integration*, pp. 262–74.

[42] Peter Fergusson, 'The Twelfth-Century Refectories at Rievaulx and Byland Abbeys', in *Cistercian Art and Architecture in the British Isles*, ed. by C. Norton and D. Park (Cambridge: Cambridge University Press, 1986), pp. 160–80.

uncovers strong memorial and historical connections between the changing design of this building and the abbots who instigated those changes.[43]

Seven years ago the general subject of monastic space formed the focus for a Cistercian conference, the proceedings of which were published as *L'Espace cistercien*.[44] This was the first time the idea of space and its various manifestations had been addressed by Cistercian scholars *en masse*, although isolated examples of interest in monastic space had existed previously.[45] Reflected in the proceedings of this conference was a wide range of interests, from Cistercian measurement of space to the ways in which heritage organizations and individuals have attempted to conserve the space of Cistercian monasteries.[46] The areas of study illustrated in *L'Espace cistercien* remained quite carefully circumscribed, however; few papers attempted a critical analysis of space itself, despite the focus of the conference. What these conference papers did reveal were the many disciplines associated with the field of Cistercian studies and the very specific interests of each.

More recently, the work of Terryl Kinder reveals the way in which architectural historians have been quicker to adopt an interdisciplinary approach to the study of medieval Cistercian life. In her *L'Europe cistercienne*, published in 1997, Kinder has chosen to explore the Cistercian world through its buildings primarily, but also through its liturgy and its local influences and contexts.[47] Kinder's book is wide-ranging in focus, dealing with Cistercian abbeys throughout Europe from the late eleventh century onwards. Yet despite the broad nature of such a focus, Kinder's work is important and instructive in demonstrating just how fundamental an interdisciplinary approach to the Cistercian landscape is in any historical evaluation of the order's nature and change. A similar interdisciplinary approach has been taken by Jens Rüffer in relation to twelfth-century Cistercian monasticism. Rüffer, whilst concentrating on issues of culture and aesthetics, has also explored Cistercian architecture both in the context of style and precedent, and in the context of context and meaning.[48] I turn now to a brief

---

[43] Peter Fergusson and Stuart Harrison, 'The Rievaulx Abbey Chapter House', *The Antiquaries Journal*, 74 (1994), 211–55. Most recently see Peter Fergusson and Stuart Harrison, *Rievaulx Abbey: Community, Architecture, Memory* (London and New Haven: Yale University Press, 1999).

[44] *L'Espace cistercien*, ed. by Leon Pressouyre (Paris: Comité des Travaux Historiques et Scientifiques, 1994).

[45] M.-A. Dimier, 'Le mot *locus* employé dans le sense de *monastère*', *Revue Mabillon*, 58 (1972), 133–54; P. Noisette, 'Usage et représentation de l'espace dans la *Regula Benedicti*', *Regulae Benedicti Studia*, 14/15 (1985/6), 69–80; Wayne Teasdale, 'A Glimpse of Paradise: Monastic Space and Inner Transformation', *Parabola*, 18 (1993), 59–62.

[46] David N. Bell, 'The Measurement of Cistercian Space', in Pressouyre, ed., *L'Espace cistercien*, pp. 253–61; Lord Montagu of Beaulieu, 'Beaulieu, a Former Monastic Estate Today', in Pressouyre, ed., *L'Espace cistercien*, pp. 483–89.

[47] Terryl Kinder, *L'Europe cistercienne* (La Pierre-qui-Vire: Editions Zodiaques, 1997).

[48] Jens Rüffer, *Orbis cisterciensis: Zur Geschichte der monastischen ästhetischen

discussion of current historiography within Cistercian studies, and some more detailed explanation of the Cistercian focus of this book.

Thirteenth-Century Cistercians

The Cistercian order, established in England with the foundation of Waverley abbey in 1128, expanded very quickly.[49] In Yorkshire, the geographical area with which this book is concerned, eight Cistercian houses existed by the end of the twelfth century. These were the abbeys of Fountains, Rievaulx, Sawley, Kirkstall, Jervaulx, Roche, Byland and Meaux. I have chosen to focus on these eight houses in this study for a number of reasons. First, these abbeys were in reasonably close proximity to each other, both topographically (Fig. i) and in terms of spiritual and administrative contact. Personnel was exchanged from house to house, while the many property disputes between these houses found in the statutes of the Cistercian General Chapter indicate that economic interests were often shared as well. Second, these abbeys represent something of a cross-section of the Cistercian endeavour, as within this group we find both extremely wealthy and large monasteries, such as Fountains or Rievaulx, and we find much smaller houses that were often financially precarious, such as Jervaulx or Roche. Close relationships between some of these particular abbeys should not be understood to indicate homogeneity among them; even within this small group, there are significant differences in the practice and expression of Cistercian spirituality. This may also be seen in differences in the architectural concerns of the houses, as will be elaborated below.

I have not included women's religious houses in this sample, a decision which requires some explanation. Women's religious houses were certainly founded in Yorkshire during the twelfth and thirteenth centuries, according to the principles of Cistercian monasticism and despite the order's initial reluctance to recognize them.[50] However, I believe that a useful study of the space of Cistercian nunneries would both require and constitute a different project to this present study. With Roberta Gilchrist, I do not believe that the study of women's houses should be a mere addition to primary studies of male monasteries. Rather, difficulties associated with the very fragmentary nature of source material in particular, calls for specific historical enquiry into these houses. I do not wish to include women's houses only as footnotes or appendices to the epistemologies created by male religious of the thirteenth century. Hence, I

Kultur im 12. Jahrhundert (Berlin, 1999).

[49] For the foundation of the Cistercian order in England, Janet Burton, 'The Foundation of the British Cistercian Houses', in Norton and Park, eds., *Cistercian Art and Architecture in the British Isles*, pp. 24–39; Janet Burton, *Monastic and Religious Orders in Britain 1000–1300* (Cambridge: Cambridge University Press, 1994).

[50] See Janet Burton, *The Yorkshire Nunneries in the Twelfth and Thirteenth Centuries*, (York: Borthwick Institute of Historical Research, 1979); S. Thompson, 'The Problem of Cistercian Nuns in the Twelfth and Early Thirteenth Centuries' in *Medieval Women*, ed. by D. Baker, (Oxford: Basil Blackwell, 1978), pp. 227–52.

Introduction

Fig. i: Map of northern England showing the eight Cistercian houses.

acknowledge the validity of a spatial reading of women's religious domains, and reserve that task for another project.

The choice of the thirteenth century as the focus of this book also deserves some explanation. Generally, post-twelfth-century Cistercian history has been under-studied by Cistercian scholars, who still privilege the initial euphoria of twelfth-century reform and the figure of Bernard of Clairvaux.[51] More specifically, scholars have traditionally written about Cistercian history in terms of 'ideals' and 'reality', two categories which have provided a popular, if not standard, narrative paradigm.[52] The supposed dichotomy of Cistercian ideals and

---

[51] For the most recent example, see Martha Newman, *The Boundaries of Charity: Cistercian Culture and Ecclesiastical Reform 1098–1180* (Stanford: Stanford University Press, 1996). The historiography on St Bernard is vast. For one introduction to the many studies devoted to Bernard, see Jean Leclercq, *Receuil d'études sur Saint Bernard et ses écrits*, 3 vols (Rome, 1962–69); for a more recent collection, Guido Hendrix, *Conspectus bibliographicus Sancti Bernardi ultimi patrum, 1989–1993* (Leuven: Peeters, 1995).

[52] Louis Lekai, *The Cistercians: Ideals and Reality* (Kent, OH: Kent State University, 1977); Louis Lekai, 'Ideals and Reality in Early Cistercian Life and Legislation', in *Cistercian Ideals and Reality*, CS 60, ed. by J. Sommerfeldt (Kalamazoo: Cistercian Publications, 1978); R. Roehl, 'Plan and Reality in a Medieval Monastic Economy', *Studies in Medieval and Renaissance History*, 9 (1972), 83–113; Constance Bouchard, 'Cistercian Ideals versus Reality: 1134 reconsidered', *Cîteaux*, 39 (1988), 217–31. The paradigm follows the style of narrative typified by Johan Huizinga, *The Waning of the Middle Ages: A Study of the Forms of Life, Thought and Art in France and the Nether-*

reality is founded on the assumption that after the first century of Cistercian expansion, the order compromised the ideals of the founders and consequently suffered from an irretrievable decline in spiritual standards. This decline is seen to have manifested itself in a number of areas: the introduction of separate abbot's lodgings in some Cistercian houses, the relaxing of early rules against the eating of meat, and the decorative nature of later Cistercian churches, are some examples. Thus early Cistercian documents, buildings, legislation, and liturgical practice are understood to represent the 'ideal', while post-twelfth-century changes to that 'ideal' are understood to signal the regrettable 'reality' of decline.

This paradigm is unsatisfactory for several reasons. Primarily, the fundamental premise on which the argument is based suffers from a serious flaw. That is, in order to set up a dichotomous relationship between an early 'ideal' and a later 'reality', it is assumed that there was, in fact, an early ideal or set of ideals that can be isolated and defined. Recent scholarship on Cistercian legislation has shown that this is not the case.[53] Work in Cistercian architecture has uncovered the diversity of early buildings—not the absolute uniformity that is often presupposed.[54] And on the most fundamental of spiritual issues, even Bernard of Clairvaux has been shown to be notoriously paradoxical.[55] The early founders of Cîteaux were not nearly as uniform and organized as some historians have concluded. Of course, part of the blame for the assumption that there was a Cistercian 'ideal' comes from the propaganda machine of the Cistercian order itself, and particularly from twelfth-century writers such as William of St Thierry, whose *Vita Prima* is a superb example of the power of representation. Yet modern historians have been too quick to replicate this convenient paradigm uncritically. One exception to this is Terryl Kinder, who laments that historians have been slow to recognize that Bernard of Clairvaux was quite aware that a 'static' monasticism could never survive, and that adaptations and changes would always have to be made. Kinder rightly points out that the equation of change and decline has been enduring both pedagogically and ideologically.[56]

The idea of decline as an historical concept may also be challenged on a more general level.[57] 'Decline' presupposes the existence of a prior period of

---

lands in the Fourteenth and Fifteenth Centuries, trans. by F. Hopman (London, E. Arnold & Co., 1924); and for the genesis of the style, Edward Gibbon, *The History of the Decline and Fall of the Roman Empire,* ed. by J. Bury, 7 vols (London: Methuen, 1896–1900).

[53] See for instance the overview of Chrysogonous Waddell, 'The Cistercian Institutions and their Early Evolution: Granges, Economy, Lay Brothers', in Pressouyre, ed., *L'Espace cistercien,* pp. 27–38.

[54] For a critique of the myth of uniformity, see Jean-Baptiste Auberger, *L'Unanimité cistercienne primitive: mythe ou réalité?* (Achel: Cîteaux. Studia et Documenta 3, 1986).

[55] I am grateful for the comments of Fr. Michael Casey at the 32nd International Congress on Medieval Studies at Kalamazoo, 1997 in his paper on 'The Meaning of Poverty for Bernard of Clairvaux'.

[56] Kinder, *L'Europe cistercienne,* p. 14.

[57] For one critique of ideas of decline in the later Middle Ages, see Peter Burke,

growth or progress that has broken down or altered in some way. As a way of explaining historical change, ideas of progress and decline are inadequate for several reasons. Initially, decline is a concept that gives historical change a negative value, while decline's antonym, progress, claims a positive value. One aim of postmodern scholarship has been to expose such claims for concepts of progress, in particular, as deriving from nineteenth-century historical traditions which charted a white, male, bourgeois path through history at the expense of other groups, cultures, and experiences.[58] Concepts of progress and decline are never uncontested categories, once the epistemological production of these categories is exposed. Ideas of historical decline also seriously undervalue the contingencies associated with historical change, sweeping together infinite experiences into one narrative. I argue that historical change may be more usefully explored in terms of its contingencies. This is another reason why I have chosen to concentrate on the study of the eight Yorkshire Cistercian houses: all of these houses were established in the twelfth-century wave of Cistercian expansion which swept across England, and all survived until the Dissolution. The thirteenth century is a time that falls between these two cataclysmic events, and is therefore a time which, in conventional narratives, heralds the start of 'decline', 'laxity', or 'decadence'. I challenge this common framework.

The field of Cistercian studies has been generally one of diversity. From Jean Leclercq's 'invention' of a monastic theology at the Dijon conference in 1953, Cistercian scholars have both sought to locate Cistercianism under Leclercq's monastic theology umbrella, and to identify the Cistercian experience as very particular and singular.[59] The establishment of the Institute of Cistercian Studies at Western Michigan University has added to the Cistercian claim for autonomy

---

'European Ideas of Decline and Revival c. 1350–1500', *Parergon*, 23 (1979), 3–8. See also Heiko A. Oberman, *The Harvest of Medieval Theology: Gabriel Biel and Late Medieval Nominalism* (Cambridge, MA: Harvard University Press, 1963).

[58] See the critique of such histories by Stephen G. Nichols, 'The New Medievalism: Tradition and Discontinuity in Medieval Culture', in *The New Medievalism*, ed. by M. S. Brownlee, K. Brownlee, and S. G. Nichols (Baltimore: Johns Hopkins University Press, 1991); Lee Patterson, 'On the Margin: Postmodernism, Ironic History and Medieval Studies', *Speculum*, 65 (1990), 87–108.

[59] For the best known expression of Leclercq's monastic theology, see Jean Leclercq, *The Love of Learning and the Desire for God: A Study of Monastic Culture*, trans. by C. Misrahi (New York: Fordham University Press, 1974; 2nd rev. edn). The autonomy of Cistercian studies has meant that the impact of the non-monastic world on the order is often ignored. Despite some standard works on the role of patrons and benefactors within the Cistercian milieu, there has been little work on other non-economic relationships between Cistercians and seculars. For some examples of scholarship on the economically driven exchanges between monks and seculars, see Bennett D. Hill, *English Cistercian Monasteries and their Patrons in the Twelfth Century* (Urbana: University of Illinois Press, 1968); Susan Wood, *English Monasteries and their Patrons in the Thirteenth Century* (Oxford: Oxford University Press, 1955); Joan Wardrop, *Fountains Abbey and its Benefactors 1132–1300* (Kalamazoo: Cistercian Publications, 1987); Constance Bouchard, *Holy Entrepreneurs: Cistercians, Knights and Economic Exchange in Twelfth-Century Burgundy* (Ithaca and New York: Cornell University Press, 1991).

as a *discrete* area of scholarly and spiritual study. Within the discipline of Cistercian studies itself areas of scholarship continue to be delineated and separated. Works attempting a synthesis of academically separate areas are uncommon. This is particularly the case with Cistercian history, which remains often separated into areas such as twelfth-century theological history, economic history, architectural history, and histories of particular houses.[60] The exception to these types of categories are broadly based survey histories, such as that of Roger Stalley for the Cistercian houses of Ireland.[61] Models such as these are useful in a number of ways yet remain very general 'survey' accounts.

The separation of Cistercian history into such fragmentary units does not mean that the field of Cistercian studies lacks any cohesion. Recent work by David Bell,[62] collections such as that edited by Christopher Norton and David Park,[63] and the plethora of edited collections of diverse essays in the Cistercian Studies series, for example, all attempt to draw from and include the achievements of a number of different disciplines. The annual Cistercian Studies conference at Western Michigan University is evidence of collective interest in a number of different fields. Scholars such as Martha Newman and Peter Fergusson have recently broken down some of the barriers between disciplines, incorporating theology and social history in the former case, and incorporating architecture, liturgy, and spirituality in the latter.[64] This book is intended to contribute to the synthesis of expressions of Cistercian devotion and spirituality, by considering particular material cultures together with the institutional, theological, and social conditions which produced and defined them.

Sources

The sources I have used exhibit the same characteristics as the majority of sources for the study of the medieval period. They are fragmentary, disparate and eclectic and tell many different stories. The way that I have chosen to

---

[60] This is especially evident in the case of the Yorkshire Cistercian houses. Fountains abbey, for instance, has enjoyed much attention on the basis of its architecture. See the preliminary bibliography included by Glyn Coppack, *English Heritage Book of Fountains Abbey* (London: English Heritage/Batsford, 1993) p. 121. For the social history of the abbey, see Wardrop, *Fountains Abbey and its Benefactors*. Yet there is still no history of this site which takes into account the many other elements of Cistercian life.

[61] Roger Stalley, *The Cistercian Monasteries of Ireland: an Account of the History, Art, and Architecture of the White Monks in Ireland from 1142–1540* (London and New Haven: Yale University Press, 1987).

[62] David N. Bell, 'The English Cistercians and the Practice of Medicine', *Cîteaux*, 40 (1989), 139–74. Bell links the building of new infirmaries to broader issues of medicine.

[63] Norton and Park, eds, *Cistercian Art and Architecture in the British Isles*.

[64] Newman, *Boundaries of Charity*; Fergusson, 'Twelfth-Century Refectories'. For recent studies on monasticism in Yorkshire, see *Yorkshire Monasticism: Archaeology, Art and Architecture from the Seventh to the Sixteenth Centuries*, ed. by Lawrence R. Hoey (London and Leeds: British Archaeological Association and W. S. Maney, 1995).

assemble these fragments forms the plan of this particular book: it does not assume to tell a complete story, or to represent the only narrative that could be written of histories of space using these sources. I shall give a brief overview here of the sorts of sources that predominate in my reading of monastic spaces. More detail on each, of course, will be given in the body of the text.

First and foremost, the remaining sites of Fountains, Rievaulx, Sawley, Kirkstall, Jervaulx, Roche, and Byland are the material sources I have considered.[65] These sites are in various states of reconstruction and disrepair, and it is extremely important to remember that the buildings and stones that we see at these abbeys today are most certainly not, for the main part, those known by thirteenth-century Cistercians. Historians, archaeologists, and architects have long lamented the difficulties in accurately reconstructing (on plan or by rebuilding) not only the character of the original sites, but even all the changes that were made to the buildings throughout the medieval period. In 1904, the archaeologist H. Astley wrote disgustedly to the *Yorkshire Archaeological Journal* that post-medieval alterations to the buildings made it almost impossible to recover the original plan of the Yorkshire Cistercian houses—'"Capability" Brown was let loose upon Roche, and it is the havoc which his hands wrought that makes it so difficult to discover the plans of the buildings'.[66] The same sort of difficulty may be found at other sites. At Byland abbey, for instance, an excavation led by Martin Stapylton in the late 1800s resulted in the loss of the twelfth-century altar slab to the nearby monks of Ampleforth (as a gift), together with the loss of stones from the monastery to decorate Stapylton's own house.[67] At Kirkstall abbey, the entire precinct is carved in two by the A65 motorway, while the buildings themselves are closed off to the public for fear of vandalism.

Nevertheless, recent research—both archaeological and historical—has allowed some clearer glimpses of the original fabric of these buildings, together with some more conclusive dating of alterations made to them. At Fountains abbey, for instance, the work of Glyn Coppack has uncovered the site of the community's first wooden church,[68] while evidence of the fire that decimated the

---

[65] Meaux abbey was completely destroyed during the Dissolution, and traces of the medieval site may only be seen from earthworks visible from the air. See David Knowles and J. K. S. St Joseph, *Monastic Sites from the Air* (Cambridge: Cambridge University Press, 1952). For some reconstruction of the buildings at Meaux, see Peter Fergusson, *Architecture of Solitude: Cistercian Abbeys in Twelfth-Century England* (Princeton: Princeton University Press, 1984), pp. 133–36.

[66] H. Astley, 'Roche Abbey', *Yorkshire Archaeological Journal*, 10 (1904), 199–220 (p. 218).

[67] For a description of the archaeological assaults on Byland over the last century, see Stuart Harrison, 'The Architecture of Byland Abbey' (unpublished MA dissertation, University of York, 1988).

[68] Coppack, *Fountains Abbey*. This contrasts with the previous finding of W. St John Hope, 'Fountains Abbey', *Yorkshire Archaeological Journal*, 15 (1899), 269–402; Roy Gilyard-Beer, 'Fountains Abbey: the Early Buildings 1152–60', *Archaeological Journal*, 125 (1968), 313–19.

church in 1146 has been found in the east and west cloister ranges and in the south transept of the church itself.[69] At Byland abbey, similar advances have been made, with the work of Stuart Harrison, who has been able to establish the chronology of the claustral building programme from formerly covered archaeological remains.[70] Although no medieval plan survives from any of these sites, such recent work has allowed the chronology of the architecture to come to light, and to place the building developments in the context of changes to other Cistercian sites, both in England and on the Continent. The thirteenth-century alterations to the earlier buildings form the major architectural focus of this book, while the background and chronology of such changes will be outlined in the relevant chapters.

Written texts from these Cistercian houses form the next group of sources to be considered. These are extremely various, ranging from devotional texts to histories, from poetry and prose to medical texts. Although Cistercian scholars have created a vast historiography of twelfth-century English Cistercians such as Aelred of Rievaulx, later writers from the abbeys of Yorkshire have yet to attract such attention. One such Cistercian is Stephen of Sawley, who has been credited with producing a *Speculum Novitii* and other devotional tracts. These texts are an important focus of chapters one through three.[71] Another Cistercian from early thirteenth-century Yorkshire, Matthew of Rievaulx is also a figure yet to make a more public appearance on the stage of monastic scholarship.[72]

---

[69] Coppack, *Fountains Abbey*, p. 32.

[70] Harrison, 'Architecture of Byland Abbey'.

[71] E. Mikkers, 'Un Speculum Novitii inédit d'Etienne de Sallai', *COCR*, 8 (1946), 17–68; an English translation is *Stephen of Sawley Treatises*, ed. by Bede K. Lackner and trans. by Jeremiah O'Sullivan, CFS 36 (Kalamazoo: Cistercian Publications, 1984), pp. 85–122. Surprisingly little work has been done on Stephen of Sawley, despite the fact that his is one of extremely few Cistercian voices to emerge from the thirteenth century in Yorkshire. Some general accounts of Stephen's work are J. McNulty, 'Stephen of Eston, abbot of Sawley, Newminster and Fountains', *Yorkshire Archaeological Journal*, 31 (1934), 49–64; H. Farmer, 'Stephen of Sawley', *The Month*, 29 (1963), 332–42; J. Holman, 'Stephen of Sawley: Man of Prayer', *Cistercian Studies*, 21:2 (1986), 109–22; Caroline Walker Bynum, *Docere Verbo et Exemplo: An Aspect of Twelfth Century Spirituality*, Harvard Theological Studies 31 (Missoula, MT: Scholars Press, 1979), pp. 157–63. For other editions of Stephen's known work, see E. Mikkers, 'Un traité inédit d'Etienne de Salley sur la psalmodie', *Cîteaux*, 23 (1972), 245–88; A. Wilmart, 'Les Méditations d'Etienne de Sallai sur les Joies de la Vierge Marie', *Revue d'Ascétique et de Mystique*, 10 (1929), 268–415; A. Wilmart, 'Le Triple Exercice d'Etienne de Sallai', *Revue d'Ascétique et de Mystique*, 11 (1930), 355–74.

[72] Paris, Bibliothèque Nationale, MS Lat. 15157. Matthew's poetry has been partially transcribed by A. Wilmart, 'Les Mélanges de Mathieu, Préchantre de Rievaulx au début du XIIIe siècle', *Revue Bénédictine*, 52 (1940), 15–84; additional transcription may be found in Jean Leclercq, 'Lettres de vocation à la vie monastique', *Studia Anselmiana*, fasc. 37, Analecta Monastica 3rd series (Rome: Orbis Catholicus, 1955) and in James Girsch, 'Matthew of Rievaulx on Select Vices: Poetry and Incidental Prose from Paris BN MS Lat. 15157' (unpublished essay, Toronto, 1983). For other mentions of Matthew, see Richard W. Southern, *The Making of the Middle Ages* (London: Cresset Library,

Better known is Hugh of Kirkstall, whose *Narratio de Fundatione Fontanis Monasterii de Comitatu Eboracensi* is an invaluable source for Fountains abbey.[73] Other Cistercian writers remain nameless, although there are several extant manuscripts containing anonymous works from the Yorkshire houses during the thirteenth-century.[74] I shall address the specificity and detail of such sources throughout.

Legislative or institutional sources are also important. The Cistercian order was bound to follow the *Regula S. Benedicti,* together with other forms of legislative and customary direction. First and foremost of these were the statutes of the Cistercian General Chapter, which met annually at Cîteaux, and which all Cistercian abbots, unless excused, were required to attend. The *Statuta* produced as a result of the meetings of the General Chapter have been edited by J. M. Canivez, and provide one important source for broader institutional direction within the order itself.[75] Legislation pertinent to the practice of daily life was

---

1987, rev. edn), p. 107; Brian Patrick McGuire, *Friendship and Community; The Monastic Experience 350–1250* (Kalamazoo: Cistercian Publications, 1988), esp. pp. 369–73; A. G. Rigg, *A History of Anglo-Norman Literature 1066–1422* (Cambridge and New York: Cambridge University Press, 1992), esp. pp. 136–38; G. Raciti, 'Une allocution familière de S. Aelred conservée dans les mélanges de Mathieu de Rievaulx', *Collectanea Cisterciensia*, 47 (1985), 267–80.

[73] Hugh's *Narratio* is transcribed by J. R. Walbran, ed., *Memorials of the Abbey of St Mary of Fountains* (London: Publications of the Surtees Society 42, 1842). An English translation may be found in A. Oxford, *The Ruins of Fountains* (Oxford: Oxford University Press, 1910). Other work on Hugh of Kirkstall's history includes D. Bethell, 'The Foundation of Fountains Abbey and the State of St Mary's York 1132', *Journal of Ecclesiastical History*, 17 (1966), 11–207; Derek Baker, 'The Genesis of Cistercian Chronicles in England: the Foundation of Fountains Abbey', *Analecta Cisterciensia*, 25 (1969), 14–41 and Derek Baker, 'The Genesis of Cistercian Chronicles in England: the Foundation of Fountains Abbey', *Analecta Cisterciensia*, 31 (1975), 179–212.

[74] For some lists of these, see C. Talbot, 'A List of Cistercian Manuscripts in Great Britain', *Traditio*, 8 (1952), 402–16; J. Morson, 'Cistercian MSS from the Collection of Sir Sydney Cockerell', *Collectanea Cisterciensia*, 21 (1959), 330–33; Neil R. Ker, *Medieval Libraries of Great Britain: A List of Surviving Books* (London: Offices of the Royal Historical Society, 1964); Neil R. Ker, *Medieval Manuscripts in British Libraries*, 3 vols (Oxford: Clarendon Press, 1969–83); David N. Bell, *An Index of Authors and Works in Cistercian Libraries in Great Britain*, CSS 130 (Kalamazoo: Cistercian Publications, 1992); David N. Bell, *The Libraries of the Cistercians, Gilbertines and Premonstratensians*, Corpus of British Medieval Library Catalogues 3 (London: British Library in association with the British Academy, 1992). These are some starting points for finding Cistercian manuscripts. Also useful are catalogues of individual libraries and collections, which I have included in the bibliography.

[75] J. M. Canivez, *Statuta Capitulorum Generalium Ordinis Cisterciensis ab anno 1116 ad annum 1786*, 8 vols (Louvain: Bibliothèque de la Revue d'Histoire Ecclésiastique, 1933–41). Canivez's edition is not without its problems and criticisms, particularly in relation to the dating of the earliest legislative documents. For a more recent edition of the so-called 'foundation' documents, see Jean de la Croix Bouton and J. B. van Damme, *Les plus anciens textes de Cîteaux: Commentaria Cisterciensis, Studia et Documenta*, II (Achel, 1974); for the need for revisions to Canivez's edition, see Chrysogonous

also codified early on in the order's history. The *Ecclesiastica Officia* is one of the fullest examples. This book of usages has been edited five times.[76] Legislative material provides one of the sources for exploring the regulation of space, and the ordering of liturgical and other practices which occur within Cistercian houses. As I have mentioned previously in relation to the work of Pierre Bourdieu, close reading of the ways in which space is *performed* may illuminate some of the symbolic meanings that particular sites may possess.

Finally, I have considered other material associated with the legal and economic status of Yorkshire Cistercians during the thirteenth century. This material includes charters and deeds relevant to the relationships between the monasteries and their patrons and benefactors;[77] legal records from the English assizes;[78] and other fragments gleaned from the Close Rolls, the Fine Rolls and similar non-monastic jurisdictions. By looking at the ways in which Cistercians encountered the secular sphere, the spaces occupied by the monks of Yorkshire can be considered more broadly.

Waddell, 'Towards a New Provisional Edition of the Statutes of the Cistercian General Chapter, c. 1119–1189', in *Studiosorum Speculum: Studies in Honour of Louis J. Lekai O. Cist.*, ed. by F. R. Swietek and J. Sommerfeldt, CSS 141 (Kalamazoo: Cistercian Publications, 1993), pp. 384–419.

[76] See Ph. Guignard, *Les Monuments Primitif de la Règle cistercienne* (Dijon: Imprimerie Darantière, 1878); C. Noschitzka, 'Codex Manuscriptus 31 Bibliothecae Universitatis Labacensis', *ASOC* 6 (1950), 1–124; Bruno Griesser, 'Die "Ecclesiastica Officia Cisterciensis Ordinis" des Cod. 1711 von Trient', *ASOC*, 12 (1956), 10–288; *Les Ecclesiastica Officia Cisterciens du XIIe siècle*, ed. and trans. by D. Choisselet and P. Vernet (Reiningue: La Documentation Cistercienne 22, 1989); Chrysogonous Waddell, *Legislative and Narrative Texts from Early Cîteaux: An Edition, Translation and Commentary* (Cîteaux: Commentarii Cistercienses, 1999). For some commentary on the *Ecclesiastica Officia*, see B. Schneider, 'Cîteaux und die Benedictinische Tradition: Die Quellenfrage des *Liber Usuum* im Lichte der *Consuetudines Monasticae*', *ASOC*, 16 (1960), 169–254 and B. Schneider, *ASOC*, 17 (1961), 73–111; Bede K. Lackner, 'Early Cistercian Life as described by the *Ecclesiastica Officia*', in *Cistercian Ideals and Reality*, ed. by J. Sommerfeldt, CSS 60 (Kalamazoo: Cistercian Publications, 1978), pp. 62–79.

[77] For instance, *Abstracts of the Charters and Other Documents Contained in the Chartulary of the Cistercian Abbey of Fountains*, ed. by W. T. Lancaster, 2 vols (Leeds: [J. Whitehead & Son, printers], 1915); *The Coucher Book of the Cistercian Abbey of Kirkstall, in the West Riding of the County of York: Printed from the Original Preserved in the Public Record Office*, ed. by W. T. Lancaster and W. Paley Baildon, Publications of the Thoresby Society vol. 8 (Leeds: [J. Whitehead & Son, printers], 1904); *Early Yorkshire Charters*, ed. by W. Farrer, C. T. Clay and E. M. Clay, 13 vols (Leeds, 1914–65); *Cartularium Abbathie de Rievalle*, ed. by J. Atkinson (London: Publications of the Surtees Society 83, 1889). Other collections and individual charters contained in unedited MSS will be referred to in more detail throughout.

[78] Among the edited material, *Rolls of the Justices in Eyre Being the Records of Pleas and Assizes for Yorkshire in 3 Henry III (1218–19)*, ed. by D. M. Stenton (London: Publications of the Selden Society 56, 1937); C. T. Clay, ed., 'Three Yorkshire Assize Rolls for the Reigns of King John and King Henry III', *Yorkshire Archaeological Journal*, 44 (1911).

This book is structured around specific monastic sites and spaces. The first three chapters deal with representations of abstract and metaphorical spaces through the use and functions of particular physical sites. Chapter one deals with the creation of boundaries in the monastic world, and concentrates particularly on the experience of the Cistercian novice. I explore the ways in which imagined or psychological divisions between the spaces of the monastic and non-monastic worlds were created and defined through the transformation and regulation of the novice's memory. Chapter two looks more closely at the ways in which the material site of the cloister was used and described metaphorically. The cloister as symbol and metaphor of paradise is the focus here. Chapter three also deals with the ways in which physical sites could be used to describe imagined or abstract spaces, this time in relation to the Cistercian church. The church was a site where the visionary played an important role, in that the dominant effect of thirteenth-century church building was to direct the devotional gaze beyond the physical space of the church, and to facilitate the imagining of the space of the other-world.

Chapters four through six focus on the regulation and order of monastic spaces through the discipline of the monastic body. The chapter house as a site for accusation, confession, and punishment is the concern of chapter four. Here, I explore manifestations and meanings of discipline, and the associations that the chapter house held for the confinement and correction of outwardly disorderly monks, whilst retaining its status as one of the principal monastic sites in which notions of community were articulated. In the following chapter I look more closely at representations of the body in the context of the monastic infirmary, particularly concentrating on the practice of blood-letting as a disciplinary tool. The infirmary (with the chapter house) was a site where the physical body became a space wherein order and disorder were resolved. Chapter six deals with a group of specific monastic bodies—those of the lay brothers or *conversi* within Cistercian houses. The spaces and sites occupied by lay brothers were very carefully demarcated, and this chapter investigates some of the reasons for and effects of that regulation.

Space or sites within the monastery were not always perceived by Cistercians in easily defined or discrete categories of material and imagined. Monastic boundaries could be challenged or violated and, as a consequence, thrown open to new meanings and implications. Chapter seven seeks to explores such circumstances through an examination of the act of apostasy, and its effect on Cistercian perceptions of space. Another such instance where topography was mutable was when death intruded into the monastery. The final chapter of this book looks at changes to commemorative practices throughout the thirteenth century. One manifestation of these changes was the creation of closer relationships between the spaces occupied by the dead and those occupied by the living. This is especially clear in the increasing number of lay burials within previously inaccessible areas of the Cistercian monastery, such as the church.

Space was topographically, institutionally and spiritually decisive in the lives and deaths of those monks who inhabited the Cisterian monastic houses of northern England. The ways in which they understood their landscapes provides the historian with important insights into not only how individuals within a particular monastic order positioned themselves in relation to geography, architecture, institution, community, and cosmos, but how we might think about the ostensible dialectic between regulation and imagination, between freedom and enclosure. Those questions form the heart of the following pages.

CHAPTER 1

Boundaries and Memories: The Space of Transition

According to the aged monk Serlo, whose memories formed the basis of Hugh of Kirkstall's *Narratio de Fundatione Fontanis Monasterii in Comitatu Eboracensi*, the establishment of Fountains abbey occurred in two stages. During the first stage, the buildings of the monastery had not yet been constructed, and the monks slept under an elm tree. Serlo marvelled:

> This thing was surely memorable: the sight of the soldiers of Christ in their first campaign [...] spending the winter in their tents, triumphing by the constancy of their faith over the world itself and the prince of the world. All slept under one elm, a brotherhood poor but mighty in the Lord.[1]

Later, the second stage of the monastery's foundation occurred, when 'the monks built huts, [and] established workshops, singing, and chanting'. Serlo remembers that 'they received with reverence the holy message, and as heated wax takes the impression of the seal, so they took the form of the holy institution'. In Serlo's view, Fountains abbey was thus created by the formation of 'a brotherhood mighty in the Lord', and by the construction of a fixed and defined monastic site that was distinctively Cistercian.[2]

Boundaries that enclosed the Cistercian abbey and excluded the outside world were carefully created and defined from the order's late eleventh-century foundation through the thirteenth century. The monastic precinct and its walls, for example, formed topographical boundaries that delineated the physical point at which the monastery was separated from the secular world. Yet the division of secular and religious realms was not only physical or geographic. Imagined or mental boundaries were also constructed to provide separation. The creation of these non-physical boundaries was an especially crucial part of the Cistercian

---

[1] *Mem. F.*, p. 34.

[2] *Mem. F.*, p. 47.

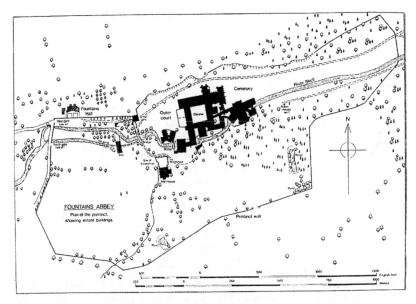

Fig. 1.1: The precinct of Fountains abbey (Courtesy, Dr Glyn Coppack, Ancient Monuments Inspector, English Heritage)

novice's shift from the non-monastic to the monastic world. In this chapter, I concentrate on those transitional spaces occupied by the Cistercian novice. During a novice's year of probation, his memory was appropriated and reformed in order to better facilitate his separation from non-monastic life. Cistercian uses and understandings of memory illuminate the ambiguous relationship between physical sites and more abstract spaces.

Physical Boundaries

Physical demarcation of the monastic precinct was a significant statement of enclosure, difference, and community, serving a range of pragmatic and conceptual requirements. Boundaries denoted legal ownership of land, provided a symbolic reminder of the separateness of secular and religious worlds, and afforded the enclosed inhabitants of the monastery some protection from the outside world.[3] More generally, Walter Horn has shown that the statement of a precise topographical boundary was not an invention of the high Middle Ages. He cites Pachomius as being the one who 'set about the architectural solution to the [...] requirements of the monastic community by surrounding his monastery with a wall—not so much as a response to brigandry [...] but rather because the wall was a symbol of monastic self-determination, shelter—a barrier against contamination by the impure and noisy world outside—and an aid in establishing a corporate morale and in supervising monastic chastity'.[4]

A simple plan of Fountains abbey demonstrates the boundary system. (Fig. 1.1) The central buildings of the abbey, which enclose the cloister, stand almost in the middle of a larger precinct, which covers an area of seventy acres. On the north side, according to the plan, this precinct is defined by a natural boundary of rock face (as it was at Roche abbey), while on the three remaining boundaries, there is a wall. This stands 3.4 metres high and was probably built during the time of abbot John of Kent, between 1220 and 1247.[5] The plan immediately reveals that the substantial precinct was home to a variety of other buildings, including the malthouse, woolhouse, millhouse and animal houses, while the area designated the 'outer court', standing just inside the main western entrance to the abbey, served as the reception area to the monastery. At the east end of the outer court was the gatehouse (not apparent on Coppack's plan), which was also a thirteenth-century construction.[6] The area covered by the abbey can therefore be seen to be divided into three main zones: the cloister, the outer

---

[3] Gilchrist, 'Community and Self'; Jane Sayers, 'Violence in the Medieval Cloister', *Journal of Ecclesiastical History* 41:4 (1990), 533–42; Paul Crossley, 'English Gothic Architecture', in *Age of Chivalry: Art in Plantagenet England 1200–1400*, ed. by J. Alexander and P. Binski (London: Royal Academy of Art in association with Weidenfeld and Nicholson, 1987), pp. 60–73.

[4] Walter Horn, 'On the Origins of the Medieval Cloister', *Gesta*, 12 (1973), 13–52 (p. 15).

[5] Coppack, *Fountains Abbey*, p. 62.

[6] R. Gilyard-Beer, *Fountains Abbey Yorkshire* (London: HMSO, 1970), p. 70.

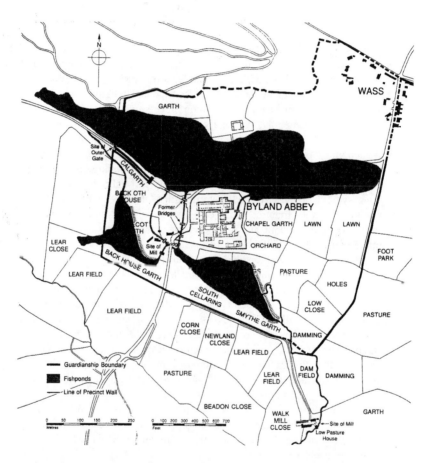

Fig. 1.2: The precinct of Byland abbey (Courtesy, English Heritage)

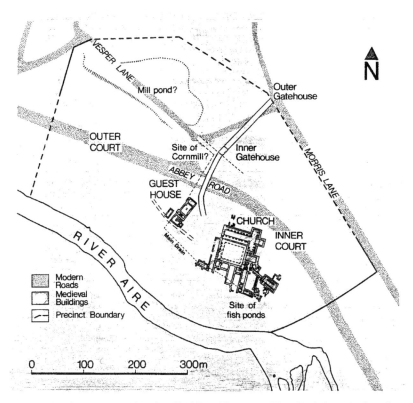

Fig. 1.3: The precinct of Kirkstall abbey (Courtesy, West Yorkshire Archaeology Service)

court with gatehouse, and the precinct, which also had its own gatehouse.

Although the precinct wall at Fountains abbey was a later addition to the monastery, there is no reason to assume that there was no physical boundary between Fountains abbey and the secular world until the thirteenth century. Rather, the periphery of the monastery seems to have been marked by the outer court itself, rather than the precinct, which housed an earlier version of the gatehouse. The boundary that marked the beginning of the monastery was therefore originally closer to the cloister and its buildings than the later precinct wall delineated. The precinct itself at Byland abbey has been mapped by Stuart Harrison (Fig. 1.2), who showed that the wall marking the monastic precinct stretched around an area of about 100 acres and stood eight feet high.[7] At Kirkstall, a similar plan was followed (Fig. 1.3), although the area enclosed was only forty acres. It is apparent from these precinct plans that the activities contained within the monastery itself were quite definitely separate from not only the farming activities of the court (where the millhouse and so forth were held), but also from the precinct boundary itself. This is not apparent from the present ruins of any of the sites, particularly that of Kirkstall, which has lost its court and precinct boundaries to the urban environment built up around it.[8]

The point of entry to the monastery was, as mentioned, marked by the gatehouse. The *Carta Caritatis* had decreed that before a new monastery was to be occupied by the monks, it must have a gatekeeper's cell or porter's lodge,[9] which indicates that control over the mingling of the secular and monastic worlds was certainly an early Cistercian priority. The Yorkshire houses, however, reveal that the remains of the substantial stone gatehouses cannot be dated to the early years of the houses' foundation. This suggests that there were temporary wooden structures in place for at least the first forty years of the Cistercian presence in the north of England. The later gatehouses, particularly those of Roche, Fountains, and Kirkstall, may be considered to represent a clear iconographic statement of Cistercian identity, on the basis of their functional qualities, their architectural similarities and their grandeur.[10] Part of the thirteenth-century stone outer gatehouse at Byland still stands (Fig. 1.4), and as the illustration reveals, was an imposing structure, allowing for both pedestrian and vehicular traffic. The outer gatehouse at Byland led to a walled lane, and then to the inner gatehouse, which doubly distanced the heart of the monastery from the outside world. The same plan was followed at Roche abbey, where

---

[7] Harrison, 'The Architecture of Byland Abbey'.

[8] The inner gatehouse, for example, has been utilized as the abbey museum, and is now on the north-east side of the A65. For the monastic precinct generally, see Stephen Moorhouse, 'Monastic Estates: Their Composition and Development' in *The Archaeology of Rural Monasteries*, ed. by R. Gilchrist and H. C. Mytum (Oxford: British Archaeological Reports, British Series 203, 1989), pp. 29–81.

[9] For some commentary on this, see Lekai, *Cistercians: Ideals and Reality*, p. 448.

[10] Peter Fergusson, 'Porta Patens Esto: Notes on Early Cistercian Gatehouses in the North of England', in *Medieval Architecture and its Intellectual Context: Studies in Honour of Peter Kidson*, ed. by E. Fernie and P. Crossley (London: Hambledon Press, 1990), pp. 47–59 (p. 51).

Fig. 1.4: Byland abbey, two views of the remains of the outer gatehouse (Photos, author)

there are substantial parts of the twelfth-century inner gatehouse remaining, and at other Cistercian houses. The interior of these double-storey structures held the porter's lodge, access for wagons, carts, and horses, and a hall.[11]

The gatekeeper was, therefore, the first person that a visitor or new member of the monastery would meet, a knowledgable older man who knew how to give and receive messages, as the *Regula S. Benedicti* had stipulated.[12] Walter Daniel's *Vita Aelredi* mentions that Aelred encountered the prior, the guestmaster, and the keeper of the gate when he arrived at Rievaulx for the first time, and that once Aelred had made his decision to enter the novitiate, he waited for four days at the guesthouse, presumably with other lay people, before he was admitted into the novices' quarters.[13] To some extent the function of the gatekeeper was to direct people to the right place, but identifying people was also important. The Cistercian book of usages, the *Ecclesiastica Officia*, reveals that the *portarius* had to be at his post after Lauds, and when guests rang the bell, was to open the door and greet them, discover their purpose, and then admit them into the monastery. He was not to speak with monks or lay brothers of the Cistercian order, and he was to be aided by a *subportarius* who would take over the gatekeeper's duties when he was in the church or the chapter house.[14]

Most significantly, however, the *portarius* played an integral role in directing and controlling the lay people who came to the monastery. This was not only evident in the greeting and identification of visitors to Cistercian houses, but also in the duty of the *portarius* to distribute bread to the poor and to travellers. The *Ecclesiastica Officia* stipulates that the gatekeeper was to have bread ready in his cell for travellers, and that bread should be given to the poor at the gate. On special occasions, such as Good Friday, the porter was to go to the church with the other brethren, but was to go back to the gate almost immediately the liturgy was over. Likewise, in summer, when the community processed from the church to the dormitory, no-one bar the sacristan, cellarer, guestmaster, abbot's cook, infirmarer, and porter (or subporter) was to leave.[15]

It was also the porter who was responsible for selecting a small number of the poor from the crowd at the gate for the *mandatum* ritual. There were to be as many poor admitted to the cloister as there were monks in the *cenobium*, while the guestmaster, assisted by both choir monks and lay brothers, washed the feet of the poor at the same time that the weekly Maundy was taking place in the

---

[11] Roland W. Morant, *The Monastic Gatehouse and Other Types of Portal of Medieval Religious Houses* (Lewes: Book Club Guild, 1995).

[12] E. Manning, 'La Règle S. Benoît selon les MSS cisterciens (texte critique)', *Studia Monastica*, 8 (1966), 213–66 (p. 263), ch. 66: 'senex sapiens qui sciant accipere responsum et reddere'.

[13] *The Life of Ailred of Rievaulx by Walter Daniel*, ed. and trans. by F. M. Powicke (Edinburgh and London: Nelson, 1950), p. 14: 'Occurrit prior, hospitalis et portarius [...]'; and pp. 15–16: 'Qui ut michi postea fassus est [...] quia dies illos quatuor quibus ibi morabatur mille annos estimabat pre desiderio quo concupivit induci festinanter in cellam noviciorum'.

[14] Griesser, 'Ecclesiastica Officia', p. 279.

[15] For the preceding see Griesser, 'Ecclesiastica Officia', pp. 199, 247, 279.

monastery.[16] The role of the *portarius* in this ritual was clearly significant in not only ushering the poor into the cloister, but also in controlling the relationship between the 'outside' world and the monastery. This was manifested in other ways as well. Women, for example, were not allowed into the monastery proper, although they were hospitably housed in the guesthouse west of the main court. The *portarius*, according to the *Ecclesiastica Officia*, should not allow children into the church, while anyone who came to the monastery accompanied by women was to remain outside, with the women, where they were given sustenance.[17] Evidence that people did flock to the gatehouses of Cistercian houses may be elicited from Hugh of Kirkstall's *Narratio*, in which Serlo of Fountains remembered that in 1133, a traveller came 'ad januam', crying and begging in Christ's name for charity.[18] Serlo also remembered that in 1134 there was a great famine, during which time many poor people gathered 'ad januam Fontanensem', hoping for relief.[19]

Violation of the physical boundaries of the Cistercian monastery could take a number of forms. Secular elements, either invited or uninvited, could intrude into the abbey, or the inhabitants of the abbey could turn to the outside world. The latter situation could be either authorized—business transactions, customary visits undertaken by abbots to their daughter houses, or to the General Chapter— or unauthorized, the main example being apostasy, which I will discuss in more detail in chapter seven of this book. An example of the transgression of the boundaries between the monastic and non-monastic worlds may be found in the comments of Serlo of Fountains, who laments that the persecution of the Cistercian Order by King John in the early thirteenth century was a disaster for Fountains abbey, not so much financially—although twelve hundred marks of silver were demanded from the abbey—but because monks who were supposed to live in the seclusion of the cloister were now forced to disperse throughout villages and towns, amongst soldiers and lay people in order to find food.[20]

---

[16] Griesser, 'Ecclesiastica Officia', pp. 197–98.

[17] Griesser, 'Ecclesiastica Offica', p. 279.

[18] *Mem. F., De Pane dato Peregrino*, pp. 49–50.

[19] *Mem. F., De Revelatione Trinitatis*, p. 123: 'Invaluit fames sub hie diebus, et erat concursus pauperum multus ad januam Fontanensem'. Serlo also remembered that during this time the monks of Fountains built little huts where the poor were received: '[...] fecerunt eis casas parvulas [...] ubi se reciperent pauperes [...]'. For the large number of grants for the poor, who gathered at the gates of Fountains abbey during the thirteenth century, see Wardrop, *Fountains Abbey and its Benefactors*.

[20] *Mem. F., De abbate Johanne*, p. 125 et seq. at p. 126. The culmination of John's contempt for the Cistercian order came in 1200 at the parliament of Lincoln, where worried Cistercian abbots gathered to request leniency in John's taxation demands. John ordered the abbots to be trodden underfoot by his horses. See *Monasticon Anglicanum*, ed. by W. Dugdale, 6 vols (London: Longman, Hurst, Rees, Orme & Browne, 1817–1830), vol. 1, p. 926. The ultimate irony came in 1203, when John founded the Cistercian house of Beaulieu 'pro salutate animae nostrae et animarum antecessorum et heredum nostrorum'. For John's *volte-face*, see Christopher J. Holdsworth, 'Royal Cistercians: Beaulieu, Her Daughters and Rewley' in *Thirteenth-Century England IV*, Proceedings of

Boundaries around a Cistercian monastery were thus constructed by walls, gates, and gatekeepers. Boundaries physically separated the outside world from the inner sanctum of the abbey, although it is clear that there was a certain amount of movement across the frontier. Some historians have sought to explain the crossing of boundaries between the monastery and the world in terms of Cistercian decline. That is, any ambiguity in the division between the two spheres is seen to indicate the failure of initial Cistercian 'ideals' and the inevitable 'reality' of compromise. This historiographical trend is particularly evident in studies of the economic relationships between Cistercian houses and the secular realm.[21] Implicit in such historical arguments for Cistercian decline is the assumption that the Cistercian claim to live 'in terra deserta, in loco horroris et vastae solitudinis'[22] established an extremely clear and ideally impenetrable *physical* boundary between the abbey and the outside world. This being the case, the argument for decline infers that Cistercian monks were not only traitors to the spiritual ideals of their founders in enjoying social and economic relationships with the non-monastic world, they were also contravening the most fundamental spatial dictum of monastic culture—withdrawal from the world.

This view has been challenged recently by Martha Newman, who concedes that tension did exist between the monastic idea of withdrawal and Cistercian involvement with the secular world. Newman contends, however, that this tension was negotiated by the creation of a particularly Cistercian culture, which drew on monks' individual and collective experiences to enable the Order to 'reject society's norms [yet] still share with the surrounding society ingrained ideas and customs'.[23] Newman also argues that 'the boundaries around Cistercian monasteries were never impermeable', and that the split between the monastic and non-monastic medieval worlds was never as absolute as historians have maintained. Newman's comments raise some pertinent questions about the nature and function of boundaries in the creation of the Cistercian cultural and spiritual environment. Extending Newman's contention, I argue that the boundaries of the Cistercian abbey were not only physical, but that they were also psychological.

Spaces of Transition

Sociologists such as A. P. Cohen have argued that boundaries are necessary for communities because they 'encapsulate the identity of the community', and that

---

the Newcastle on Tyne Conference 1991, ed. by P. R. Coss and S. D. Lloyd (Woodbridge, Sussex: Boydell and Brewer, 1992), pp. 139–50.

[21] See for instance, Roehl, 'Plan and Reality'; Hill, *English Cistercian Monasteries and their Patrons*.

[22] Deuteronomy 32. 10. This phrase appears in many of the Cistercian foundation histories. See for example William of St Thierry's *Vita Prima*, *PL* 185, col. 241, which describes the abbey of Clairvaux in these terms.

[23] Newman, *Boundaries of Charity*, p. 5.

'boundaries are marked because communities interact in some way or other with entities from which they are, or wish to be, distinguished'.[24] In other words, the boundary is a meaningful and, as Victor Turner might say, 'multivocal' symbol of both identity and difference and both separation and inclusion.[25] This view differs from functionalist theories drawn from Durkheim, whose emphasis on the separateness of the sacred and the profane implies a consensus of meanings for both domains.[26] In Cohen and Turner's view of the function of boundaries, it is important to remember that the boundary is not just a physical manifestation: it is also an idea, expressed and understood in different ways. Cohen goes on to argue that people become aware of a culture when they stand at its boundaries, and that boundaries are 'symbolic receptacles filled with the meanings that people impute to and perceive in them'.[27]

Anthropologists have understood the boundaries of social systems as being where interactions between the inhabitants of that system contract.[28] So, for example, the end of a village is one sort of boundary. Where the village is part of a larger system, for example in trade networks, the boundary of the community may be described as relative to social contexts and sets of activities. Such anthropological readings have also been used to describe the different sets of frontiers that marked the life of the medieval hermit, who lived on the margins of his/her world, and existed on at least three frontiers, 'geographical, social and temporal'.[29] In the Cistercian monastery, too, we can see 'concentric' boundaries in the physical relationship between precinct, court, and cloister, boundaries within boundaries. In terms of interaction, these demarcated areas serve to describe the sorts of activity that the topography might represent—devotion in the cloister, domestic labour in the court, and the end of the earthly monastic world at the precinct walls.

Yet, in the Cistercian monastery, as with other communities, a boundary could also be understood to express ideas of transition. This is particularly apparent in the ways in which a novice was received into the monastery. During his year of probation, a novice remained technically at the frontier of Cistercian monastic life. Until he had made his formal profession, a novice did not have the status of a secular person, nor that of a Cistercian religious; he occupied a liminal space in which the monastery and the outside world intersected. The example of the novice demonstrates that the division between the cloister and

---

[24] Anthony P. Cohen, *The Symbolic Construction of Community* (Chichester: E. Horwood; London and New York: Tavistock Publications, 1985), p. 12.

[25] Victor Turner, *The Ritual Process: Structure and Antistructure* (London: Routledge & Kegan Paul, 1969).

[26] Emile Durkheim, *The Elementary Forms of the Religious Life*, trans. by J. W. Swain (London: Allen and Unwin, 1976).

[27] Cohen, *The Symbolic Construction of Community*, p. 19.

[28] Thomas H. Eriksen, *Small Places, Large Issues: An Introduction to Social and Cultural Anthropology* (London and Chicago: Pluto Press, 1995), p. 67.

[29] Christopher J. Holdsworth, 'Hermits and the Powers of the Frontier', *Reading Medieval Studies*, 16 (1990), 55–76.

the world was not easily or sharply created. Until the formal and ritualistic adoption of the Cistercian life had occurred, the novice remained both within and without the monastic community, in a space of transition.

The *Ecclesiastica Officia* detailed the process by which a new member was admitted to the Cistercian abbey.[30] The first condition was that whoever wished to become a monk would only be admitted to the chapter house (where he would formally request entrance into the novitiate) four days after having made his initial request. During the chapter meeting, the potential monk was interrogated by the abbot, at whose feet the candidate was prostrated and to whom the candidate was to indicate that he wished to serve. This routine, which took place after the reading of the *Regula S. Benedicti* in the chapter house, was repeated each day for three days. The novice lived in the guesthouse during this period, until the third day, when he was taken to the novitiate. It was at this point that the novice's year of probation was deemed to begin. Certain situations precluded men from entering the novitiate, one such example being if a man was currently married. This was also addressed in the *Ecclesiastica Officia:* if it was suspected that a married man had somehow managed to enter the novitiate, he would be ejected. Similarly, if someone was found to have already been admitted as a monk (presumably to another order), he would also be expelled.[31]

During the novice's year of probation, several regulations applied which served to distinguish this new member of the monastery from the rest of the Cistercian community. Not only did the novices occupy a separate building or sets of rooms within the monastic precinct, but they were also physically distinguished from more senior members of the community by remaining untonsured. A novice was also not permitted to use monastic sign language to communicate with other monks, he was to leave the church before the choir monks after the Divine Office, and he was to wear a different habit.[32] In other respects, a novice was expected to live in the same way as the rest of the community, reading, eating, and working at the same times as other monks. If the novice were to die during this year, he would be buried in the same way as a monk.

After the probationary year was over, the novice returned to the chapter house where he had made his initial commitment to the Cistercian Order. He revoked all his possessions, and was then led to the church, where he was blessed by the abbot before being tonsured. If the novice was literate, he would then read his profession standing at the presbytery steps. The novice master would read the profession for an illiterate monk. After a number of prayers had been recited by the community, the novice then processed around the choir, prostrating himself at the feet of the entire community. Returning to the presbytery steps through the middle of the choir, the novice would then prostrate himself once more and finally receive the monastic cowl, which had been asperged and blessed. At the

---

[30] For the following, see Griesser, 'Ecclesiastica Officia', pp. 263–65.

[31] Griesser, 'Ecclesiastica Officia', p. 264.

[32] Griesser, 'Ecclesiastica Officia', p. 263.

end of this service, the novice would take his place in the choir with the rest of the monks.[33]

The novice's period of transition from the secular realm to the monastic world was marked by several features. First, a division or boundary was created between the world outside the abbey and the world within it. This was signified by both the physical relocation of the novice's body from one material space to another and by ceremony or ritual, as described above. The period of transition was also a time of learning, in which the novice was expected to divest himself of the trappings of secular life, such as personal property, and work toward the vocation of the Cistercian monk. Walter Daniel, for example, described the novitiate as a testing place, where 'a novice finds it so hard to stamp out the old and endure the present and take precaution against future vices'.[34] However, this time of transition was also a time in which the novice's status within the monastic world was not yet fixed. Until the novice had formally made his profession in the church, had been tonsured and presented with the clothing of a monk, he remained a figure still connected with the non-monastic realm. The novice occupied two spaces simultaneously—as a semi-secular figure, the novice might still be associated with the outer world, while as a figure moving toward unity with God, the novice belonged to the inner world of the monastery.

The ways in which the novice negotiated the move from the secular world to the monastery were regulated by legislative means, such as those included within the *Ecclesiastica Officia*. However, the novice was also advised to utilize other strategies to move through this period of transition. One of the most significant was the reformation of the novice's memory.

Memory and Rethinking the Personal Past

Towards the middle of the thirteenth century, an instructional tract for new members of the Cistercian community was composed, possibly by Stephen of Sawley, abbot of Newminster and Fountains. This work was entitled a *Speculum Novitii*—a mirror for novices. The mirror in the title of this text referred to the novice's conscience, which was to be examined daily, both during confession and in times of private meditation. 'When you go to confession', the novice is advised, 'you may use words such as [...] "My mind wanders through such diverse places such as castles, schools, gatherings [...] At times things I saw and heard in the past come into my memory [...]"'.[35] The novice is also urged to

---

[33] Griesser, 'Ecclesiastica Officia', p. 264. For the ceremonial and ritual aspects of the novice's transition from the secular to monastic world, see Giles Constable, 'The Ceremonies and Symbolism of Entering the Religious Life and Taking the Monastic Habit from the Fourth to the Twelfth Century', in *Signi e riti nella chiese altomediovale occidentale* (Spoleto: Settimane di studio del Centro Italiano dei Studi sull' alto Medioevi 33, 1987), pp. 771–834.

[34] *Life of Ailred*, p. 17.

[35] Mikkers, 'Speculum', p. 45. For the following see M. Cassidy-Welch, 'Confessing to Remembrance: Stephen of Sawley's *Speculum Novitii* and Cistercian Uses of

confess thoughts that are not necessarily drawn from his own experience, but from his imagination—'idle thoughts' (*multa otiosa*). Mainly, however, the author concentrated on the sorts of thoughts that the novice might find in his memory, things that the novice may actively remember from personal experience. These thoughts are listed for the purpose of providing prompts for confession:

> I have thought long on building a church, writing books, managing the house; or on hunting, horse-racing and other such things. At times the image of the coupling of man and woman comes to my mind; at other times my memory alone is occupied with such things.[36]

Subsequent lines of the *Speculum* were devoted to the listing of other subjects for confession. These seem wide-ranging. The novice must confess should he complain about the vicissitudes of Cistercian life ('fasts, vigils, the coarse food, the hard and menial labour'), or speak words conducive to laughter; he must confess if he has felt anger, has been boastful, and has failed to set a good example to others. If he has been hypocritical or malicious, unkind or judgemental toward others, he should confess these faults as well. And if the novice did not feel or show sufficient devotion, if he bowed 'absentmindedly' or ran more slowly to the Divine Office than to the table, then he was a 'stumbling block' to the others, and needed to confess his behaviour.[37]

The author of the *Speculum Novitii* immediately positioned remembrance of the secular past as a sin that needed to be confessed, together with other forms of transgressive behaviour. The relationship between these transgressions may be seen in the very personal or individual focus of each. Complaining about the food in the monastery, for example, is a result of undue care for the pleasures of the body. Remembering the secular world was included with these subjects for confession, because the non-monastic past is also centred on the self—memory's point of reference is the singular, bodily experience of the individual. Another example of this may be seen in Stephen's advice to the novice to correct himself if he remembers and takes pride in his familial connections: 'At times I extol myself on account of my [...] noble ancestry [...] I think I am worth something when I am worth nothing',[38] the novice was encouraged to say. Remembering and taking pride in ancestry is antithetical to the monastic ethos in that it privileged the individual over the community, and provided a very personal, historical link with the secular world the novice is to have left behind. As the *Speculum* said, even more seriously, these matters were against the will of God or against the order.

Memory', *Cistercian Studies Quarterly*, 35:1 (2000), 13–27.

[36] Mikkers, 'Speculum', p. 45.

[37] Mikkers, 'Speculum', pp. 45–46.

[38] Mikkers, 'Speculum', pp. 45–46. For similar ideas regarding ancestry, see Aelred of Rievaulx's *De Institutione Inclusarum*, *PL* 32, col. 1462, where Aelred says that it is a question of vanity if Christ's handmaid prides herself on the fact that she was born of noble parents: 'Vanitas est si ancilla Christi intus in animo suo glorietur se nobilis ortam natalibus'.

It is significant that the act of remembrance did not need to be a voluntary one, according to the *Speculum*. A novice was still guilty of wrongful thinking even if the memories he brought to mind were intrusions or distractions.[39] Explicitly, these memories were said to be evil. The novice is advised that 'if an evil thought should intrude [...] react at once and say "whose image is this?" On hearing that it is the devil's, put an end to it'.[40] Remembrance of the personal past was seen as sinful or dangerous for a number of reasons. On a practical level, it could be argued that with an individual's belief that he has individual worth comes a potential threat to the acceptance and stability of the communal nature of the monastic institution. This may be manifested in apostasy or disruption. The potential instability of the Cistercian novice was certainly of concern to all abbeys. In Walter Daniel's *Vita Aelredi*, a story is told of a secular clerk, who came to the monastery at Rievaulx as a novice, but rapidly found life there too difficult and attempted to leave the abbey. The novice revealed to Aelred that he could not endure 'the daily tasks', or the food and clothing, and that he was 'tormented and cast down by the length of the vigils'.[41]

The tale of the recalcitrant novice shows not only the practical difficulties experienced by some novices in assimilating into the monastery, but it also suggests that the lure of the outside world may still be strong even after the decision to remain within the confines of the abbey had initially been made. Most importantly, Walter Daniel's description of the secular clerk reveals the destructive nature of what the storyteller describes as 'mental instability'.[42] The reason that the clerk found it so impossible to live at Rievaulx was not because the food was poor, the work was difficult, and the clothing was rough. It was, according to Walter Daniel, because the clerk's will was changeable, because he wavered between one life and another, and because his temperament was inconstant.[43] The memory, which crosses the secular/ monastic divide and is based on the personal experience of the individual, can be seen as an issue of will, representing stability once personal memories have been expunged. When Stephen of Sawley encouraged his novice to restructure his memory at the very start of his monastic career, he was attempting to train the novice's will, to eradicate the temptations of the secular world, and to minimize potential times

---

[39] Mikkers, 'Speculum', p. 46: 'Sed his et de omnibus [...] feci scienter vel nescienter aut alii per me, quorum memini vel non memini, me reum confiteor Deo et promitto emendationem et veniam postulo'.

[40] Mikkers, 'Speculum', p. 60.

[41] *Life of Ailred*, p. 30. For apostasy, see F. Donald Logan, *Runaway Religious in Medieval England c.1240–1540* (Cambridge: Cambridge University Press, 1996); for disruption see, M. Cassidy, '*Non Conversi Sed Perversi*: the Use and Marginalisation of the Cistercian Lay Brother', in *Deviance and Textual Control: New Perspectives in Medieval Studies*, ed. by M. Cassidy, H. Hickey, and M. Street (Melbourne: Department of History, University of Melbourne, 1997), pp. 34–55.

[42] *Life of Ailred*, p. 24: 'instabilis animo'.

[43] *Life of Ailred*, pp. 24–25. See also Michael Casey, 'The Dialectic of Solitude and Communion in Cistercian Communities', *Cistercian Studies*, 23 (1988), 273–309 (p. 279).

of instability. Stephen evidently realized that the monastery was occasionally a difficult place in which to live, and he urged the novice to heed the advice of St Jerome, who had advised that 'if the solitude of the desert is upsetting you, take an imaginary stroll through paradise. For, as long as you are strolling in paradise, you will not be in the cloister'.[44]

The *Speculum Novitii*'s representation of personal memory as a sin was based on St Bernard's view that 'all past events are to be censured'.[45] Janet Coleman has suggested that unlike St Augustine who saw the memory as a 'treasure house' of images that could be used as religious analogy, St Bernard was concerned with the purification or 'blanching' of the memory, which had been tainted and polluted by non-monastic experience. Coleman claims that for St Bernard, memory was the surviving connection between the world and the monastery, a connection that needed to be severed. This view certainly holds true for the text of the *Speculum*. However, there are also other more complex factors present in Cistercian understandings of *memoria*, which indicate that the retraining of the novice's memory represents spiritual progress, rather than a simple severance of the secular world. In relation to Cistercian theology in a more general context, I would suggest that memory was closely related to the first step in St Bernard's path to unity with God, the acquisition of humility or an inner movement toward God.

Humility

Monastic doctrines of humility derived from the mystical theology of St Augustine, who had seen pride as 'the cause of all our sickness'.[46] St Bernard expressed the need for humility in terms of not only the acquisition and practice of a virtue, but also as part of a process of self-knowledge, that would lead the monk to union with God.[47] Humility was thus a virtue, and its acquisition is a crucial stage in the quest for what John Sommerfeldt describes as 'the ecstasy of contemplation'.[48] Bernard saw the learning of humility as a logical and

---

[44] Mikkers, 'Speculum', pp. 67–68: 'Si te molestat solitudo eremi, paradisum mente deambula: quoties in paradisa deambulaveris, toties in claustro non eris'. See Jerome, *Ep. 14 ad Heliodorum monachum, PL* 22, col. 354: 'Infiniti eremi vastitas te terret? sed tu paradisum mente deambula. Quotiescumque illuc cogitatione conscenderis, toties in eremo non eris'.

[45] Janet Coleman, *Ancient and Medieval Memories: Studies in the Reconstruction of the Past* (Cambridge: Cambridge University Press, 1992), p. 181.

[46] See Andrew Louth, *The Origins of the Christian Mystical Tradition From Plato to Denys* (Oxford: Clarendon Press, 1981), who quotes Augustine's *Tractatus in Joannem*, XXV.16.

[47] *De Gradibus Humilitatis et Superbiae*; *De Diligendo Deo*. For accounts of Bernard's theology, see Etienne Gilson's classic, *Le théologie mystique de St Bernard*, 4th edn (Paris: J. Vrin, 1980).

[48] John R. Sommerfeldt, *The Spiritual Teaching of Bernard of Clairvaux: An Intellectual History of the Early Cistercian Order*, CFS 125 (Kalamazoo: Cistercian

progressive sequence, like 'notes in a scale from which harmony is produced'.[49] The idea that the learning of humility was an element of knowing the self appears in Aelred of Rievaulx's *Speculum Caritatis* in two forms. Aelred prefaced the *Speculum* with a fairly standard apologetic for presuming to write on the subject of charity at all:

> Gentle and discreet humility is indeed the virtue of the saints, whereas mine and that of others like me is a lack of virtue. Of this kind the prophet said: Notice my humility and rescue me. He was not asking to be rescued from a virtue, nor was he boasting about his humility. Rather, in his dejection, he was imploring aid [...] and so I undertake an impossible, inescapable task worthy of criticism: impossible because of my faint heartedness, inescapable because of your command, and deserving of the criticism of anyone who looks at it closely.[50]

Later, Aelred returned to the subject of pride and humility, this time concentrating on theological and spiritual meanings of humility. Aelred understood that it is certainly 'pleasant' and 'glorious' to 'perceive oneself loftier than the world, and by standing on the peak of a good conscience to have the whole world at one's feet'.[51] Yet, as Charles Dumont points out, Aelred regarded this sort of *superbia* as symptomatic of 'loving oneself wrongly' and forgetting God.[52] According to Aelred, it is this forgetfulness that must be overcome in order to begin the return to union with God. In the *De Institutione Inclusarum*, Aelred expanded on humility as 'the sure and safe foundation of all the virtues',[53] again in the context of beginning an ascetic spiritual life. The significance of humility in the *De Institutione Inclusarum* is related to the virtue of chastity, and Aelred described two types of pride that may undermine this virtue—carnal (divided again into boasting and vanity), and spiritual.

Twelfth-century Cistercians saw humility as the absolute precondition to the possibility of transcendence. Humility was, according to Bernard and Aelred, especially necessary at the beginning of the monastic life, although maintaining humility was certainly an ongoing and difficult process that would continue to determine a monk's experience within the monastery. The acquisition of humility is also stressed throughout the *Speculum Novitii*. The relationship between humility and memory is close. In a broad theological sense, *memoria* was one of the divisions of the soul, particularly in Augustinian theology. The

---

Publications, 1991), p. 53.

[49] Southern, *Making of the Middle Ages*, p. 219. The music metaphor used by Southern is particularly appropriate in the case of Stephen of Sawley, who suggests to his novice that in finding suitable thoughts for meditation, a theme should be selected and a melody created, the chosen subjects for contemplation being the notes and the chords, arranged with frequent interchanges. See Mikkers, 'Speculum', p. 53.

[50] Aelred of Rievaulx, *Speculum Caritatis*, PL 195, cols 503–04; Aelred of Rievaulx, *The Mirror of Charity*, trans. by E. Connor and intro. by C. Dumont, CFS 17 (Kalamazoo: Cistercian Publications, 1990), p. 73.

[51] *Speculum Caritatis*, col. 554.

[52] *Speculum Caritatis*, col. 508.

[53] *De Institutione Inclusarum*, PL 32, cols 1451–74.

other divisions were, of course, *intelligentia* or reason and *voluntas*, or the will.[54] The acquisition of humility related to the exercise of the will; in the monastery, the first stage of acquiring humility is leaving the secular world to enter the religious life—the start of self-knowledge. This was the stage at which the novice should begin to overcome his forgetfulness of God and begin to live out the knowledge of truth. As Patrick Geary has pointed out, the Middle Ages inherited two classical memory traditions: the Platonic, which emphasized memory as the way of acquiring true knowledge, and the 'prosaic', that is the art of training the memory. The exercise of the will and the use of the memory in order to attain humility relate to the former Platonic epistemology whose purpose it is to 'lead the mind to recover knowledge that had been lost'. Remembering and the use of the memory was thus not so much a technique or art of recollection, but a psycho-spiritual capacity and a way of knowing.[55]

Memoria as a division of the soul reveals that memory was not always confined in meaning to commemorative ideas, but could be understood as a tool for transcendence. Memory was also a significant part of monastic life in other ways. The eulogizing of Cistercian founders and past abbots and the perpetuation of particular remembrances in particular parts of the monastery, such as the use of the chapter house for burial and the commemoration of the dead and the living in liturgy, indicate the construction of sites of memory—as Pierre Nora would describe—that continued to play an authoritative role in the mediating of institutional memories.[56] These forms of remembrance or commemoration will be discussed later in this book. What needs more focused discussion at this point are the ways in which memory was reformed, and the ways in which we may understand this reformation in terms of the creation of another boundary, marking the division between the monastery and the secular world.

Meditation and the Reformation of the Memory

Throughout the *Speculum Novitii*, it is clear that the novice's memory was not to be eradicated altogether. Rather, the author of the text argued that the novice's capacity for remembrance should be utilized not only to train the mind to forget the experienced past, but also to create a new past, full of new memories. The strategies used to construct this new past were essentially mnemonic, and were described in terms of spiritual progress. Spiritual progress, according to the

---

[54] See Sommerfeldt, *Spiritual Teaching*, p. 7, who notes that Bernard generally accepts the Augustinian division (for example, Sermon 11 on The Song of Songs), but that at other times, this division is altered to reason, will, and emotion (for example, Sermon 4 for the Feast of all Saints). For another view, see E. Von Ivanka, 'La structure de l'âme selon S. Bernard', *ASOC*, 9 (1953) 202–08, who rightly emphasizes the essential indivisibility of the soul's function; although the soul may, like the Trinity, be notionally compartmentalized, its ultimate nature is unificatory.

[55] Patrick Geary, *Phantoms of Remembrance: Memory and Oblivion at the End of the First Millennium* (Princeton: Princeton University Press, 1994), p. 16.

[56] Nora, 'Les Lieux de Mémoire'.

Speculum, was made by both the acquisition of humility, as outlined above, and the enacting of that humility in the environment of the monastery. This translation of inward feeling to outward behaviour hinges, in this text, on the mnemonic strategy called *meditatio*. Meditation was not explicitly defined by the author of the *Speculum Novitii*, although he was clear that its effect would be a 'more sublime and higher contemplation of God himself'.[57] Meditation may be understood as a process of enlightenment, achieved by specific and disciplined concentration and reflection, or what I argue is a series of essentially mnemonic practices. In the fourth chapter of the *Speculum Novitii*, seven ways of meditation are listed: admiration, praise, comparison, longing, formulation of mental pictures of Christ, rejoicing, and thankfulness. These strategies are used to dictate to the novice exactly how, in all circumstances, the novice was to reflect on his actions and monitor his thoughts.[58]

For the most part, *meditatio* is seen as the technical skill the novice should use in order to formulate 'mental pictures' of specific events in the Christian past. For example, in formulating the various forms of meditation, the novice ought to 'bring to mind the events of salvation history'.[59] The novice was urged to consider the events of the Christian past and to feel that he had 'actually been present at each'.[60] The language that is used here to describe these events is the language of remembrance, and the memories to which the author primarily referred in this context are the memories of Christ's birth, passion, and death. For example:

> Recall his Passion [...] the derision, spitting, cuffing, chaining of his hands, blindfolding, the scourging ordered by Pilate, the mockery of Herod, the jeering voices which demanded his crucifixion and the mocking soldiers intoxicated with sour wine who genuflected before him and hit him on the head with rods. See him on the cross, his hands elevated like the evening sacrifice—his limbs stretched, his face pale, his bones in pain, his joints broken—and the cry of his great sadness is the sadness brought on by the separation of his soul and spirit.[61]

Even more specifically, the novice was told to visualize the joyfulness of Mary when the angel appeared to her while she was reading the prophet Isaiah, to imagine the expression on Christ's face as he appeared to the Magi, to ponder on

---

[57] Mikkers, 'Speculum', p. 50.

[58] See Leclercq, *Love of Learning*, p. 73, who says that *meditatio* is the reflecting on and awakening of ideas within the self. However, Leclercq also says that *meditatio* can refer to the reading of a text and the learning of it by heart, using the whole body (that is, the memory, the will, and the intelligence). I would follow Leclercq's broad definition, rather than that of, for example, Paul Gehl, who describes *meditatio* as merely meditative reading. Although processes of *meditatio* were certainly present in reading, Stephen of Sawley shows that this practice was present in other areas of the monastic life. See Paul F. Gehl, 'Competens Silentium: Varieties of Monastic Silence in the Medieval West', *Viator*, 18 (1987), 125–60 (p. 138).

[59] Mikkers, 'Speculum', p. 51.

[60] Mikkers, 'Speculum', p. 53.

[61] Mikkers, 'Speculum', p. 52.

the amazing sight of Christ washing the feet of fishermen and tax collectors and to think on 'his tenderness in receiving Mary Magdalen when, in tears, he proceeded to raise Lazarus from the dead'. The humanity of Christ is the focus of these images, but the evocation of a form of interior remembrance within the novice is what gives life to these visions.[62]

These scenes are framed in memorial and historical terms that immersed the novice in the experience of the events. Stephen of Sawley's use of images that are not always written in the Gospels reinforced the experiential and personal quality of the novice's visualization. Although the novice was given guidelines for the sorts of images he was to imagine, the precise nature of these 'mental pictures' was left to the novice to create. In this way, interaction between the novice and his thoughts was defined by the Christian experience. However, it was not only the recollection of events that was important but also the novice's involvement in those events. This was described in the *Speculum* as becoming 'one with Him'.[63] In this sense, the act of meditation was essentially receptive; the novice was filled with the knowledge and shared experience of Christ's suffering and death to the extent that it became his own. Imagining the crucifixion, for instance, the novice was told that 'the weakness of hanging there will be yours, the pallor of shaking limbs yours, the shedding of blood yours and the last breath of the crucified yours'.[64]

The process of meditation involved responses that were verbal and emotional. At Vespers, for example, the novice should 'run back to the Lord's Cross [...] be completely filled with anxiety'[65] and for a spoken affirmation, announce that 'the door is closed' to Satan.[66] Although the act of meditation was a fundamentally private one—there must always be a time to be fully alone with God, usually between dawn and the third hour—the fact that it was also incorporated into the actions of the novice's monastic existence made it a public manifestation at the same time. Meditation must also be practised at all times. In the chapter 'Meditation Day and Night' it is indicated that at each stage of the day, the novice should engage with this way of thinking. On waking, the novice ought to think of Christ's resurrection, and on going to bed, he ought to think of Christ's burial.[67]

Meditation was also a highly sensory practice, the performance of which relied on the use and training of the whole body. Hearing is one example. In the

---

[62] For the preceding see Mikkers, 'Speculum', pp. 51–52. See also Jean Leclercq, 'The Imitation of Christ and the Sacraments in the Teaching of St Bernard', *Cistercian Studies*, 9 (1974), 36–54 (p. 41) for *memoria* as a form of interior remembrance and as liturgical commemoration.

[63] Mikkers, 'Speculum', p. 53: '[u]nus fias cum eo spiritus in aeternum'.

[64] Mikkers, 'Speculum', p. 53.

[65] Mikkers, 'Speculum', p. 64: 'Cum vespera canitur, recurre ad crucem Domini et Nicodemo et Joseph satage cum omni sollicitudine et roga, ut deponatur cum omni lenitate et quiete'.

[66] Mikkers, 'Speculum', p. 60: 'Vade retro satanas. Iam ostium clausum est [...]'.

[67] Mikkers, 'Speculum', p. 64.

fourth chapter of the *Speculum Novitii*, the novice is encouraged to hear words of praise, for instance, 'Blessed be the name of the Lord'. Hearing words of admiration, too, would help the novice to 'grow in health until you wax feathered under the wings of the Mother of Grace'. Speaking was encouraged too—the 'voice of longing' should be used to say 'When shall I go and behold the face of God'. 'Do not hesitate', said Stephen of Sawley, 'to give form in your heart to anything that can be activated within you by God or the angels, or that can be said or that can happen in conjunction with the various events and circumstances of our redemption'.[68]

The practice of meditation served to give the novice a new past, one that was filled with collective Christian memories familiar to all members of the Cistercian community. Having confessed to remembrance of his personal, non-monastic past, the Cistercian novice was encouraged to construct an entirely new history for himself using the memories of Christ's birth, passion and death. These memories replaced the secular and experiential memories of a novice's individual past. The creation of a new past for the novice also created another sort of boundary or division between the monastery and the outside world. Having severed his mental attachment to the life he had experienced before joining the Cistercian Order, a novice could progress as a monk and as a Christian toward unity with God. Meditation was thus a tool with which the spirit was enriched just as much as it was a way of casting off the past. The discipline of meditation was practised every day within the monastery, not only in the formulation of mental pictures, as Stephen of Sawley's *Speculum Novitii* described, but also in the mnemonic strategies used in the monks' daily *lectio divina*.

Mnemonic Strategies

A significant aspect of the technique of meditation was its structure and organization, and its relationship to other medieval practices of memory. The act of *meditatio* is strikingly similar to typical methods of visualization and memorization with which the monk was familiar in his daily reading. Medieval memory was understood to have a materiality *sui generis*, and hence was thought to occupy a particular space in the mind. This space was able to be filled and emptied, altered and maintained, inscribed and deleted, according to the discipline of mnemonic practice. Information was compartmentalized into regulated segments recalled by memorial cues. In the case of reading, an aid to memory might be the written word itself and the shape of the letters, or the division, ordering, counting, and alphabetizing of segments of text.[69] In relation to *meditatio*, disciplinary techniques based on memorial 'prompts' were also important. Stephen of Sawley's use of the hours of the Divine Office as memorial cues follows these pre-established norms of mnemonically based

---

[68] Mikkers, 'Speculum', pp. 50–51.

[69] Mary Carruthers, *The Book of Memory: A Study of Memory in Medieval Culture* (Cambridge: Cambridge University Press, 1990).

reading.[70] For example, the novice is to think on scenes from the life of Christ:

> At Lauds, think of the apprehended Christ. At Prime, think of Christ standing before Pilate [...] During Terce, think of Christ raised on the cross [...] At Sext, think of the darkness which fell upon the earth up to the ninth hour [...] At None, think of Christ dying [...] At Vespers, run back to the Lord's cross [...At] Compline, think how you are [...] watching the Lord's tomb so that when he arises you can run and [...] hold his feet.[71]

The similarity between this process and monastic reading techniques is clear. The mnemonic practice of *meditatio* is, as outlined previously, a highly visual one, but it is also a highly technical one. That is, the novice was instructed to train his mind to think on particular things at particular times of the day, or to think of them in very particular ways.

The Cistercian order certainly availed itself of these didactic and mnemonic practices derived from the classical tradition. As I have shown in relation to Stephen of Sawley's *Speculum Novitii*, the strategy of forming mental pictures in order to remember served the reformation of the novice's memory, and helped to establish an alternative past for the individual. Such mnemonic strategies were also present in the daily reading of the Cistercian monk. One small example is found in Cambridge, Jesus College MS QB 17,[72] a late-twelfth-century manuscript from Rievaulx abbey, usually cited for the library catalogue it contains. At fol. 117b, a set of memorial verses appears on the number of chapters in each book of the Bible. These verses are severely abbreviated, and as such, create a rhythmic 'song' when read aloud.[73] The mnemonic strategy in this text is the combination of the sound of the words together with the visual prompt of their abbreviation. Again, the senses of hearing and seeing are used to activate the memory. Stephen of Sawley was also most emphatic that the novice should memorize what he has read, and quoted William of St Thierry's letter to the brothers at Mont-Dieu, saying: 'Deposit a portion of the daily reading in the

---

[70] For medieval memory and reading, see Michael T. Clanchy, *From Memory to Written Record. England 1066–1307* (Oxford: Blackwell, 1993), esp. p. 191 et seq., 'What Reading Meant'; Michael Camille, 'Seeing and Reading', *Art History*, 8 (1985), 26–49.

[71] Mikkers, 'Speculum', p. 64.

[72] Jesus College MS QB 17 is listed in M. R. James's catalogue as no. 34. See M. R. James, *A Descriptive Catalogue of the Manuscripts in the Library of Jesus College, Cambridge* (London: Clay, 1895), pp. 43–56 (with addenda to the description of this MS in a separate sheet inside the catalogue held at Jesus College).

[73] Cambridge, Jesus College MS QB 17, fol. 117b. A transcription of the first couple of letters reads: 'Ge.l.Ex.g.Le.x.una.dua'. Spoken aloud, the rhyme may be rewritten as 'Ge/lex/G/lex/una/dua'. The meaning of these letters can then be retranscribed again to expand into 'Genesis 50 (the 'L' represents the Roman numeral for 50), Exodus 40 (the lower case 'g' represents the 'q' for *quadraginta*), Leviticus 20 (the 'x' refers to the number ten and the 'una dua' refers to 'one ten times two', i.e., 20). The rest of the mnemonic makes up the remaining seven books in Leviticus and completes the remainder of the Bible.

depths of your memory. There it is better digested and more abundantly ruminated on upon recall'.[74]

With respect to Martha Newman's arguments relating to the formation of Cistercian culture, outlined earlier, we can see that memory was an important part of this process.[75] Newman contends that the transmission of monastic culture through words and texts was achieved primarily 'through the memories of [the] monks'.[76] Newman suggests that the internalization of information, 'ideals' and 'assumptions' played an extremely important role in defining the Cistercian world, and distinguishing it from the secular realm. This process of internalizing information and ideas may also be found in the encouragement of monks to read and remember the written word. Internalization was also practised in the novice's creation of a new Christian history through the discipline of meditation. The similarity of mnemonic techniques used in *meditatio* and in remembering the written word can both be interpreted as contributing to the formation of a distinctive Cistercian cultural space.

Twelfth- and thirteenth-century Cistercians were aware that the boundaries of monastic life were not so clearly defined as the walls and gatehouses of the enclosure might imply. Divisions between the secular world and the monastery were imagined, understood, negotiated, and controlled, while the monastic novice faced an especially careful transition from the secular world of his past to the Cistercian abbey. Other medieval Cistercians, such as the author of the *Speculum Novitii*, recognized that boundaries between the monastery and the non-monastic world needed to be created mentally as well as physically. The reformation of the novice's memory and the fabrication of a new past discussed in the *Speculum Novitii* represent one way in which the division between the monastery and the world could be negotiated. This was to be done by the practice of meditation, or the practice of visualization combined with mnemonic strategies familiar to the monk in his daily *lectio divina*. Remembering and forgetting were significant aspects of the novice's progress from the worldly to the religious just as they were fundamental tools in the formulation and determination of boundaries between the geography of the monastic world and that of the secular sphere.

Monastic boundaries certainly created spatial divisions, yet boundaries also created meaningful spaces within the monastery itself, spaces which reveal that medieval Cistercians were not simply concerned with topographical separation, but also with imagined space. The material site of the cloister is one example. Cistercian representations of the cloister reveal that although boundaries were created to enclose this space and to separate it from the non-monastic milieu, the cloister was also a space frequently equated with the possibility of transcendence. In this way, the cloister was both material and metaphorical, physically located and mentally imagined.

---

[74] Mikkers, 'Speculum', p. 59.

[75] See Newman, *Boundaries of Charity*, esp. pp. 1–19.

[76] Newman, *Boundaries of Charity*, p. 19.

CHAPTER 2

Metaphorical and Material Space: The Cistercian Cloister

Hic est tuus paradisus, ut vere ducit te ad paradisum.[1]

Conrad of Eberbach's *Exordium Magnum* tells of a vision experienced by Christian, a Cistercian monk. In his vision Christian, who was convinced that the French monastery of Cîteaux was not to be found on earth but rather in heaven, was taken to the site of Cîteaux. Here, Christian witnessed the monks of the earthly Cîteaux chanting the Divine Office surrounded by a great light. Above the monks, Christian also saw angels singing the same Office, housed in a choir that imitated the earthly one below them. Christian wondered at this 'blessed and holy assembly', where Cistercians were equated with angels and where the site of Cîteaux was equated with heaven on earth. In this story, the monk Christian is witness to the divine presence, not only in that he sees the angels, but also in that he sees the Cistercian monks. More significantly, the material site of a Cistercian abbey is represented as the earthly manifestation of heavenly space.[2]

Christian's vision took place in the early thirteenth century, a period when the Cistercian order was still expanding, but when new brands of itinerant monk were claiming real reform and visible poverty throughout Europe. These wandering preachers were not confined by the principles of spatial segregation or withdrawal from the world that bound the Cistercian order. Yet despite the popularity—and even threat—of the Franciscans and Dominicans, and indeed unorthodox or heretical itinerants, Cistercians of the thirteenth century were still concerned to demarcate their own fixed topographical spaces within traditional discourses of withdrawal from the world. What *does* change in Cistercian understandings of space from the twelfth century, and which is most forcefully

---

[1] Matthew of Rievaulx, *Ad Ricardum Epistula*, ed. by Wilmart, 'Mélanges', p. 79.

[2] Brian Patrick McGuire, 'An Introduction to the *Exordium Magnum Cisterciense*', *Cistercian Studies*, 27:4 (1992), 277–97.

articulated in the literature of the thirteenth century, is the notion of metaphorical space.

From the middle of the twelfth century, Cistercians were more interested in the relationship between earthly and heavenly topographies, like that described in Christian's vision of the heavenly Cîteaux. Cistercian sites themselves became metaphors for the spaces of paradise. One site in particular that served to reiterate the Cistercian claim to direct entry to heaven was the monastic cloister, which was described in Cistercian literature as signifying the space of the heavenly paradise. This metaphor of cloister as paradise was also extended to refer to the entire monastic site as well as the 'cloistered life'. The cloister was thus not only a material site, but it was also an imagined or abstract space. Performative practices associated with the cloister, such as processions and liturgical rites, were also important in creating and continuing such meanings for the site, while visions such as that recorded by Conrad of Eberbach circulated to remind the changing world of the early thirteenth century that the Cistercian order was perhaps fixed in material site, but remained nonetheless the space of eternity.

The Sites

The origins of the cloister and the development of this 'standard' feature of the monastic plan has been described as 'one of the great mysteries of medieval architecture'.[3] The earliest illustration of what we may describe as a 'traditional' cloister plan is generally held to be the plan of St Gall (*c.* 820) (Fig. 2.1). Although in some fifth-century Syriac religious houses there is evidence of courts or *atria* around which buildings were symmetrically, neither earlier Egyptian monastic sites nor Celtic monasteries appear to have used a cloister as the central feature of the monastic site. The early northern English Benedictine houses of Whitby, Monkwearmouth, and Jarrow in the ninth century reveal no cloister either. It is not, it seems, until the later importation of Cluniac plans of the St Gall type that English monastic houses began to utilize the cloister model.[4] The increasing popularity of the cloister may be attributed to the need for a site for communal rituals, and the need for a way of segregating choir

---

[3] Walter Horn, 'On the Origins of the Medieval Cloister', *Gesta*, 12 (1973), 13–52 (p. 13). See also the other essays devoted to the monastic cloister in the same issue, including Paul Meyvaert, 'The Medieval Monastic Claustrum'; Leon Pressouyre, 'St Bernard to St Francis: Monastic Ideals and Iconographic Programmes in the Cloister', 71–92 and W. Dynes, 'The Medieval Cloister as Portico of Solomon', 61–69. See also Christopher Brooke, 'Reflections on the Monastic Cloister', in *Romanesque and Gothic: Essays for George Zarnecki*, ed. by N. Stratford, 2 vols (Woodbridge: Boydell and Brewer, 1987), 1, pp. 19–25; L. Seidel, 'Medieval Cloister Carving and the Monastic Mentalité', in *The Medieval Monastery*, ed. by A. MacLeish (St Cloud, MN: North Star Press of St Cloud, 1988), pp. 1–16.

[4] This had not been a progressive development on the continent either; at St Riquier, for example, as late as *c.* 790, the cloister was triangular.

monks from lay workers. In Cistercian houses, both these features were certainly present in the use of the cloister site.

The Cistercian cloister was an open space situated on the south side of the church (Fig. 2.2). This space was not always perfectly square—at Sawley abbey, for example, the cloister was oblong. In all of the Yorkshire houses, the cloister was enclosed on four sides by arcades. These were not usually glassed in during the thirteenth century, although they were roofed. A reconstruction of part of the cloister arcade at Rievaulx abbey shows the open windows of the arcading (Fig. 2.3). Importantly, the cloister was not always present in Cistercian houses at their foundation, or even as part of the first stage of monastic building programmes. At Fountains, for instance, the cloister did not appear as a central element until 1144, twelve years after the abbey had been founded.[5] At Byland, the cloister was partially demarcated at the same time as the church (between 1160–70) as part of the original plan, but was not completed until around 1180.[6] At Kirkstall, it was not until thirty years after the monastery's foundation that the cloister was finished.[7] Work on the boundaries of the cloister, or the ranges, often continued well after the claustral site itself was fixed. At Meaux, the claustral buildings were being altered as late as 1210–1220,[8] as at Fountains.[9] Rievaulx abbey had two cloister sites: one in the normal position which was even larger than the cloister at Fountains,[10] and another which was attached to the infirmary (Fig. 2.4). Despite the many stages of building which took place in the Yorkshire houses throughout the medieval period, the cloister's traditional position south of the church remained the same.

As a plan of Rievaulx abbey shows, the open space of the cloister was enclosed on the north side by the church, on the south side by the refectory, warming house, and kitchen, on the west side by the lay brothers' range, and on the eastern side by the library, chapter house, parlour, and treasury (Fig. 2.5). The arcades on the northern side of the cloister were those utilized for *lectio divina* and were furnished with benches. This north arcade was also the place where the wax tablet was stored on which was written the names of monks with particular duties for the week.[11] The principal monastic buildings were, therefore, grouped around the central claustral space, while the doors facing outward to the

---

[5] The cloister was included under the instruction of Henry Murdac. See Coppack, *Fountains Abbey*, p. 26.

[6] Fergusson, *Architecture of Solitude*.

[7] 'Fundacie abbathie Kyrkestall', in *Monasticon Anglicanum*, ed. by W. Dugdale, 8 vols (London: Printed for Longman, Hurst, Rees, Orme, & Browne, 1817–30), p. 181.

[8] *Chronicon Monasterii de Melsa*, ed. by E. A. Bond, 3 vols (London: HMSO, 1866–88), 1, p. 286.

[9] The covered arcades were rebuilt by Abbot John of Kent [1220–47]. See 'Chronicon Abbatem Fontanensium', *Mem. F.*, p. 136.

[10] For the Rievaulx cloister, which measured 140 square feet, see Fergusson and Harrison, *Rievaulx Abbey: Community, Architecture, Memory*, passim.

[11] The recess in which this *tabula* was housed survives at Fountains, See Gilyard-Beer, *Fountains Abbey Yorkshire*, p. 41.

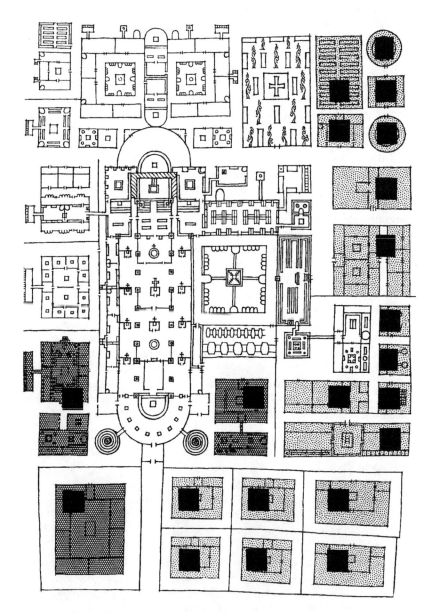

Fig. 2.1: Plan of St Gall, showing the cloister south of the church (Courtesy, English Heritage)

Metaphorical and Material Space 51

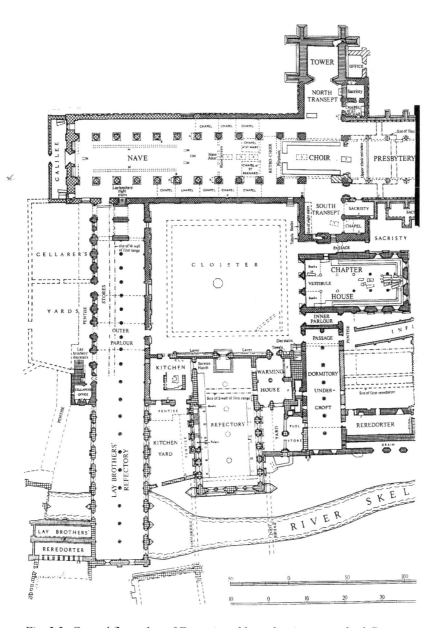

Fig. 2.2: Ground floor plan of Fountains abbey, showing a standard Cistercian claustral design (Courtesy, English Heritage)

Fig. 2.3: Reconstruction of part of the cloister arcade at Rievaulx abbey (Photo, author)

cloister from these buildings were the sole means of access to the cloister garth. Access was clearly regulated in other ways as well. At Byland and Kirkstall, for instance, in place of the usual door to the west range, there is a 'lane' leading to the lay brothers' quarters (Fig. 2.6). This replaced the traditional cloister walk, and had benches on its east side at Byland. The lane itself was shut off from the view of the cloister by a doorway, while the lay brothers' day stairs opened onto the cloister side of the range.[12] The 'lane' effectively pushed the lay brothers' quarters away from the cloister proper, minimizing the contact that this segment of the monastic community may have had with the cloister. At Rievaulx abbey, the original refectory on the south range of the cloister was realigned from east/west to north/south during the last decade of the twelfth century.[13] Peter Fergusson argues that this change in position should be related to the Cistercian desire that a monk should not leave the cloister. At Rievaulx, the original position of the south range refectory did not allow direct access to the cloister

---

[12] At Fountains, for example, there was no 'lane', so the lay brothers' day stairs opened onto the west side of the range.

[13] Peter Fergusson, 'The Twelfth Century Refectories at Rievaulx and Byland Abbeys', in Norton and Park, eds., *Cistercian Art and Architecture,* pp. 160–80. North/south aligned refectories are found at Kirkstall, Byland, Rievaulx, Fountains, Waverley, and Furness.

Fig. 2.4: The main cloister at Rievaulx abbey and the infirmary cloister at Rievaulx abbey (Photos, author)

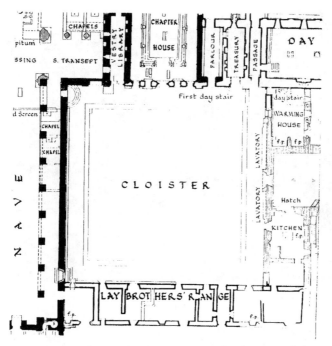

Fig. 2.5: Plan of Rievaulx abbey, showing the cloister and the ranges (Courtesy, English Heritage)

from the kitchen or warming house, as these buildings lay essentially behind the refectory. With a north/south alignment, all these domestic buildings to the south could open onto the cloister space itself and a monk would never technically leave the cloister area. The rebuilding of the south range at Rievaulx draws attention to the fact that early Cistercian legislation understood *claustrum* to mean the buildings attached to the monastery—not simply the open space in the centre of the inner monastic precinct. The injunction against monks leaving the cloister should, therefore, be broadly interpreted to mean that *claustrum* referred to the monastic precinct as a whole, as well as the immediate site of the cloister.[14]

The cloister site may be seen to serve a number of spatial functions within the monastic precinct, including demarcation and enclosure. Primarily, the four sides of the cloister garth demarcated the boundaries of the main sites of liturgy, discipline, domesticity, and labour within the monastery: the northern cloister arcade adjoined the church, the eastern arcade was dominated by the entrance to the chapter house (Fig. 2.7), the southern arcade was associated with the refectory (Fig. 2.8), while the western arcade marked the domain of the *conversi*, whose work was predominantly manual. The demarcation of these particular areas of monastic activity reflects the integral status of each in the

---

[14] *Statuta* 1134, 6, t. 1.

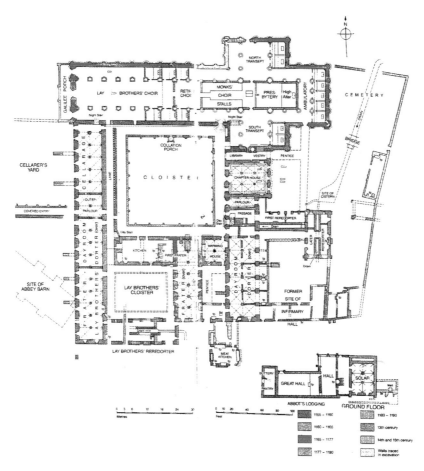

Fig. 2.6: Plan of Byland abbey, showing the lay brothers' lane west of the cloister (Courtesy, English Heritage)

Fig. 2.7: Jervaulx abbey, entrance to the chapter house; Kirkstall abbey, entrance to the chapter house; Fountains abbey, entrance to the chapter house (Photos, author)

Metaphorical and Material Space 57

Fig. 2.8: Rievaulx abbey, south range, entrance to the refectory (Photo, author)

monastic world. This is also evident in processional uses of the cloister, where the principal stations of liturgical processions reiterated these areas of activity within the monastic site. I shall return to the processional and ritual uses of the cloister later in this chapter. Segregation and demarcation may also be perceived in the construction of the closed cloister arcades. These arcades, which ran around the perimeter of the cloister, served as walls and as ambulatories, at once providing a means of access from one part of the cloister to another, whilst maintaining the separateness of each monastic area. In the context of these spatial functions of enclosure and separation, the openness of the claustral colonnade itself may seem ambiguous. However, as Robert Venturi has argued, walls in architecture represent the resolution of open and closed spaces, and that a seemingly paradoxical combination of solid wall and open void does not necessarily undermine this resolution.[15] In other words, an open space may still represent a wall. In the context of the open cloister arcades, the linear outline of the form of the colonnade arches mirrors the linear direction of the arcade path, which continues to designate the accessible area.
The open spaces between the colonnade arches, therefore, delineate *in absentia* the enclosing wall.

Finally, the cloister replicated the enclosure of the monastic precinct itself.

---

[15] Robert Venturi, *Complexity and Contradiction in Architecture*, 2nd edn (New York: Museum of Modern Art, 1977).

Material manifestations of enclosure within the Cistercian monastery began with the boundary walls which skirted the precinct, included the outer court of the abbey, and ended with the site of cloister—a system of boundaries within boundaries. The cloister site, as the innermost monastic enclosure, was therefore an area generally inaccessible to secular people, but highly accessible to choir monks. The position of the monastic cloister south of the church may therefore be explained in terms of both architectural precedent and very general spatial functions such as segregation and enclosure.

More specifically, the cloister was one of the central sites for the expression and practice of communal rites. These rites were both liturgical, such as processions, and domestic, such as shaving. Another significant ritual that was performed within the cloister was the *mandatum* rite, conducted each week for the monastic community, and once a year for the poor, who were admitted into the usually inaccessible cloister to have their feet washed by the monks on Maundy Thursday. Such rituals may be seen to reflect the communal emphasis of monastic life as it was practised in the cloister itself.

Uses of the Cloister

Unlike Benedictine liturgy, the use of processions in Cistercian houses was minimal, an innovation that was strongly criticized by non-Cistercians during the twelfth century, including Abelard.[16] The earliest Cistercian rite allowed only two major processions per annum—one on Palm Sunday and the other at Candlemas, while a third was added in 1151 before the conventual Mass on Ascension Day. One more procession was included in 1223 for Assumption Day, the last addition for the thirteenth century. There were three principal stations demarcated for processions in the *Ecclesiastica Officia*, where an antiphon was to be sung.[17] Processions began in the church, and moved around the cloister, stopping first at the eastern range, then the refectory, then the west range, finally ending next to the church.[18]

The whole Cistercian community participated in these major processions. The procession was generally led by the subdeacon, who sprinkled holy water on the ground in front of the community, then the deacon, who carried the cross. The rest of the choir monks followed in the same order in which they stood in the church, but in pairs. The abbot or daily priest followed the choir monks, then came the novices and finally the lay brethren.[19] Visitors and guests to Cistercian

---

[16] See Abelard's *Ep. 10* in *PL* 178, cols 335–40.

[17] Griesser, 'Ecclesiastica Officia', p. 195 for the Palm Sunday processions.

[18] Griesser, 'Ecclesiastica Officia', p. 195.

[19] Griesser, 'Ecclesiastica Officia', p. 195. The solidarity exemplified in a procession being carried out in pairs had been stressed by St Bernard, who said that two by two was the right way to travel—'if someone wants to travel alone [*solitarius*] he upsets the procession'. See Michael Casey, 'In communi vita fratrum: St Bernard's Teaching on Cenobitic Solitude', *Analecta Cisterciensia*, 46 (1990), 243–61 (p. 250).

houses were also a part of these processions—at Candlemas for instance, guests were given a candle like the rest of the monastic community, and took part in the procession around the cloister arcades.[20] The inclusion of non-monastic participants in liturgical events such as the Candlemas procession must qualify some of the more prohibitive elements of the cloister space, which I have discussed above. Although enclosure and inaccessibility were two fundamental features of the cloister, the *Ecclesiastica Officia*'s incorporation of guests and visitors into the liturgical performance of the cloister site shows that these features could be negotiated.

Closer examination of one of these four principal processions around the cloister arcades demonstrates how the space of the cloister together with the movement of the processions was redolent with complex meaning for Cistercians. The Palm Sunday procession provides an example. This procession, like the others, began in the church. Palm leaves were spread on the steps of the presbytery and asperged with holy water before the service began. The first antiphon to be sung was the *Pueri Hebreorum*, during which the palm fronds were distributed to the choir monks and to the novices. Lay brothers and visitors received these leaves only if there were some left after the first distribution. While the cantor began the second antiphon, the *Occurrunt Turbe*, the subdeacon left the church. He took with him the holy water, which was to be used during the subsequent procession, as I have already noted. He was followed by the deacon, who brought the cross, and then by the rest of the congregation.

Walking out of the church on the north-eastern side of the cloister, the procession continued straight ahead, along the east claustral range. As the procession approached the first station, the antiphon *Cum Appropinquaret* was intoned by the cantor. When this antiphon was finished, the procession moved toward the south range, while the next antiphon began, the *Collegerunt*. Leaving the cloister arcade outside the refectory, the community processed to the west range, where the cantor sang *Unus Autem Ex Eis*. The procession then walked toward the church to the north of the cloister, while the antiphon *Ne Forte* was sung. At the last station, in front of the church, the cantor began the antiphon *Ave Rex Noster,* while the community inclined toward the cross. The *Gloria Laus* was chanted before the community ultimately returned inside the church. Immediately prior to the congregation re-entering the church, the Gospel was read out by the deacon, who turned to the east while the community faced one another. Two choir monks entered the church before the rest of the procession, singing the *Gloria Laus* again while the congregation then entered the church to finally deposit the palm fronds which they had been holding onto the steps of the altar.[21]

The Palm Sunday procession may be interpreted in various ways. The event which is described in the movement around the cloister arcades is, of course, Christ's triumphal entry into Jerusalem, while this procession has generally been identified with Cistercian monastic emphasis on *imitatio Christi*.[22] The travelling

---

[20] Griesser, 'Ecclesiastica Officia', p. 212.

[21] Griesser, 'Ecclesiastica Officia', pp. 195–96.

[22] Jean Leclercq, 'The Imitation of Christ and the Sacraments in the Teaching of St

of the community that takes place during this procession thus imitates the travelling Christ. Even more broadly, the moving of the procession itself reiterates the idea of the Christian life as a journey, or as a process of travelling, described by Augustine as the *peregrinatio ad patriam*. In the case of the Palm Sunday procession, the immediate *patria* is the church, while the ultimate *patria*, for the processing monks and for all Christians, is heaven. Leclercq has also argued that the Palm Sunday procession's symbolism is made all the more forceful with the reading of the Passion.[23] The procession's symbolic travelling around the cloister is therefore clearly explained by the reading of the Gospel at its conclusion.

The relationship between the monastic procession and Christ's own journey is further accentuated by the presence of the cross, which preceded the procession and faced the community throughout the entire ritual. The cross may be understood to serve as a way of directing the procession, and it may be seen in terms of mnemonic function. In other words, the constant presence of the cross (as a representation of Christ) could stir the Christian memories of the Cistercian community, emphasizing with every step of the procession the journey of Christ. The cross thus served as a visual reminder of the similarity between the procession and the life of Christ. It is also worth noting at this point that other rites used the cross at the head of a procession as a way of warding off the devil, while in the English ritual, the Eucharist was sometimes used as well as the cross at the head of the procession.[24] In the Cistercian rite, I suggest that the dominant function of the cross was to underline Christian identification with Christ's journey.

The procession thus signifies a precise moment in salvation history, imitating and representing what is past. Yet, by the performance or playing out of that historical episode, the events are relocated to the present. Further, the journey that takes place around the arcades replicates one that is to come in the future. Historical time is collapsed in the performance of this procession, while the division between eschatological and earthly time is blurred. In this way, the Palm Sunday rite is at once commemorative, affirming and aspirational. The cloister, as the site for this temporal transformation, becomes 'holy ground', where the liturgical life of the monks is infused with memory and with hope. This may also be seen in a number of visions which involve processions, one of which is found in the thirteenth-century *Exordium Magnum*. In this collection, a monk is said to have seen in a vision a glorious procession of acolytes, priests, and subdeacons, all of whom held candles and all of whom were dressed in white. The procession was led by Mary, who was said to have illuminated the whole church with her light, while the group moved toward the infirmary to collect a dying monk. The monastic procession in this vision has been

Bernard', *Cistercian Studies*, 9 (1974), 36–54 (pp. 49–50).

[23] Leclercq, 'Imitation', p. 49.

[24] V. D'Avino, ed., *Guillelmi Duranti Rationale divinorum officiorum I–IV*, ed. by A. Davril and T. M. Thibodeau, Corpus Christianorum. Continuatio mediaevalis 140 (Turnhout: Brepols, 1995); Miri Rubin, *Corpus Christi: The Eucharist in Late Medieval Culture* (Cambridge: Cambridge University Press, 1991), esp. pp. 243–71.

transformed into an other-worldly and powerful event, in which the boundary between the living and the dead is obscured.[25]

Another liturgical episode, the *Benedictus Aquae* rite, illustrates similar principles, while emphasizing the purity of the claustral landscape. The rite was primarily one of exorcism, and involved asperging of the altar and the cloister arcades with a mixture of holy water and salt.[26] This rite was semi-processional, but unlike the Palm Sunday procession, the *Benedictus Aquae* rite took place every Sunday in Cistercian houses after Terce and before Mass. During this rite, it seems that the community remained within the church building, while an obedientary scattered the salt and water around the cloister. The processional aspect of the *Benedictus Aquae* ritual was limited to the church, where the community would approach the presbytery steps for the blessing after the exorcism had been completed.[27] This rite was therefore predominantly protective and cleansing. The asperging of the cloister at the same time as the blessings were taking place in the church underscores the symbolic importance of the cloister as the centre of monastic activity. As we have seen, processional use of the cloister site is one strong example of the imposition or fabrication of symbolic and referential meaning for the space. In the context of the *Benedictus Aquae* ritual, the cloister also assumes a symbolic significance. Taking a liturgical activity outside the church establishes the sacred nature of the claustral landscape, and the *Benedictus Aquae* sprinkling of water around the arcades sacralizes this topography. With this weekly sacralization, the *terra sancta* of the cloister was thus renewed and fortified.

The cloister was also the site for a number of other ritual events. One of the most important of these was the *mandatum* rite, or the washing of feet. The *mandatum* rite was included in the *Regula S. Benedicti*, and involved both the monastic community and lay people on Maundy Thursday.[28] On this day, a group of poor people (*pauperes*) was led into the cloister. Here they sat down and removed their shoes while the water and other necessary accoutrements for the *mandatum* were organized. The monks themselves were responsible for the washing of the feet. They came from the church into the cloister and were assisted in the rite by lay brothers. After each person had his feet washed and kissed, he was given a piece of money by the choir monk who had performed the *mandatum*. This gift was given while the choir monk knelt down and kissed the hand of the recipient. When all the *pauperes* had received the *mandatum* and been given their money, they were then led out of the cloister into the guest house.[29]

Although the *mandatum* rite for the poor took place only once each year, the ritual occurred every Saturday within the Cistercian monastery for the

---

[25] For the preceding see Brian Patrick McGuire, 'Purgatory, the Communion of Saints and Medieval Change', *Viator*, 20 (1989), 61–84.

[26] Griesser, 'Ecclesiastica Officia', pp. 223–24.

[27] Griesser, 'Ecclesiastica Officia', p. 224.

[28] Manning, 'Règle de S. Benoît', p. 244, ch. 35, and p. 253, ch. 53.

[29] Griesser, 'Ecclesiastica Officia', p. 198.

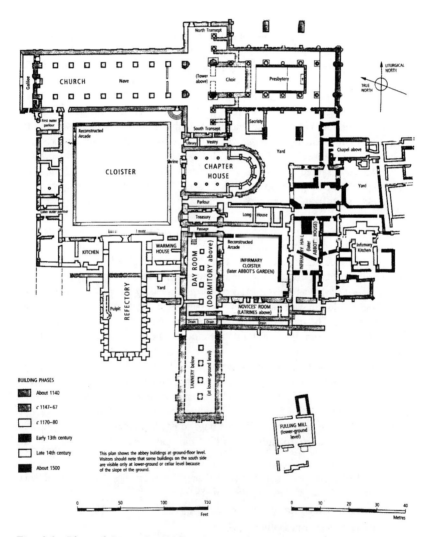

Fig. 2.9: Plan of Rievaulx abbey, showing the lavers (basins) used for the mandatum *against the refectory wall (Courtesy, English Heritage)*

community, and each week for guests. Again, the *Ecclesiastica Officia* provided the instruction for this rite, outlining the procedures for the monks, abbot, and designated helpers who performed the *mandatum* for the community. The abbot himself only washed the feet of twelve members of the community—that is, four monks, four *conversi*, and four novices.[30] If there were not enough novices in the abbey, then more lay brothers received the *mandatum*. The names of those to conduct the *mandatum* were written on the *tabula* in the cloister, while the rite itself took place on the south side of the cloister, adjacent to the wall of the refectory (Fig. 2.9). Peter Fergusson has found that the association of the *mandatum* rite with the refectory was an innovation in Cistercian houses, where the south wall of the cloister at houses such as Fountains, Rievaulx, and Byland included a bench above the basin. Monks could therefore sit on this bench, with their feet in the basin for the *mandatum* rite.[31]

More generally, the association of *mandatum* and refectory was important in giving 'architectural definition' to a liturgical event. In other words, the transcendent meaning of the ritual act was given physical form. The *mandatum* rite, done in imitation of Christ, who had washed the feet of the apostles prior to the Last Supper,[32] carried with it meanings of humility, intimacy, and community. The washing of the feet was also symbolic of baptism, where water washes away sin, and provides both blessing and absolution. The Cistercian *mandatum* certainly included these meanings. The site of the cloister for the performance of this rite could be associated with the chamber in which Christ had washed the feet of the twelve apostles, and it could be associated with the *templum dei*, mentioned in the liturgy for the *mandatum hospitum*. The immediate location of the cloister, therefore, was symbolically transformed in this ritual to signify biblical and heavenly spaces.

Community within the cloister was also signified by the use of that site for domestic activity, such as shaving and cutting hair. The community congregated in the cloister to be shaved, and to shave each other. The abbot decided who was to do the shaving, and it was directed that haircuts should not be too severe, but should be above the ears.[33] Most interestingly, the *Ecclesiastica Officia* specified that no-one should presume to shave another unless invited to do so. By the same token, once a monk had asked another to shave him, his request could not be refused. The act of tending the body was not, therefore, an individual concern, but was also a communal act of intimacy in this context. Like the *mandatum* rite, shaving may be understood as a collective act of humility. It should be remembered, however, that lay brothers were not a part of this routine—their beards were a mark of their status. The bodily contact between one monk and another was thus confined to choir monks.

The use of the cloister for symbolic, ritual purposes is one example of the

---

[30] Griesser, 'Ecclesiastica Officia', p. 199.

[31] Fergusson, 'Twelfth-Century Refectories', p. 178.

[32] John 13. 4–15.

[33] For the rite see Griesser, 'Ecclesiastica Officia', p. 249.

ways in which referential meanings were attributed to the material site.[34] The Palm Sunday, *Benedictus Aquae*, and *mandatum* rites show that the immediate topography of the claustral landscape could be transformed and metaphorically transcended through symbolic practices. The symbolic nature of the cloister was also the subject for some analysis by non-Cistercians during the twelfth century. Sicard of Cremona, for instance, believed that the four sides of the cloister represented contempt of the self, contempt of the world, love of one's neighbour and love of God,[35] while Honorius Augustodunensis saw the site of the cloister itself as replicating the portico of Solomon constructed next to the holy temple in Jerusalem.[36] During the eleventh and twelfth centuries, representations of the cloister as portico of Solomon were the subject of new interest and it has been suggested that the equation of cloister/portico was a way of providing 'a mantle of dignity for an indispensable functional type, which [...] would otherwise lack the necessary cachet of tradition'.[37] In other words, one of the reasons for the popularity of the cloister plan during the Middle Ages was that this physical site could be metaphorically identified with the site where the apostles were said to have gathered. The portico of Solomon was a place of communion,[38] and for the monastery, provided a blueprint for the expression of apostolic community.

Other symbolic meanings were also attributed to the cloister during the medieval period. One of the most prevalent of these was the identification of the cloister with paradise. William Durandus, for instance, described the cloister as a symbol of the heavenly paradise, while a similar motif was common among Benedictine preachers of the eleventh and twelfth centuries.[39] Cistercians, too, used comparable language to depict the monastic life and the monastic landscape. Uses of the paradisal metaphors in Cistercian literature, however, reveal that the identification of cloister and paradise extended beyond the fixed location of the cloister site itself. The entire monastic precinct and the concept of apostolic, monastic life was represented in paradisal terms. This is especially the

---

[34] I have concentrated on the use of the cloister for liturgical and processional purposes in this chapter, but it should also be remembered that the cloister served as the venue for education and reading. For the Cistercian legislation on these activities, see *inter alia* Griesser, 'Ecclesiastica Officia', pp. 245–46 for *De Collatione*.

[35] Sicard of Cremona, *Mitrale seu De Officiis Ecclesiasticis Summa*, PL 213, cols 25–26: 'In hoc claustro quatuor sunt latera, contemptus sui, contemptus mundi, amor proximi, et amor Dei'. Sicard goes on to identify the buildings surrounding the cloister in terms of their metaphorical identifications: '[...] Capitulum, secretum cordis; refectorium, delectatio sanctae meditationis; cellarium, sancta Scriptura; dormitorium, munda conscientia; oratorium, vita immaculata; hortus arborum et herbarum, congeries virtutum [...] '.

[36] Honorius Augustodunensis, *De Claustro*, PL 172, col. 590: 'Claustralis constructio juxta monasterium est sumpta porticu Salomonis constructa juxta templum'.

[37] Dynes, 'Portico', p. 68.

[38] Acts 4. 11; Acts 5. 20–21, 42.

[39] Rationale Divinorum Officiorum, I, 1, 42 et seq.; Jean Leclercq, *La vie parfaite. Points de vue sur l'essence de l'état religieux* (Paris/Turnhout: Brepols, 1948), pp. 161–69.

case from the late twelfth and early thirteenth centuries, when as mentioned previously, the Cistercian order was faced with the rise of new monastic groups who decried enclosure as indulgent, and who claimed the entire world as their monastery. During this period, representations of the cloister and the cloistered life as symbols and earthly images of heaven redefined the space of the Cistercian monastery in terms of its transcendent potential. One example from Yorkshire is found in the letters and poetry of Matthew of Rievaulx.

The Cloister as Paradise

During the early part of the thirteenth century, Matthew, precentor of Rievaulx abbey, wrote a series of letters to the prior of Beverly Minster.[40] In these letters, Matthew exhorted his friend to join the Cistercian order without delay. 'Flee', says Matthew to the prior, 'from the midst of depravity and perversity, from danger to security, from work to quietude, from shadow to light, from the corruption of the flesh to the delight of the spirit'. The Cistercian abbey is the antithesis of all that is corrupt and worldly, in Matthew's view, and if his friend hastens to join the monastic life, the gates of paradise will be open to him. In this letter the monastery is a garden of delights, but unlike the garden of Eden, the monastery is a garden where women do not practise seduction, where the beauty of the forbidden tree does not cause one to fall into error, and where the serpent does not deceive. 'Paradise', says Matthew, 'is among us here, in spiritual exercise, simple prayer and holy meditation'.[41]

In other letters, Matthew reiterates similar themes. He suggests the nearby monastery of Meaux as the place where his friend should join the Cistercian order, describing this site as the door to heaven, and as a place that abounds in sweet spirituality. God himself is in this place, according to the precentor.[42] In another letter, Matthew describes his own abbey of Rievaulx as holy ground,[43] and although he recognizes that the vocation of the Cistercian monk may be difficult to live out, he promises that it is a sweet state in which to die.[44] Meaux abbey, like other Cistercian houses, is equated with a paradisal garden, where lilies of the valley and roses grow and where the scent of cinnamon, myrrh, and

---

[40] Wilmart, 'Mélanges', pp. 72–74, Jean Leclercq, 'Lettres de vocation à la vie monastique', *Studia Anselmiana*, fasc. 37, Analecta Monastica 3rd series (Rome: Orbis Catholicus, 1955), pp. 173–79. There are three letters to the prior of Beverly dealing primarily with the attractions of the Cistercian order, although Matthew wrote others to his friend.

[41] For the preceding, see Leclercq, 'Lettres', p. 178–79. At p. 179: 'Hic paradisus intra nos est, scilicet spiritualis exercitatio, pura oratio et sancta meditatio'.

[42] Leclercq, 'Lettres', p. 177.

[43] Wilmart, 'Mélanges', p. 74.

[44] Leclercq, 'Lettres', p. 178.

other spices sweetens the air.⁴⁵ The monastery is a refuge, a haven and an open door.⁴⁶

These general images of the monastery as sacred ground and heavenly garden are repeated to describe the cloister in one of Matthew's poems on the vice of ambition. In this poem, *De ambitio qui est radix omnium malorum*, Matthew reiterates the idea that the Cistercian way of life will lead the monk to the gates of heaven.⁴⁷ Matthew then goes on to describe the destructive effects of ambition, arguing that this vice will preclude the monk from living his life in the paradisal cloister. In Matthew's view, the desire for the world is antithetical to the inner 'paradise' of life in the cloister. Here, the term *claustrum* signifies the living out of the monastic experience, and in particular, the inner life of the monastery. The equation of cloister and paradise in this case is the equation between the contemplative, enclosed life of the monk, and the heavenly peace that will result from the pursuit of this life. Matthew uses the architectural image of the cloister in order to represent the enclosure of monastic life in general.

It has been argued that the medieval cloister symbolized a 'compromised' paradise, where elements of both Eden and Jerusalem were present.⁴⁸ In this view, the cloister was a synthesis of two types of paradise—the earthly paradise (equated with Jerusalem), and the heavenly paradise (equated with a garden). The cloister included both earthly elements, in that it was created by man-made architecture, and heavenly elements, in that its central focus was a garden. The medieval cloister was thus a reminder that 'Eden survives by compromise in a fallen world'.⁴⁹ A twelfth-century example of the duality of city and garden in the imagined paradise may be found in the vision of a monk from Rievaulx abbey, who was taken to heaven by St Benedict. The paradise seen by this monk included a garden with fruit, herbs, and birds, as well as a golden castle and a city.⁵⁰

Medieval Cistercians, however, concentrated primarily on the images of the garden in their representations of the claustral landscape as paradise, as Matthew of Rievaulx's letter and poems show. To some extent, the metaphor of cloister as paradisal garden can be traced to the image of the enclosed garden in the Song of Songs; like this garden, the space south of the church was associated with

---

⁴⁵ Leclercq, 'Lettres', p. 178: 'Hic fragrat odor inaestimabilis suavitatis, hic redolent flores rosarum et lilia convallium [...] cinnamum et balsamum, myrrha et aloe [...] '.

⁴⁶ Wilmart, 'Mélanges', p. 72: 'Ordo Cistercii [est] singulare refugium, felix portus egenorum, porta patens [...] '.

⁴⁷ Paris, Bibliothèque Nationale, MS Lat. 15157, fol. 54.

⁴⁸ W. A. McClung, *The Architecture of Paradise: Survivals of Eden and Jersualem* (University of California Press, Berkeley, 1983).

⁴⁹ McClung, *Paradise*, p. 24.

⁵⁰ See Giles Constable, 'The Vision of a Cistercian Novice', *Studia Anselmiana*, 40 (1956), 95–98; Giles Constable, 'The Vision of Gunthelm and Other Visions Attributed to Peter the Venerable', *Revue Bénédictine*, 66 (1956), 92–114. A slightly abbreviated version appears in Helinand of Froidmont's *Chronicon*, PL 212, cols 1060–63.

warmth and light.[51] The centrality of the natural world and images of paradisal landscapes may also be traced in other Cistercian representations of their monasteries and of gardens in general. In Hugh of Kirkstall's account of the foundation and early history of Fountains abbey, the monk Serlo described the importance of the elm tree under which the Fountains monks sheltered until the monastery buildings were constructed. This elm tree, which 'still lives unhurt with its green leaves and thick foliage', was a natural protector for the first monks of Fountains. Serlo recalls:

> There was an elm tree in the middle of the valley, with luxuriant foliage after the manner of that kind of tree, which with its leafage tempered for the beast which lay beneath it the cold in winter and the heat in summer. Hither the holy men went to seek lodging beneath its shelter.[52]

The elm tree, representative of the natural world, was not only protective of the intrepid monks of Fountains, but it also signified the first, central site in which community was expressed. As Serlo notes, 'all slept under one elm, all lay between one elm'.[53] The elm tree was, metaphorically, the first cloister at Fountains abbey.

A late thirteenth-century poem on Susanna and the Elders by the Yorkshire Cistercian Alan of Meaux includes a description of Joachim's garden in which similar paradisal images of a *locus amenus* are utilized.[54] The garden is filled with lilies and roses, which give out a delicious scent, and the fountain in the garden's centre is 'clearer than glass and more precious than gold'.[55] There are trees of all varieties, maples, beeches, and pine trees, while apple trees and palm trees grow as well. The images here are ones of growth, abundance, and fertility, while the language used by Alan of Meaux is typically 'tactile', as Caroline Walker Bynum has described.[56] This 'tactile' language reappears in the

---

[51] Song of Songs 4. 12 and 16. For images of the cloister and the monastic life as paradise, see *inter alia*, Jean Leclercq, 'Le cloître est-il un paradis?', in *Le message des moines à notre temps*, ed. by His Em. Cardinal Fumasoni Biondi et al. (Paris: A. Fayard, 1958), pp. 141–59; Stanley Stewart, *The Enclosed Garden*, (Madison: University of Wisconsin Press, 1966); Terry Comito, 'Sacred Space and the Image of Paradise: The Cloister Garden' in *The Idea of Garden in the Renaissance*, ed. by T. Comito (Hassocks: Harvester Press, 1979); Joan M. Ferrante, 'Images of the Cloister—Haven or Prison?', *Medievalia*, 12 (1989), 57–66; J. T. Rhodes and Clifford Davidson, 'The Garden of Paradise', in *The Iconography of Heaven*, ed. by C. Davidson (Kalamazoo: Medieval Institute Publications, 1994), pp. 69–109; George Ferguson, *Signs and Symbols in Christian Art* (Oxford: Oxford University Press, 1966), pp. 43–44.

[52] *Mem. F.*, pp. 34–35.

[53] *Mem. F.*, pp. 34–35.

[54] J. H. Mozely, 'Three Medieval Poems on Susanna', *Studi Medievali*, n.s. 3 (1930), 27–52 (pp. 41–50).

[55] Mozely, 'Three Medieval Poems on Susanna', pp. 44–45: 'Ortus aroma dabat [...] / [...] Lilia nempe rosis'; 'Fons ibi lucidior vitro, preciosior auro'.

[56] Caroline Walker Bynum, *Jesus as Mother: Studies in the Spirituality of the High Middle Ages* (Los Angeles and London: University of California Press, 1982), p. 79.

identification of Cistercian landscapes with heavenly gardens and with the heavenly paradise.

Similarities between the topography of Cistercian monasteries and the topography of heaven may be found in other literature from the Yorkshire Cistercian houses. Walter Daniel tells us that Rievaulx abbey lay in the Rye valley, a 'kind of second paradise of wooded delight', where the river's waters 'give out a gentle murmur of soft sound and join together in the sweet notes of a delicious melody'.[57] Fountains abbey was initially a *locus horroris*, but was transformed into a place of healing and a gateway to heaven.[58] In the *Speculum Novitii* attributed by Mikkers to Stephen of Sawley, the novice was reminded to thank God that he had been given access to the 'long closed entrance to Paradise', the angels' table filled with the chanting of heaven and the music and harps of paradise. The world outside, according to Stephen of Sawley, was a land of drought and the symbol of aridity and death, while the paradisal cloister is a place of sweet delights.[59] In Germany around the same time, Caesarius of Heisterbach told of the abbot who lamented the loss of one of his novices to the knights who had persuaded the novice to leave: 'You are this day casting your brother out of Paradise and lodging him in Hell'.[60]

The relationship between the cloistered life and paradise is also extended to include the inhabitants of Cistercian monasteries. Walter Daniel's *Vita Aelredi* portrays the monks of Rievaulx abbey as angels and as angels might be, they are clothed in undyed wool: 'pure white in vesture as they are white in name'. Walter Daniel also reports that Aelred of Rievaulx was transfixed by the story of these angels on earth, exclaiming: 'And where, oh where is the way to these angelic men, to these heavenly places?'[61] Likewise, Serlo of Fountains described the arrival of the Cistercians in England as the arrival of 'men of extraordinary holiness and perfect piety, who spoke on earth with the tongues of angels'.[62] The most angelic of these men, in the eyes of Matthew of Rievaulx, were abbots William and Aelred of Rievaulx, and abbot Silvan of Byland. William, the founder and architect of Rievaulx, is described as a flower without a thorn,[63] while Aelred is seen as a mirror, and the embodiment of the 'way of peace'. Abbot Silvan is like a precious sapphire, as are Cistercians generally, who 'shine like jewels' and are precious, the gold against which Benedictine excesses are measured as lightweight, or silver.[64]

---

[57] *Life of Ailred*, pp. 12–13.

[58] *Epistulare Carmen de Fontibus*, in Wilmart, 'Mélanges', p. 69.

[59] Mikkers, 'Speculum', p. 68.

[60] *Caesarii Heisterbacensis monachi ordinis Cisterciensis dialogus miraculorum*, ed. by Joseph Strange, 2 vols (Coloniae: Sumptibus J. M. Heberle, 1851), 1, ch. 14. For the preceding see M. Cassidy-Welch, 'Incarceration and Liberation: Prisons in the Cistercian Monastery', *Viator*, 32 (2001), forthcoming.

[61] *Life of Ailred*, p. 13.

[62] *Mem. F.*, p. 5.

[63] Wilmart, 'Mélanges', p. 55.

[64] Wilmart, 'Mélanges', p. 63: 'Ordo niger est velud argentum, Cystercius aurum'.

The physical site of the cloister was not, therefore, the only part of the monastery that was described as a paradisal or heavenly space. The whole monastic site and the Cistercian way of life could also be depicted in these terms. The cloister itself, already defined as the site in which the earthly world could be metaphorically transcended in the performance of liturgical ritual and processions, provided the model for representations of the monastery as paradise. In this way, we might see that the cloister is a space in which a finite locality is connected to a theology of the infinite.[65] Within the cloister, spaces of theological abstraction have been anchored to the earth. This is achieved through the performance of symbolic rituals, or the practices of a defined group. Not only is the cloister a 'place of promenade and assembly', but it is also an architectural microcosm of a larger symbolic system. Such ideas certainly hold true for the Cistercian cloister. The signification of infinite space by a 'finite locality', however, should be extended for the Cistercian environment. Not only were performative practices important in creating referential meanings for the cloister, but narrative descriptions of the claustral landscape reinforced these meanings. Furthermore, these meanings were expanded by Cistercians to describe the whole monastic site.

The infinite nature of eschatological space was also signified in Cistercian theological discourse, and specifically in ideas of ascension. Bernard of Clairvaux for example, had argued that the path to God had three stages. First, a monk needed to feel an inward movement toward God. Then he needed to practise his devotion in the environment of an abbey. Then, lastly, unity with God, or ascension would ensue.[66] Bernard described the final stage as becoming free. Man, Bernard had said, is already free in his will, but he is only ultimately truly free by the redeeming work of Christ the liberator. The *Speculum Novitii*, too, described the Cistercian's journey toward God as ascension and salvation. As was said to the novice: 'You will ascend to God with your whole heart, your whole being will be drawn to him, you will be united to him with all your strength. You will become one with him for all eternity'.[67] This could only happen, according to this text, in the Cistercian monastery. That is to say, the monk can transcend all that is earthly, but only if he first locates himself in a specific and confined place, the monastery.[68]

Visions and miracles that are said to take place in the cloister reinforce its spiritual and symbolic significance. One example from Jocelin of Furness's *Life of Waldef* describes a vision in which the cloister played a central role. Jocelin tells that the abbot of Melrose visited Rievaulx abbey, and sat down in the

---

[65] Following Lefebvre, *Production of Space*, pp. 216–17.

[66] For Bernard's theology of ascension, see for instance, *Sermo super Cantico Canticorum* 27 and 69.

[67] Mikkers, 'Speculum', p. 53.

[68] For a modern monastic view of the idea of freedom in enclosure, see Teasdale, 'A Glimpse of Paradise: Monastic Space'. Teasdale says that 'the monastic enclosure protects [...] it liberates because it confines. It allows for an inner discipline by providing an outer boundary. It liberates by narrowing our options and our focus, thus creating the psychological space for prayer and contemplation'.

cloister to sleep. Unable to doze, he started to recite his psalms. As he did so, a 'bright figure of angelic aspect' appeared, 'leading a great white-robed multitude in regular order'. The figure introduced himself as William, first abbot of Rievaulx, and 'those who follow are monks and lay brothers under my charge [...] the gems you see shining in my crown and vestments are for the souls I have acquired for God [...] and since we earned eternal rest in this place, we visit it three times a year'.[69] The *Exordium Magnum* replicates the type of vision described by Jocelin of Furness, reporting that a monk of Clairvaux witnessed a glorious procession passing through cemetery, cloister, and church. There were clergy of every type, saints dressed like Cistercians all in white, the apostles Peter and John, and lastly Mary. They moved through the cloister to the infirmary, where they collected a monk who had been ill for some time.[70] The *Dialogus Miraculorum* of Caesarius of Heisterbach posits the cloister as the place where the devil may strike, introducing disruption and noise to a place that is otherwise represented as a place of quietude and order. The fiend, says Caesarius, hurled a flaming arrow at a monk of Heisterbach and when this failed to disrupt his prayers: 'the fiend stirred up about him so great a noise, that the whole floor of the cloister [...] seemed to resound with the clatter of boots of the monks running hither and thither'.[71]

A more prosaic example tells of a monk of Fountains abbey, who had just laid out the newly illustrated pages of a manuscript (of the *Life of Godric of Finchdale*) to dry around the cloister arcades, when he was called into the church for Vespers. While inside the church, a violent storm broke and the monk, forbidden by the rule to interrupt the Divine Office, listened in horror to the sound of the wind and the rain whipping around the cloister.[72] After church, the monk had to go to bed and, again, the rule forbade him to return to the cloister. He spent a wretched night tossing and turning and worrying about the loss of his precious pages. Toward morning, St Godric himself visited the monk in a vision and assured him that his manuscript would survive the tempest.

---

[69] G. J. McFadden, 'An Edition and Translation of the Life of Waldef, Abbot of Melrose by Jocelin of Furness' (unpublished PhD dissertation, Columbia University, 1952), at ch. 19: William, as the founding abbot of Rievaulx, receives attention from other Cistercian writers as well. See for example Matthew of Rievaulx's *De Willelmo Primo Abbate Rievallis*, discussed above. William's epitaph appears in London, British Library, Cotton MS Titus D XXIV, fol. 81:

Dormit in hoc tumulo quondam celeberrimus ille
Ordinis interpres, religionis odor;
Sol patriae, pater ecclesiae, lux fusa per orbem;
Cuius fundator et patriarcha domus;
Insignis virtute, fide, spectabilis ortu,
Abbas Willelmus totus apostolicus.

[70] See McGuire, 'An Introduction to the *Exordium Magnum Cisterciense*'.

[71] *Dialogus Miraculorum*, book 4, ch. 96.

[72] Otto Lehmann-Brockhaus, *Lateinische Schriftquellen zur Kunst in England, Wales und Schottland vom Jahre 901 bis zum Jahre 1307* (Munich: Prestel Verlag, 1956), vol. 3, no. 5935–38.

When morning came, the monk rushed downstairs to discover that Godric had been right; his manuscript was dry and exactly where he had left it.

In these visions, as with other narrative representations, the cloister is not only the topographical site for witnessing the extraordinary; it is also the site in which the distinction between the earthly and the otherworldly is obfuscated. In the cloister, the space of the monastery and the space of the afterlife were continually associated and combined. Not only did the cloister signify other spaces, but those spaces—especially paradise—were perceived to be present in the fixed cloister site.

Medieval Cistercians recognized the cloister as the centre of their monasteries, as an area which served to demarcate architecturally the principal areas of labour, liturgy, prayer, and domesticity within Cistercian houses. The cloister was also a site for liturgical expressions of community. However, the cloister was also perceived to be a space abundant with symbolic meanings and was associated in particular with the heavenly paradise. This identification of an earthly topographical site with the imagined space of heaven was primarily expressed in terms of symbolic transcendence, described in processions and in literary representations of the cloister and monastery. These representations, although not necessarily Cistercian innovations, were certainly utilized from the late twelfth century by Cistercians like Matthew of Rievaulx to describe the Cistercian cloistered life.

Ideas of transcending an earthly space, however, were also present in other areas of monastic life where devotional practices were conducted. The performance of prayer and meditation in the church, together with other communal liturgies, also stressed imagined space within a fixed material site. Like Christian's vision of the heavenly and earthly Cîteaux in which monks and angels mirrored each other in a heavenly landscape, the Cistercian church of the thirteenth century reflected otherwordly imaginings and the possibility of transcendence. In particular the reorganization of Cistercian church buildings during the thirteenth century indicates that material and imagined spaces were—in the north of England—becoming more and more entwined.

CHAPTER 3

The Cistercian Church

Seeing and Believing

The medieval Cistercian church was a space devoted to visualization and imagination. This claim may seem surprising, given that one of the adjectives most frequently used to describe Cistercian church building is not 'creative', but rather 'austere'. Even in the medieval period itself, Cistercians were (initially, at least) renowned for the simplicity of their church design. Peter Cantor, for example, reported that St Bernard burst into tears when he saw shepherds' huts, because they reminded him of the uncluttered Cistercian buildings of old.[1] And for the thirteenth-century French architect, Villard de Honnecourt, Cistercian churches were so simply formulated that their proportions and dimensions could all be expressed in a basic *ad quadratum* plan (Fig. 3.1). More importantly, St Bernard had been extremely forthright in expressing his contempt for churches crowded with ornament, sculpture, and visual distraction in his famous *Apologia ad Guillelmum Abbatem*.[2] Aelred of Rievaulx, too, remarked critically of non-Cistercian monasteries that: '[...] even in cloisters of monks you find cranes and hares, does and stags, magpies and ravens—which are certainly [...] effeminate amusements. None of these things are at all expedient for the poverty of monks, but feed the eyes of the curious'.[3] It would seem

---

[1] Peter Cantor, *Verbum Abbreviatum*, PL 205, col. 257: 'Item: Exemplum sancti Bernardi flentis eo quod videret tuguriola pastorum lecta culmo similia casulis pristinis Cisterciensium tunc habitare incipientium in palatiis stellatis et muratis'.

[2] *Sancti Bernardi Opera*, 8 vols (Rome: Editiones Cistercienses, 1963), 3, pp. 5–108. Translated by M. Casey in The Works of St Bernard of Clairvaux, vol. 1, Treatises 1, Cistercian Fathers Series 1 (Shannon: Cistercian Publications, 1970), pp. 33–69.

[3] Aelred of Rievaulx, *Speculum Caritatis*, book 1, ch. 24, PL 195 col. 572.

unequivocal that Cistercian churches were not sites in which the visual was given predominance.

Paradoxically, however, other Cistercian sources indicate that the church was perceived as a space in which seeing, visualizing and imagining were, in fact, all highly important. The instructions on the recitation of the Divine Office attributed to Stephen of Sawley, for example, encourage the visualization of mental pictures to heighten the monk's experience of Christ,[4] while Matthew of Rievaulx's poetry on the discipline of chanting the Office urges the reader to imagine celestial choirs as the office is sung.[5] Most importantly, the Cistercian church building itself relies on varieties of seeing and visualizing to describe both its own space and the spaces of the other world anticipated by the monk, as he performs liturgical duties within it. The Cistercian church was therefore a space in which a distinction between physical and mental seeing was created and defined.

The complexity of the Cistercian visual world has only recently been explored by historians and art historians, who have long sought to represent St Bernard as a 'theoretician of imageless devotion'. Although the idea of 'intellectual or spiritual vision' within medieval Cistercian houses has been examined, little work has been done on the implications of this way of seeing in an unadorned physical environment. Indeed, Michael Camille believes that the idea of transcending beyond the visual, 'itself a rare idea in the mystical tradition, was practically impossible for most people whose devotional frameworks were constructed through images'.[6] Thirteenth-century Cistercians would not agree. The visual in the Cistercian monastic milieu was extremely important in the practice of devotion and that the association between images and seeing was just as important in Cistercian monasteries as it was in other medieval Christian environments. The difference between Cistercian and other monastic houses lies in Cistercian understandings of vision, which are manifested in the construction and representation of the Cistercian church.

The height, breadth, and light of the Cistercian church, together with its eastward direction of the gaze, creates a spatial narrative that does not rely on the conventional visual stimuli of stained glass, sculpture, and images to draw attention to the profundity of the building and its meanings. The church is emptied of worldly visual cues and the monk's attention is focussed on space itself. The sound of the Divine Office, which reverberated around the building seven times each day, also reinforces the depth of space within the church. The Cistercian church, like other medieval basilicas, draws the gaze to the east end of the building, which from the thirteenth century, became the site for chapels devoted to the Virgin and the power of intercessionary prayer. These chapels

---

[4] Mikkers, 'La psalmodie'.

[5] Paris, Bibliothèque Nationale, MS Lat. 15157, fols 45^v-46.

[6] Jeffrey M. Hamburger, 'The Visual and the Visionary: The Image in Late Medieval Monastic Devotions', *Viator*, 20 (1989), 161–82 (p. 180); Michael Camille, 'Visionary Perception and Images of the Apocalypse in the late Middle Ages', in *The Apocalypse in the Middle Ages*, ed. by R. K. Emmerson and B. McGinn (Ithaca and London: Cornell University Press, 1992), pp. 276–89 (pp. 287–88).

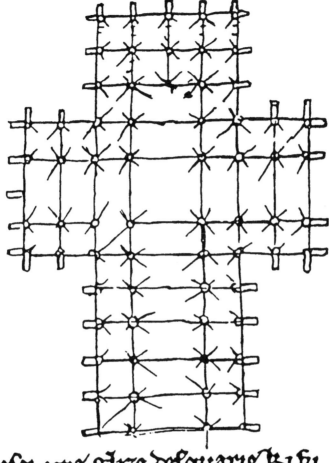

Vesci une glize desquarie ki fu
esgardee a faire en l'ordene des cistiaus

Fig. 3.1: Villard de Honnecourt, plan of a paradigmatic Cistercian church. From E. Stiegman, 'Analogues of the Cistercian Abbey Church', in A. MacLeish, ed., The Medieval Monastery, Medieval Studies in Minnesota 2 (St Cloud, MN: North Star Press of St Cloud, 1988), pp. 17–33 at p. 22.

provide a sharp visual cue which serves to remind the monk of the relationship between the earthly site of the church and the paradisal and purgatorial spaces that follow this temporal world. Placed below the strongest of the building's sources of light, intercessionary chapels indicate the possibility of negotiating the trials of purgatory to rise with the *lux mundi*. Thus, the space of heaven may be the object of visualization in the church, just as much as it is the object of metaphor in the cloister.

From the novelty of the finished church at Byland abbey in the late twelfth century to the completion of the Chapel of Nine Altars at Fountains in the mid-thirteenth, Cistercian church architecture of this period demonstrates a growing interest in the relationship between space and the visual, an interest that we may pragmatically link to changing emphases in liturgical practices together with obligations to patrons and benefactors. Conceptually also, a discernible iconography of light and sound came to dominate the Cistercian churches of northern England from the later twelfth century, an iconography that may be related to new understandings of space within traditional discourses of Cistercian devotion.

The East End of the Church

The Building Programmes

Byland abbey's church had always been radical. Not only was it the largest Cistercian oratory in Britain when it was built, but its similarity to French cathedral architecture emphasized light and space in a way that was significantly different from other twelfth-century Cistercian churches in the north of England.[7] The fact that the high altar was surrounded by arcading meant that the eastern end of the church was able to accommodate an extra five chapels with windows above them on the easternmost wall (Fig. 3.2). This innovation effectively pushed the high altar out into the body of the presbytery, encouraging the eyes of the monk to focus on this altar 'island' and then to look past it to the light and space behind. In the thirteenth century, the choir at Byland was enclosed by a stone wall and the whole church was repaved with green and yellow geometric tiles. The church seems to have been fairly colourful—the interior was limewashed with a masonry pattern picked out in red paint and some of the piers in the presbytery show vine-leaf designs.[8] Surviving tile pavements reveal an interest in geometric patterning, too (Fig. 3.3). Work at Byland was completed in a number of stages, but it is generally agreed that the

---

[7] The Cistercian precedent for the east end of the church at Byland has been suggested as Morimond by Christopher Wilson, 'The Cistercians as "Missionaries of Gothic"' in Norton and Park, eds., *Cistercian Art and Architecture in the British Isles*, pp. 86–116 (p. 112, note 90). For the influence of the architecture of the Somme Valley at Byland and an overview of French Gothic style in England during the early part of the thirteenth century, see L. Grant, 'Gothic Architecture in Southern England and the French Connection in the Early Thirteenth Century', in *Thirteenth Century England III, Proceedings of the Newcastle upon Tyne Conference 1989*, ed. by P. R. Coss and S. D. Lloyd (Woodbridge: Boydell and Brewer, 1991), pp. 113–26 (p. 117). See also Fergusson, *Architecture of Solitude*; Peter Fergusson, 'The South Transept Elevation of Byland Abbey', *Journal of the British Archaeological Association*, 3rd series 38 (1975), 155–76.

[8] Harrison, 'Architecture of Byland Abbey'.

The Cistercian Church

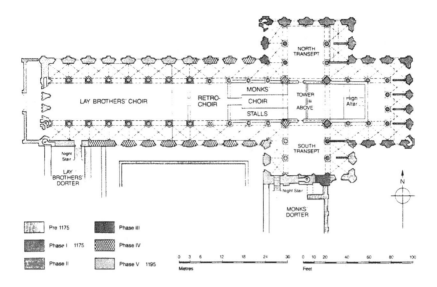

Fig. 3.2: Byland abbey, plan of church (Courtesy, English Heritage); Byland abbey, view to east end of church (Photo, author)

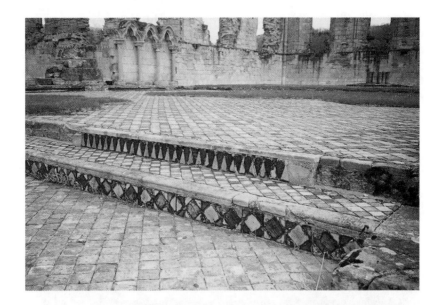

Fig. 3.3: Byland abbey, altar, detail of tiled steps; Byland abbey, cloister tiles, detail from west cloister arcade (Photos, author)

Fig. 3.4: Byland abbey, west façade of church, remains of rose window (Photo, author)

work was finished during the 1190s,[9] with the western end of the nave and the galilee porch. Above the western doorway was a rose window, the surround of which survives in part (Fig. 3.4).

---

[9] Harrison, 'Architecture of Byland Abbey'; Fergusson, *Architecture of Solitude*. For a different view, see Charles Peers, *Byland Abbey* (London: HMSO, 1952) and Nikolaus Pevsner, 'Byland Abbey' in his *Yorkshire the West Riding* (Harmondsworth: Penguin Books, 1960). Both Pevsner and Peers dated the final stage of the Byland church to the first quarter of the thirteenth century.

At Jervaulx abbey, the church is mainly late-twelfth/early-thirteenth century—slightly later in date, therefore, than the enormous structure at Byland. The Jervaulx church also includes an aisled presbytery with a flat eastern end. According to Nicola Coldstream, this aisled presbytery represents a 'decisive break with the immediate past',[10] that is, with continental influence, although the overall plan has been considered to be fairly traditional in contrast to other thirteenth-century choir plans in England. The church at Jervaulx is one of the first Cistercian ones in England to use this design for its east end, and it differs from Byland in that the Jervaulx main arcade follows right through to the east wall (Fig. 3.5). This layout still leaves room for chapels, but visually includes them in the *presbyterium* area, rather than creating a separate zone divided by an ambulatory.

The whole eastern end of the church at Rievaulx began to be rebuilt during the 1220s, with the main alterations taking place around the transepts and the presbytery. Although the presbytery remained the same shape—like that of Jervaulx—it was elongated and expanded. Five chapels were included at the end of the extension, with an ambulatory all around the end. One of these chapels was dedicated to Aelred. These replaced the three chapels originally at the end of each transept, while other chapels were introduced to the lay brothers' part of the choir during the thirteenth century (Fig. 3.6). The decorative elements of the new extension to the east end of the Rievaulx church are significantly 'un-Cistercian' in nature, with moulded arches and clustered piers, although it is clear that the style was locally influenced rather than French Gothic in character. Emphasis is given to light and lightness in construction in this building, reflecting a wider architectural shift throughout England toward the use of French Gothic forms; the heavier—and so-called more austere—elements of a twelfth-century presbytery like that of Kirkstall are noticeably absent from the elevated and decorated style of the Rievaulx church (Fig. 3.7). Other thirteenth-century decorative additions to the Cistercian church at Rievaulx included sculpture; a stone figure of Christ in Majesty from the *pulpitum* still survives in part.[11]

The most spectacular remodelling of the eastern end of any English Cistercian church during the thirteenth century occurred at Fountains abbey, where there had already been several alterations to the church from the time the abbey was founded.[12] Two plans indicate the radical change to the size and nature of the existing church building (Figs 3.8 & 3.9). The thirteenth century

---

[10] Nicola Coldstream, 'Cistercian Architecture from Beaulieu to the Dissolution', in Norton and Park, eds., *Cistercian Art and Architecture in the British Isles*, pp. 139–59 (p. 145).

[11] Now in the possession of English Heritage. For a detailed architectural study of Rievaulx, see Fergusson and Harrison, *Rievaulx Abbey: Community, Architecture, Memory*.

[12] For the chronology of the building work on the church, see Coppack, *Fountains Abbey*. For the most detailed study of the Chapel of Nine Altars, see Peter Draper, 'The Nine Altars at Durham and Fountains', in *Medieval Art and Architecture at Durham Cathedral, The British Archaeological Association Conference Transactions for the Year 1977* (London: British Archaeological Association, 1980), pp. 74–86.

first saw the rebuilding of the eastern end of the church in a Gothic style (c. 1203–), with an extended choir arm of five bays, which followed the design at Jervaulx. New aisles were formed from the presbytery by the introduction of a row of alternating pier forms. Lancet windows with grisaille glass were set above the arcade, while the usual distinction between the triforium and clerestory was abandoned in favour of the appearance of one upper storey. Like Rievaulx, the presbytery was extended eastwards (c. 1211),[13] but in contrast to Rievaulx—and all other Cistercian churches—the new eastern arm of the church was nine bays in length (Fig. 3.10). This Chapel of Nine Altars is the most prominent of the thirteenth-century additions to the Fountains church and it has been shown that it was part of the original plan to extend the presbytery and was completed by 1247. The nine altars positioned along the east wall were for monks who were priests and therefore obliged to say a personal mass. The most central of the nine altars seems to have been dedicated to the Virgin. Peter Draper argues that the nine altars were built at Fountains, not for any strong liturgical reason, but because Clairvaux had nine altars. There were lancet windows in each of the bays and a rose window dominated, the design of which was repeated in the multi-coloured tiled floor. The emphasis in the Chapel of Nine Altars is, again, on elevation and on light, despite the separation of each chapel by low walls. The scale of this eastern elevation was rivalled only by the later eastern chapel of nine altars at Durham, a non-Cistercian building.[14]

The east end of the Yorkshire Cistercian church was thus the focus of much attention during the thirteenth century.[15] Of course, other modifications were made to the west end of the church too, such as the introduction of lay burials. I will discuss this particular development later. However, it was the eastern part of the church, and especially the space beyond the monks' choir and behind the presbytery that was expanded and developed the most during this period. There are a number of possible reasons for the interest in this area. East had always been an important direction in church and liturgical design. The traditional location of the earthly paradise was in the east, and with the advent of crusading culture, eastward also indicated the Holy Land, the earthly paradise realized. Some years ago, Louis Gougaud provided a still useful résumé of the meanings behind the importance of east in the context of private prayer.[16] In this résumé, Gougaud stressed the association of the east with the source of light—in the natural world from the sun and symbolically from Christ, the light of the world. It is another of the associations described by Gougaud, that of the east representing the blurring of earthly and heavenly space, that may be most clearly traced in the expansion of the east end of Cistercian churches.

---

[13] *Mem. F.*, p. 128 says that the building work started under abbot John of York.

[14] Draper, 'Chapel of Nine Altars'.

[15] There was also a new church started at Meaux in 1207, which was completed in 1235–40. *Chronicon Monasterii de Melsa*, ed. by E. A. Bond, 3 vols (London: Rolls Series, 1866–68), 1, p. 326.

[16] Louis Gougaud, *Devotional and Ascetic Practices in the Middle Ages*, trans. by G. C. Bateman (London: Burns, Oates, and Washbourne, 1927).

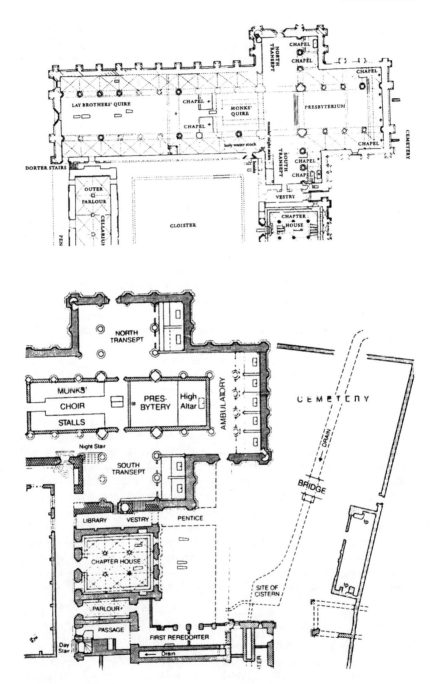

Fig. 3.5: Jervaulx abbey, plan of church (From S. Davies, Jervaulx Abbey, *London [n.d.]; Byland abbey, east end of church (Courtesy, English Heritage)*

Fig. 3.6: Rievaulx abbey, church, thirteenth-century chapels in the west end of the nave; Rievaulx abbey, church, detail of chapel and altar in the west end of the nave (Photos, author)

Fig. 3.7: Kirkstall abbey, twelfth-century presbytery; Rievaulx abbey, thirteenth-century east end of the church (Photos, author)

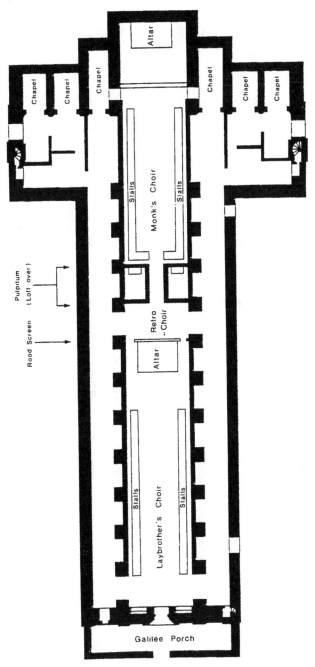

Fig. 3.8: Fountains abbey, plan of the church prior to the building of the Chapel of Nine Altars (Courtesy, Dr Glyn Coppack, Ancient Monuments Inspector, English Heritage)

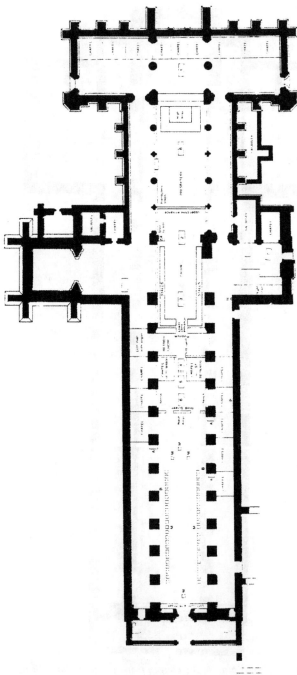

Fig. 3.9: Fountains abbey, plan of church showing the Chapel of Nine Altars (Courtesy, Dr Glyn Coppack, Ancient Monuments Inspector, English Heritage)

Fig. 3.10: Fountains abbey, Chapel of Nine Altars, exterior; Fountains abbey, detail of interior of the Chapel of Nine Altars (Photos, author)

Intercessionary Spaces

The building of chapels in the Nine Altars at Fountains abbey coincides with a period of theological preoccupation with the idea of intercession and intercessionary prayer. In the Cistercian world, this preoccupation was manifested in the centrality of Mary as a mediating figure and in the related development of ideas and representations of purgatory. Cistercian churches were, from the earliest times, dedicated to the Virgin and by 1281, Mary is mentioned as the special patroness of the order.[17] The function of the Virgin as a particularly effective agent of intercessionary prayer may be discerned in Stephen of Sawley's *Meditations on the Virgin*.[18] This text describes the space of this world as a 'laborious journey', a 'valley of tears', and a 'place of exile', in which Mary acts as a mediatrix for the salvation of her devoted people.[19] The text is structured around a series of 'petitions' and 'joys' dedicated to the Virgin. The first joy acknowledges that Mary was the liberator of souls in purgatory, as does the fourth petition which cites St Bernard: 'As all things look to their respective center [sic], so the souls in purgatory look to Mary for deliverance'. In the tenth petition Stephen implores Mary to 'prepare my impure heart so that upon your glorious intercession, the Holy Spirit may in his kindness descend into it', a sentiment repeated in the twelfth petition, where the monk begs that 'rising from the death of my soul through your holy intercession, I may also rise from the grave of evil habits in which I had lain weighed down for a long time, to the newness of holy life so that I may attain heaven at the end of my life'. The final division of the text talks of Mary's clemency and describes that even on behalf of the very wicked, she will still plead with her Son in glory for mercy. Stephen himself says to Mary that he will be completely satisfied 'if, by way of intercession you will just once turn your beautiful face on the eyes of majesty—God the father—on my behalf'.[20]

Other sources from the Yorkshire houses show similar preoccupations with Mary and her power to rescue. An early thirteenth-century collection from Rievaulx abbey, for example, includes the story of a monk who had died without receiving the *viaticum*, but who was nevertheless admitted to heaven because of Mary's influence.[21] A late-thirteenth-century/early-fourteenth-century collection

---

[17] For a general overview of this preoccupation, see Brian Patrick McGuire, 'Purgatory, The Communion of Saints and Medieval Change', *Viator*, 20 (1989), 61–84. For the legislation on dedication and patronage of the Virgin, see *Statuta* 1134, 18, t. 1 and *Statuta* 1281, 6, t. 3.

[18] A. Wilmart, *Auteurs spirituels et textes dévots du moyen âge latin* (Paris: Etudes Augustiniennes, 1932), pp. 317–60. Reprinted from his 'Les Méditations d'Etienne de Sallai sur les Joies de la Vierge Marie', *Revue d'Ascétique et de la Mystique*, 10 (1929), 268–415.

[19] Wilmart, *Auteurs spirituels*, pp. 356–57.

[20] Wilmart, *Auteurs spirituels*, pp. 341–56 for the preceding.

[21] London, British Library, Arundel MS 346, fols 69–70.

from Kirkstall contains a number of hymns dedicated to the Virgin, one of which is the well-known *Stabat Mater*, in English.[22] A story from Fountains abbey tells of a monk of an English Cistercian abbey who saw the Virgin with St Benedict and an abbot releasing the souls of dead Cistercians (which were in the form of boys), as was her nightly routine.[23] There are also many sermons from the Yorkshire houses dedicated to the Virgin,[24] and innumerable other minor tracts devoted to Mary among other collections.[25] These collections, although not peculiar to the Cistercian order, demonstrate a clear interest in the intercessionary characteristics of the Virgin.

In the context of the eastern end of the Yorkshire churches, the intercessionary emphases of Cistercian devotion may be extended. Once in the monk's choir, visual attention is drawn to the east by the source of light and by the effect of space behind the presbytery. At Fountains abbey, the eyes are also drawn to the Chapel of Nine Altars, whose central chapel was dedicated to the Virgin.[26] This chapel, below the new rose window, signified the meeting of the earthly and the divine in Mary, the agent to Christ. The fact that these chapels were also functional physical sites meant that the intercession of the Virgin could be activated through formal liturgical functions, such as the saying of a personal Mass. The addition of altars to the eastern end of Cistercian churches such as Fountains during the thirteenth century suggests that the demand for intercessionary prayer was increasing, while the act of intercession itself could be tied to a particular site.[27]

Closely related to the central role of Mary in the relationship between the space of this world and the next, is the development of the doctrine of purgatory in the Cistercian milieu. This blurring of the division between heaven/earth and hell/earth was understood to signify both tribulation and redemption and became the subject of numerous *exempla* tales throughout the thirteenth century. This is

---

[22] London, British Library, Arundel MS 248, fols 154^v-55.

[23] London British Library, Add. MS 15723, fol. 88b. See also Paris, Bibliothèque Nationale, MS Lat. 15912, fols 60–61^v for a similar vision in which the Virgin circles a cloister seven times, removing drops of molten metal from boys, before explaining that the cloister is purgatory and that the boys represent the souls of dead Cistercians.

[24] For example, a thirteenth-century collection of sermons from Fountains abbey, now Oxford, Bodleian MS Lyell 8; another MS from Fountains, now Dublin, Trinity College, MS 167; the *Mariale* from Byland now Manchester, John Rylands Library, MS 153.

[25] One example is a poem from Byland dated *c.* 1269 in Manchester, John Rylands Library, MS 153, fol. 130 and transcribed in M. R. James, *A Descriptive Catalogue of the Latin MSS in the John Ryland Library at Manchester* (Manchester: Manchester University Press, 1921; repr. Munich: Kraus, 1980), 1, p. 263.

[26] See W. St John Hope, 'Fountains Abbey', *Yorkshire Archeological Journal*, 15 (1899), 269–402.

[27] For the demands on Cistercian monasteries by patrons and benefactors, see chapter 8. For the increase in petitions for private masses, see Wardrop, *Fountains Abbey and its Benefactors*. The addition of altars to the east end of the Fountains abbey church may thus be linked to liturgical necessity, rather than merely to stylistic emulation of Clairvaux, as is suggested by Draper, 'Chapel of Nine Altars'.

especially the case in the Cistercian environment, where Brian McGuire has identified what he describes as a 'special Cistercian purgatory' in the *exemplum* literature of the late-twelfth and early-thirteenth centuries.[28] One example of the linking of purgatory with the sacred space of the monastery appears in the story of an English monk. In a dream, the monk was taken to a cloister by St Benedict, in which a number of boys and a lay brother were being consumed with fire or showered with molten metal. The Virgin Mary circled the cloister, dispensing mercy in the form of cool breezes. Benedict said to the monk: 'This cloister is the purgatory of the Cistercian order. Here are its members who sinned while they lived and are now being tortured'. McGuire rightly indicates that this is not an especially novel view of purgatory in doctrinal terms. But it is a particularly Cistercian space, and it is the praying of the Cistercian order that allows Mary to provide the soothing breeze of relief.[29]

As with the general association of eastern chapels with intercessionary prayer, the development of those areas of the Cistercian church may be linked with the interest in the space of purgatory during the thirteenth century. Purgatory indicated a moving together of the spiritual and material worlds, and emphasized the agency of the individual in mediating the duration of the purgatorial experience.[30] The action of devotion was certainly emphasized in buildings such as the Chapel of Nine Altars at Fountains, where the east end of the church was demarcated as a site for individual supplication.[31]

In the brief résumé of the main building programmes given above, I have considered some of the meanings that may be signified in the expansion of the east end of Cistercian churches, and considered some elements shared by these oratories. One is the relationship between each building. We can certainly discern familial influence from Byland to Fountains, via Jervaulx and Rievaulx, although there are individual and unique features in each. Broader theological reasons for Cistercian interest in the east end of the church were also increasingly more important from the late twelfth century. These new spaces had a range of symbolic meanings for the monks who used them, particularly in terms of sight and sound. These sensory features of the eastern end of Cistercian

---

[28] McGuire, 'Purgatory', p. 76.

[29] Paris, Bibliothèque Nationale, MS Lat. 15912, fols 60–60ᵛ. The story is translated in Brian Patrick McGuire, 'The Cistercians and the Rise of the Exemplum in Early Thirteenth Century France: A Reevaluation of Paris BN MS Lat. 15912', *Classica et Medievalia*, 34 (1983), 211–67. The same manuscript tells of the fires of purgatory that afflicted a monk of Fountains abbey. See Paris, Bibliothèque Nationale, MS Lat. 15912, fol. 126ʳ. A brother called Benedict saw his own death and told Herbert, the infirmarian. After Benedict had died as he foresaw, he appeared to Herbert and told him of his experiences. Herbert asked to feel the fires of Purgatory, and when he did, his ensuing screams of pain woke all the other brethren from their sleep.

[30] McGuire, 'Purgatory', pp. 81–82.

[31] A parallel may be drawn here between the mariological emphasis of the Cistercian churches, and the eastern ends of secular cathedrals such as Durham, where the expansion of the east end of the church allowed for more expansive veneration of St Cuthbert. See Draper, 'Chapel of Nine Altars', *passim*.

churches are especially perceptible in the very Gothic features of light and elevation.

Church Iconography

Some Trends in the Interpretation of Medieval Buildings

Scholars of iconography believe that within the texts of art and architecture are sets of symbolic meanings and references. These meanings are culturally specific, although similar forms and elements are subject to diffusion and re-interpretation from one place to another and from one country to another. The intentions of patron and architect in the case of buildings have traditionally been thought to underpin the precise nature of forms—whether 'original' or imitated—although the enormous difficulties of ever being able to establish 'intention' have long been acknowledged. This much has been consistent in iconographic approaches. However, with the acceptance of many referential meanings for the same site and space and the absolute subjectivity involved in the interpreting of architectural styles and elements, iconographers have needed to rethink some of the assumptions inherited from their earlier proponents.[32]

The middle years of this century saw a flurry of activity in the field of architectural iconography, a field then devoted to the systematizing of symbolic and referential meanings in (mostly) church architecture. In a recent article on the history of architectural iconography and the interpretative limits of this discipline, Paul Crossley concluded that many of the foundational works on the subject had fallen into the trap of attempting to harness disparate and perennially shifting cultural meanings embedded in medieval church buildings, in order to 'provide an overall ideological explanation' for these structures. In other words, there is danger in attempting to discover one purpose and one meaning for buildings which, as Crossley so eloquently enthuses, are 'heaped up from untidy masses of stone, with their myriad figures and their endlessly unfolding vistas, [and which] speak to us of truths ramified, disruptive and many-layered'.[33]

The abandonment of broad and systematic analyses of the symbolic meanings of medieval architecture has been so widespread since von Simson's work on

---

[32] For some recent views on this field, see *Iconography at the Crossroads: Papers from the Colloquium Sponsored by the Index of Christian Art*, ed. by Brendan Cassidy (Princeton: Index of Christian Art, Dept. of Art and Archaeology, Princeton University, 1993).

[33] Paul Crossley, 'Medieval Architecture and Meaning: the Limits of Iconography', *Burlington Magazine*, 130: 1019 (1988), 116–21, at p. 121. For a similar view, see Eric Fernie, 'Archaeology and Architecture: Recent Developments in the Study of English Medieval Architecture', *Architectural History*, 32 (1989), 18–29. Fernie says that there is a place for iconography in the otherwise empirical studies and vice versa—'I submit that it is not useful when one or other of these modes becomes exclusive and denies the validity of the others', pp. 28–29.

the Gothic cathedral, that a recent attempt to resurrect this methodology has been hailed as original.[34] Scholarship has more recently tended to focus on the particular. This has not always involved the exclusion of iconographic methodologies altogether; rather, architectural historians and archaeologists have sought to deal with sets of meanings exclusive to particular and local contexts. Some examples of this are the interest in the relationship between liturgy and architecture, such as Peter Draper's work on Durham cited previously, or Roberta Gilchrist's study of gender and material culture in medieval nunneries.[35]

Paul Crossley considers the whole medieval world view to have been characterized by symbol and allegory, and that medieval artefacts are hence imbued with 'crucial symbolic status'.[36] Iconography, thus, is an acute critical tool for interpreting these symbols, whether or not we agree with Crossley's implication that the medieval world was unusual in its pervasive interest in the symbolic. What constitutes the symbolic is often confined to decorative elements in church buildings—the stained glass at abbot Suger's church at St Denis, or relief sculpture on cathedral facades. The reason that I focus on the trends in this discipline is that architectural iconography has traditionally and consistently utilized the medieval church as its primary focus of analysis. Our conceptions of the symbolism and form in church buildings in the thirteenth century and onwards has therefore been heavily and irrevocably influenced by the workings and assumptions of this discipline. Cistercian churches are no exception to this, especially those in Yorkshire which are so heavily indebted to nascent Gothic forms.[37]

The conventional iconography of the Christian church has been dominated by two main ideas, which are not always mutually exclusive. The first is that the church building is symbolic of the heavenly Jerusalem, or heaven on earth—with all of its attendant hierarchies. Cistercians, too, were receptive to this traditional iconography, as a thirteenth-century manuscript from Fountains abbey shows.[38] On folio 30^r of this manuscript, we find a description of the meaning of the church building which says, 'The chuch in which the people congregate to worship God signifies the holy church which is built in heaven'. This is followed by brief expositions on the symbolism of other areas of the

---

[34] J. Wirth, *L'Image médiévale. Naissance et développements (VIe–XVe siècles)* (Paris: Méridiens Klincksieck, 1989). For the response, see J. Baschet, 'Fécondité et limites d'une approche systématique de l'iconographie médiévale', *Annales ESC*, 46:2 (1991), 375–80.

[35] And more recently, Peter Draper, 'Architecture and Liturgy', pp. 83–91. See also Gilchrist, *Gender and Material Culture*.

[36] Crossley, 'Medieval Architecture and Meaning', p. 116.

[37] See Wilson, 'Cistercians as "Missionaries of Gothic"' in Norton and Park, *Cistercian Art and Architecture in the British Isles*, pp. 86–116.

[38] Cambridge, Trinity College, MS 1054. L. H. Stookey, 'The Gothic Cathdral as the Heavenly Jerusalem: Liturgical and Literary Sources', *Gesta*, 8 (1969), 35–41 (pp. 36–37) notes a hymn included in the Roman ritual for the dedication of a church that is similar to the description of the building in the Trinity MS. This would indicate that the Trinity MS is a shortened version of the well-known rite complete with its symbolism.

church such as the windows, light, columns, and other elements of the church.[39] The church as heavenly Jerusalem is traditionally thought to have derived from the book of Revelation, where the description of the city as a building has been widely quoted. At one point in the history of architectural iconography, it was even suggested that the image of the heavenly Jerusalem was so significant and pervasive during the Middle Ages that we should not interpret church buildings as mere representations or symbols of St John's revelation, but as actual copies of it.[40] The Gothic cathedral in particular has been heralded as the most representational of the church forms seeking to illustrate the heavenly city, given its soaring vaults and jewel-like windows.[41]

The second of the dominant iconographies is that of the church as the body of Christ. This derives from the common cruciform plan of western Christian churches, which are said to emulate the cross on which Christ died. Recently, Roberta Gilchrist has extended this iconographic reading of the site and shape of the church to propose that gender played an important role in architectural decisions and iconographies. In medieval English nunnery architecture, Gilchrist identifies a tendency for women to be associated with or located in the northern areas of churches in Saxon tradition and in the context of later liturgies. Women received the Eucharist, for example, in the northern part of the church. The church as metaphor for the body of the crucified Christ consolidates women's association with the north, or the right-hand side, in invoking the memory of the Virgin tending Christ's wounds, blood from which flowed from the right side of Christ's body. Gilchrist suggests a link between the northern cloister common in women's religious houses and the duality of gendered spaces in all churches, 'between north/south, female/male, moon/sun and Old Testament/New Testament'. Gilchrist bases this argument not so much on a universal model of binary opposites, but on 'an iconography of many meanings', most specific to the religious house itself.[42]

Iconographies of the Cistercian Church

So how do iconographic approaches help us to 'read' the text of the Cistercian

---

[39] Cambridge, Trinity College, MS 1054, fol. 31^v: 'Columpnae doctores sunt qui templum dei per doctrinam sicut cronum dei evangeliste spiritualiter sublevant'. And 'crux triumphalis in medio ecclesie ponitur eo quid de medio corde redemptorem suum [...] Descriptio alphabeti in pavimento est simplex doctrina fidei in cordis humanis'; fol. 32: 'Aqua est populus, sal doctrina, cinis memoria passionis Christi, vinum atque mixtum spiritum deus et homo [...]'.

[40] Rev. 21. 15–27. H. Sedlmayr, *Die Entstehung der Kathedrale* (Graz: Akademische Druck- u. Verlagsanstalt, 1976). For the critique of this view, see Crossley, 'Medieval Architecture and Meaning', pp. 118–19.

[41] E. Mâle, *The Gothic Image. Religious Art in France of the Thirteenth Century*, trans. by D. Nussey (London: Collins, 1961) and more recently, see Barbara Nolan, *The Gothic Visionary Perspective* (Princeton: Princeton University Press, 1977), esp. p. 45 et seq.

[42] Gilchrist, *Gender and Material Culture*, p. 148.

church? It is clear that the Yorkshire churches contain strong elements of conventional church design, and that they were obviously receptive to non-Cistercian stylistic innovations as well, even though the execution of those features was 'peculiarly Cistercian'. As with interpretations of other medieval church architecture, the study of Cistercian churches has been subject to certain persistent representations and readings. For a long time, the most pervasive of these was the charge of 'anti-aestheticism' levelled at Cistercian buildings.[43] More seriously, the central position given to Bernard's *Apologia* as 'the Cistercian view on art' meant that Cistercian churches were considered to be devoid of ornament as 'an ascetical reaction against beautiful things'.[44] This view has long been challenged by those who have seen the 'austerity' of Bernardine churches as being aesthetic proof of the order's purity,[45] and by those who lament that traditional approaches to the study of Cistercian church architecture have concentrated too much on the 'Cluny-Cîteaux dichotomy' at the expense of analysing the aesthetics and meanings of Cistercian church buildings.[46] The idea of austerity in Cistercian church architecture has thus been used by scholars who see the buildings as archaic examples of 'Romanesque puritanism' in an era of Gothic perfection,[47] and by scholars who consider that the simplicity of Cistercian forms is the 'art of a utopian society'.[48] Both views concentrate mainly on the decorative elements (or their absence) in Cistercian churches.

Closely related to the tradtional focus on simplicity and austerity is the understanding that Cistercian oratories were intended to be places in which the eyes could not be distracted from the practice of devotion. Even Stiegman is

---

[43] For a résumé of these critiques, see Conrad Rudolph, 'Bernard of Clairvaux's Apologia as a Description of Cluny, and the Controversy over Monastic Art', *Gesta*, 27 (1988), 125–32. For a new reading of Cistercian aesthetics, see Rüffer, *Orbis cisterciensis*.

[44] E. Stiegman, 'Analogues of the Cistercian Abbey Church', in *Medieval Studies at Minnesota 2*, ed. by A. MacLeish (St Cloud, MN: North Star Press of St Cloud, 1988), pp. 17–33. For the place of the *Apologia* in Cistercian art and architecture, see Conrad Rudolph, *The Things of Greater Importance: Bernard of Clairvaux's Apologia and the Medieval Attitude Toward Art* (Philadelphia: University of Pennsylvania Press, 1990).

[45] F. Bucher, 'Cistercian Architectural Purism', *Comparative Studies in Society and History*, 3 (1960–61), 89–105; M.-A. Dimier, 'Architecture et spiritualité cisterciennes', *Revue du Moyen Age Latin*, 3 (1947), 255–74. For a useful summary of scholarship of Bernardine aesthestics, see C. Rudolph, 'The Scholarship on Bernard's *Apologia*', *Cîteaux* 40:1–4 (1989), 69–111.

[46] Stiegman, 'Analogues'.

[47] Even medieval commentators were entranced by the new Gothic style. Gervase of Canterbury, for example, said that Romanesque buildings were 'hewn by axes, and Gothic carved with chisels'. Cited by N. Ramsay, 'Artists, Craftsmen and Design in England 1200–1400', in Alexander and Binski, eds, *Age of Chivalry. Art in Plantagenet England 1200–1400*, pp. 49–54 (p. 49.)

[48] Meredith Parsons Lillich, 'Constructing Utopia', in *Studies in Cistercian Art and Architecture II*, ed. by M. Parsons Lillich (Kalamazoo: Cistercian Publications, 1984), pp. xi–xiv at p. xiv.

careful to point out that as early as the desert fathers, austerity in the architecture of oratories was seen as desirable—both to promote 'interiorization' and to curb 'visual curiosity [...] *concupiscientia oculorum*'.[49] The moral basis on which this assumption rests is the idea of *curiositas*. The most well-known explication of the idea of *curiositas* for the Yorkshire monasteries is described by Aelred of Rievaulx in his *Speculum Caritatis*. In this tract, Aelred considers that concupiscence of the eyes 'concerns all the superfluous beauty which the eyes like in various forms'; the monk who desires these things of beauty has not learnt sufficient interior contemplation. As the apostle Paul said, 'the things that are seen are for a time, but those which are not seen are eternal.' Example of superfluous beauty, according to Aelred, are 'buildings of extravagant size and unnecessary height', 'paintings', 'sculpture', 'walls covered with purple hangings decorated with folk sagas or battles of kings', an 'awe-inspiring glow of candles' and the 'splendour of gleaming metal'.[50]

Historiographical emphasis has therefore established that Cistercian churches were non-visual in emphasis and austere in intended design. Recent scholars have challenged this notion in a number of ways. First, such a reading of Cistercian church building reiterates the old paradigm of Cistercian decline. That is, attention is given to the aesthetic intention of the building, or its 'ideal'. Where the style moves away from 'Cistercian architectural purism', then it must follow that the style is assumed to be in some way corrupted, impure and symptomatic of Cistercian decline.[51] To a large degree, the view that describes 'architectural purism' does not take into account changes in priority, function, or even sets of meanings within Cistercian buildings; it is accepted that these are static, or should be. This, then, does not account for radical changes in the fabric of Cistercian churches, changes such as those we see at Byland or Fountains. Second, historians and art historians have only recently stressed Cistercian interest in the visual. I do not agree that visual elements were unimportant in Cistercian churches, although twelfth-century sources such as Aelred do not prioritize the decorative. But even Aelred, in quoting Paul, was aware that seeing is not always done with the eyes. Seeing with the mind was one of the cornerstones of Cistercian devotion, as Stephen of Sawley's description of mental pictures to his novice shows.[52] Terryl Kinder, too, argues that Cistercian monasticism itself was broadly founded on the principle of interior knowledge, the process by which a monk needed to know himself to know the image of

---

[49] Stiegman, 'Analogues', p. 20. Stiegman also quotes Hugh of Fouilly's *De Claustro Animae*, which made a scathing attack on the superfluity of church decoration, particularly church painting: 'One should read Genesis in the book [...] not on the walls'. (p. 23).

[50] For Paul, see 2 Cor. 4. 18. For Aelred, see *Speculum Caritatis,* book 2, ch. 24, *PL* 195, col. 572.

[51] This has been noted by M. Parsons Lillich, 'Constructing Utopia', p. xi.

[52] Mikkers, 'Speculum Novitii'. See chapter 1 for more discussion of the process of *meditatio* and the formulation of mental images. See also James France, *The Cistercians in Medieval Art* (Stroud: Sutton, 1998) for the Cistercians and visual schema.

God.[53] This interior 'way of seeing' certainly holds true for the monk's experience of the church as well. Within the context of seeing, some of the functions and uses of Cistercian church space may be related to 'new' Gothic elements, which in themselves facilitated emphasis on the visual.

Within the Church

The Divine Office and the Psalmody

'When you arrive at the church', instructs Stephen of Sawley, 'place your hand on the door and say "Depart, evil thoughts, cares, intentions, affections of the heart and appetites of the body. But you, my soul, enter into the joy of your Lord that you may gaze on the loveliness of God and contemplate his holy temple"'.[54] Immediately, the novice is expectant of experiencing God in a visual way—through the gaze of the soul. The theme of sight is discussed again later in the same chapter of the *Speculum Novitii*. It is noted that 'wandering eyes are most harmful to the mind's stability', and the novice is told to focus his eyes on one place in front of himself. Yet in the same paragraph, the novice is advised to use his eyes to visualize Christ and to form a mental picture of Christ suspended on the cross: 'Move your eyes mentally all the way up to his divine heart, which houses all the treasures of his wisdom and knowledge'.[55]

In a more specific tract on the recitation of the Divine Office, in which Stephen of Sawley concentrates on the singing of the psalms at Matins, Lauds, and Prime, the themes of seeing and visualizing occur again.[56] In this treatise, Stephen responds to the request of a monk to draw up a formula for the successful recitation of the psalms. Stephen begins by pointing out that 'different methods of psalmody affect different people in different ways' and that there are a number of meanings to be found in the psalms—'mystical, moral or anagogical'. Nevertheless, there are certain themes in the psalms that anyone should recognize, such as creation, redemption, justification from mortal sin, and glorification.[57] Like the *Speculum Novitii*, the instrument advocated by Stephen to practice reflection is meditation, which he defines as a 'reflection on the facts and individual circumstances of God's saving deeds, for the purpose of arousing the mind to experience the deep affections of joy and love or—on the other hand—of fear or wonderment, longing, praise, and similar sentiments'.[58]

The formulation of mental pictures is important in this practice, and Stephen gives a long and detailed discourse on the life and passion of Christ as an

---

[53] Kinder, *L'Europe cistercienne*, p. 15.

[54] Mikkers, 'Speculum', p. 48.

[55] Mikkers, 'Speculum', p. 49.

[56] Mikkers, 'La psalmodie'.

[57] Mikkers, 'La psalmodie', p. 259.

[58] Mikkers, 'La psalmodie', p. 267.

example of what the monk should reflect on during the recitation of the Divine Office. 'Picture in your mind these events and circumstances', says Stephen. During the night office, for example, the theme of darkness is stressed. Christ was born in the middle of the night, thus the monk ought to visualize the light of the world being hidden in the darkness. Christ, according to Stephen, is 'clothed with light as with a garment'; he casts his light all the way to heaven despite the darkness of the world. Darkness in this meditation is described also in terms of the events of Christ's life—the flight into Egypt and the poverty that Christ's family was allegedly forced to suffer, the insults of the Jews 'as they taunted him about his parents' poverty, about dining with sinners, about eating meat, about loving harlots'. Stephen stresses that Jesus himself prayed at night, receiving solace from the angels, and offers incentives for the monk to do likewise, reporting that a man who had thought on Christ's life in this way while reciting psalm 88 had been so overcome with 'the wonder of unaccustomed sweetness' that 'he felt almost drunk with joy'. Stephen is clear about the significance of this event: 'I have given you this example because it is of recent origin, experienced while this was being written'.[59]

Dark deeds are also emphasized at Lauds, where the monk who is feeling 'lazy, apathetic or sleepy at that hour should ponder on the soporific apostles at Gethsemane. 'Simon, are you asleep? Why are you so lukewarm? Could you not watch with me one hour?' are the words the monk must remember along with the subsequent awfulness of Christ's arrest. Peter's denial of Jesus, the betrayal of Judas and the fleeing of John are all mentioned here. Again, visual techniques are recommended to conjure these scenes more effectively:

> When you try out these suggestions, you should visualize—as I said a long time back—the circumstances of the actual events. For instance: stare at the rising sea, that is, at the carefully and maliciously conceived persecutions inflicted by faithless people on him who never did anything evil but instead did all things well for a people pecul[i]arly his own.[60]

The pattern of visualizing the darkest hours of the life of Christ continues until Nones when the monk should consider how, at the death of Jesus, the gates of heaven opened and the 'bodies of those who slept were raised; they came into the holy city and appeared to many.' Stephen talks of the centurion who gazed at his miracle, and remarks: 'If a centurion who was still a heathen discerned this with his physical vision, how should the Christian given to spiritual meditation be affected?' The wonder of Christ's death should be related to the formation of the Church as the monk remembers that one of the soldiers 'laid open [H]is side with a lance and from it blood and water gushed forth—a living fountain in which martyrdom, baptism and the Sacrament of the Altar have their origin'. As Stephen says, 'this is how the Church was formed from the side of Christ as he died on the Cross'.[61]

It is at Compline that the relationship between the visualizing process of

---

[59] Mikkers, 'La psalmodie', p. 282–83.
[60] Mikkers, 'La psalmodie', pp. 284–85.
[61] Mikkers, 'La psalmodie', p. 287.

meditation and light is most explicitly described. First, Stephen concentrates on the enclosing of the body of Christ in 'the narrowest of tombs'. This allows the monk to reflect once more on Jesus's patience in allowing himself—'on whom principalities, powers and angels wait in heaven'—to be guarded and entombed. In the context of darkness and constriction, Stephen considers the descent into hell.

> In your mind go with his soul into hell [...] Visualize the stupor which the brilliant and unaccustomed light caused the unclean spirits [...] See what joy he flooded over the elect who had been sitting in the darkness and in the shadow of death when he brought them light so that they might see him [...] See how gloriously he led them forth [...] all that way into his wonderful light.[62]

Thus is the sight of Christ the end of darkness. And, says Stephen, 'this is why it is good to be present at the Divine Office. Even if there are times when the mind wanders, the body is always occupied with such representations'.[63] The practice of meditation and the formulation of mental pictures is inextricably tied to the experience of seeing, the ultimate reward for which is the sight of God.

Stephen of Sawley follows Bernard of Clairvaux in stressing the importance of seeing. However, where Stephen's didactic tract is concerned with the relationship between vision and visualizing in the context of monastic discipline, St Bernard's ideas on seeing are linked to the resurrection of the body. St Bernard believed that the risen body was the material body one had on earth, and that the eyes with which one sees God in heaven must be the eyes used to see on earth. In other words, it is the actual physical process of seeing that allows the Beatific vision. Stephen of Sawley's text encourages the monk to ponder on seeing God, too, but the overall idea of seeing in his treatise is more abstract. In the recitation of the Divine Office, seeing is done with the mind more than it is done with eyes, although the metaphors of light and darkness still apply.[64]

Within the Cistercian church itself, light also played an extremely significant visual role. As mentioned previously, the monks at Fountains had complained that their twelfth-century church was too dark and crowded, which is assumed to be one of the reasons for the construction of the Chapel of Nine Altars.[65] However, a more abstract relationship may be ascertained between the use of the church space for the recitation of the Divine Office and the effect of the new light-drenched eastern extensions to the church buildings of the thirteenth century. The illumination of the east end of Cistercian churches was ensured by both the size of the windows, and by the fact that light from those windows was not obscured by the darkening propensities of stained glass. Cambridge Trinity College MS 1054 tells us that the windows of the church not only prevent wind and rain from entering the building, but also illuminate the inhabitants of the

---

[62] Mikkers, 'La psalmodie', p. 287.

[63] Mikkers, 'La psalmodie', p. 288.

[64] Caroline Walker Bynum, *The Resurrection of the Body in Western Christianity 200–1336* (New York: Columbia University Press, 1995), pp. 175–76.

[65] *Mem. F*, p. 128.

church.[66] It is quite apparent that dark Yorkshire mornings would have precluded such illumination in the first Offices of the day, nonetheless, the daylight Offices—and perhaps the sight of the source of the coming light in the church—served to remind the monk of Christ, and to reflect on the words 'you are the light of the world'.[67]

Light was also a concern of the Cistercian General Chapter throughout the thirteenth century. It was decreed in 1202 that the glass in windows should be white and without pictures. Even as early as 1152, the General Chapter had legislated that there could be a lamp burning continuously in the oratory,[68] while there are several mentions of the lights to be placed on the altar or carried in processions. In 1226, for example, it was said that in processions, there were to be two lights carried before the cross. There were to be two lights on either side of the altar, but otherwise, there was not to be excessive light unless directed otherwise by the General Chapter.[69] Thus, the main source of light entering the church was natural.

It was Erwin Panofsky who most famously argued that there is a direct connection to be made between theological principles and architecture, particularly between sources of light and Christian theology. Panofsky pays specific attention to the windows and light of St Denis and their relationship to abbot Suger's interest in the Pseudo-Dionysius' 'metaphysics of light'.[70] The methodology behind this view has been challenged by those who believe that art and architecture is influenced by more than abstract theology and the intellectual elite, and by those who suspect that there is precious little real evidence to suggest that Suger himself was interested in the Pseudo-Dionysius at all.[71] Yet, in the Cistercian church, the relationship between practices within the space of the church (such as meditation on the psalms and the recitation of the Divine Office) and more complex theological ideas can certainly be established, albeit rather differently to Panofsky's model. First, the relationship between the light of the church and Cistercian theology is to be found in the use of that space by individual monks, rather than in the uncovering of theological motives in the

---

[66] Fol. 31ᵛ: 'Fenestrae ecclesie vitrii sunt scripture divine que ventum et pluviam repelletur. Id est lasciviam prohibent et dum claritatem inde solis in ecclesiam transmittunt inhabitantes illuminant'. On the same folio, the scribe reminds us that 'luminaria ecclesie sunt illa quorum doctrina fulget ecclesia [...]'.

[67] Cambridge Trinity College MS 1054, fol. 30ᵛ. It should be noted, however, that in the late thirteenth century, there is evidence that coloured glass was occasionally in place in Cistercian houses—particularly in Germany, but to a lesser extent in England. See R. Marks, 'Cistercian Window Glass in England and Wales', in Norton and Park, eds., *Cistercian Art and Architecture in the British Isles*, pp. 211–27.

[68] B. Lucet, *La codification cistercienne de 1202 et son évolution ultérieure* (Rome: Editiones cistercienses, 1964): Dist. 1.5; *Statuta* 1152, 5, t. 1.

[69] *Statuta* 1226, 1, t. 2; *Statuta* 1270, 11, t. 3.

[70] Panofsky, *Gothic Architecture and Scholasticism*.

[71] Crossley, 'Medieval Architecture and Meaning', p. 120; Peter Kidson, 'Panofsky, Suger and St Denis', *Journal of the Warburg and Courtauld Institutes*, 50 (1987), 1–17.

building programmes themselves, as Panofsky's analysis had implied. Stephen of Sawley's text, for example, establishes that the practice of devotion within the church should be directed at themes of light and vision and that being illuminated by the light of the church is a bodily and experiential reminder of these themes. Images of Christ's life and Passion created in the novice's mind are therefore reinforced by the effect of the space around him. This relationship between the place of devotion and the mental spaces created by the monk—as expressed by Stephen of Sawley—cannot be solely attributed to the intentions of architects under the influence of theological ideas. Rather, the relationship only exists if it is constructed by the people who use the physical space—in this case, Cistercian monks. In the case of thirteenth-century English Cistercians, the light and space of the church was a place in which images of Christ could be inscribed and made present.

Singing and the Effect of Sound

These devotional and imaginative practices were not only visual, however, but were also effected by sound. Stephen of Sawley recognizes that 'your mind must be in harmony with your voice' when reciting the Divine Office.[72] This suggests that the monk is to be aware of external factors in the creation of his mental pictures. Sound, and especially singing, were important in reinforcing the interior vision of Christ—as St Bernard had said, music 'should caress the ear so that it might move the heart'.[73] As with the unadorned nature of Cistercian windows and the uninterrupted white light they allowed into the church, Cistercian singing was also to be plain and simple. The General Chapter legislated in 1202 that men ought to sing in manly voices and not in the shrill way of women, with 'false voices'.[74] This view was upheld at Rievaulx abbey by Aelred during the twelfth century, in the context of '*curiositas* of the ears':

> Where do all these organs in the church come from, all these chimes? To what purpose, I ask you, is the terrible snorting of bellows, more like a clap of thunder than the sweetness of a voice? Why that swelling and swooping of the voice? One person sings bass, another sings alto, yet another sings soprano. Still another ornaments and trills up and down on the melody. At one moment the voice strains, the next it wanes. First it speeds up, then it slows down with all manner of sounds [...] Sometimes you see a man with his mouth open as if he were breathing out his last breath, not singing but threatening silence, as it were, by ridiculous interruption of the melody into snatches. Now he imitates the agonies of the dying or the swooning of persons in pain. In the meantime, his whole body is violently agitated by histrionic gesticulations—contorted lips, rolling eyes, hunching shoulders—and drumming fingers keep time with every single note. And this ridiculous dissipation is called religious observance![75]

---

[72] Mikkers, 'La psalmodie', p. 260.

[73] Cited by Chadd, 'Liturgical Music and the Limits of Uniformity', p. 305.

[74] Lucet, *La codification cistercienne de 1202*.

[75] Aelred Of Rievaulx, *Speculum Caritatis*, book 2, ch. 24, *PL* 195, col. 571.

Aelred's opinions on music indicate the gravity with which the monk is expected to use his voice, as well as the importance of singing as part of the practice of Cistercian devotion. These ideas are to be found in some later comments on music and singing from the same abbey, this time from the monk in charge of the music there in the early thirteenth century—the precentor, Matthew of Rievaulx.

The sole recension of Matthew of Rievaulx's poems and letters attests among other things to the author's musical career and position of precentor.[76] That Matthew often felt weighed down by the responsibilities associated with this office is clear from a number of his writings.[77] Nevertheless, Matthew also managed to write a number of tracts devoted to music—a poem *De disciplina psallendi*, a hymn to St Stephen, a hymn to John the Evangelist, a hymn to John the Baptist, and a hymn to the apostles Peter and Paul. I shall concentrate on the first of these, the *De disciplina Psallendi*. The poem stresses both the importance of singing and the importance of harmony. The title refers to singing in general, but may also have the specific meaning of chanting the psalms.[78]

The poem begins by immediately relating the sound of singing in the church to the celestial choirs of heaven, describing the heavenly chorus of cherubim and seraphim as instructing monks on earth.[79] Matthew cites Moses as the instigator of singing the Lord's praises, and also mentions Aurelian, whose *Musica disciplina* contained a number of strictures relating to psalm singing.[80] The sound and effect of singing is important to Matthew; he mentions sweet melody and stresses the choral elements of the music. This latter point is important in that Matthew relates harmony not only to the tune or melody that is being sung, but also to the result of that singing. If everyone sings together, there can be no discord and there will be jubilation if the choir is in complete harmony.[81] More significantly, singing will help the monk move from the shadows into the kingdom of heaven. Psalm singing in general, according to Matthew, should be the sound accompanying all stages of life: 'habet in psalmis quod lactet puer, quod laudet adolescens, quod corrigat vitam iuvenis, quod sequatur senex'.

---

[76] Paris, Bibliothèque Nationale, MS Lat. 15157.

[77] See Wilmart, 'Mélanges', pp. 27–28 for three examples of Matthew's weariness.

[78] R. E. Latham, *Revised Medieval Latin Word-List* (London, The British Academy: Oxford University Press, 1965), p. 380 gives *psallo* as 'to make music, hymn or chant'; J. F. Niermeyer, *Mediae Latinitatis Lexicon Minus: Abbreviationes et Index Fontium* (Leiden: Brill, 1993) at p. 869 gives *psallere* as 'to sing psalms'.

[79] For other examples of this simile, see C. E. Schilla, 'Meaning and the Cluny Capitals: Music as Metaphor', *Gesta*, 27 (1988), 133–48 (pp. 133–34).

[80] 'Aurelius dicet quod psalmorum melodia/ Balsama spirat odor thus fragrans [...]'. For Aurelian on the psalms, see J. Dyer, 'Monastic Psalmody of the Middle Ages', *Revue Bénédictine*, 99 (1989), 41–74 (p. 57). Bernard of Clairvaux is also noted in the margin. The Aurelian tract is edited in M. Gerbert, *Scriptores Ecclesiastici de Musica*, (Hildesheim, 1963), I, p. 130.

[81] 'Sic aliis concors, ne sis simphonia discors'; 'Consona tota choors et sit jubilatio concors/ Laus est grata Deo superis iocunda choreis'.

Matthew of Rievaulx's poem is thus conventional in its iconography, stressing the necessity of togetherness in singing and relating the choir of monks to the angelic choirs in heaven. It is also a celebration of sound in an otherwise quiet world. However, some more general conclusions may be drawn from Matthew's emphasis on the aural merging of the space of heaven and the singing in church. The relationship between architecture and music in a Cistercian context has been explored mainly by Otto von Simson, whose interest was in Augustinian principles of modulation and their effect in establishing the proportions of Cistercian churches such as Fontenay.[82] Simson argues that St Bernard's musicality and interest in Augustine were highly influential in the geometrically 'pristine' architecture of the twelfth century. In relation to the Yorkshire Cistercian churches, it is attractive although speculative to contend that the elevation of the eastern parts of the buildings from the choir and beyond added to the already sharp acoustics of the church.[83] Whether or not the raising of voices to heaven was a factor in the design of the eastern elevations of Rievaulx and Fountains is unknown. However, the amplification of sound must certainly have been an effect of the new buildings.

The Cistercian church was one site in which ideas of transcending the earthly were expressed. Although historians and art historians traditionally tended to view the non-decorative nature of Cistercian churches as evidence of an antivisual aesthetic, it is clear that Cistercians themselves experienced the church environment in terms of seeing, visualizing and imagining. There is a range of possible interpretations of Cistercian church space during the late-twelfth and early-thirteenth century, especially in relation to the eastern extensions to some of those churches in this period. I have stressed both the use of the space and the sets of theological and symbolic meanings inscribed in its use and potential definitions. The effects and meanings of the light within the church and the sound of the devotions conducted there, are especially significant features of Cistercian texts from the Yorkshire houses.

However, some points still remain. First, it must be remembered that the east end of the Cistercian church was occupied solely by choir monks; lay brothers were segregated from the choir by a screen. This meant that the visual effect of the light and symbolic eastern end of the church was not as immediately apparent to the inhabitants of the western end of the building. The hierarchical nature of the demarcated church layout is reflected in the hierarchical nature of the Church itself. Second, the possibility of transcending earthly space, as represented in the eastern end of the church, was not a clear and inevitable progression. Rather, it was a progression to be negotiated, as the threat of purgatory indicates. Although the emphasis in devotional uses of the Cistercian church may be perceived in terms of transcendence, it is also important to remember that themes of hierarchy, trial and judgement were also present in the

---

[82] Von Simson, 'The Cistercian Contribution'.

[83] There has been some suggestion that the monks used a sound enhancing system comprising pottery urns set into the base of the choir stalls to amplify the monks' voices. See A. Stock, 'A Sounding Vase at Fountains Abbey?', *Cistercian Studies*, 23 (1988), 190–91.

practice of Cistercian devotion. Ideas of transcendence and freedom from the confines of physical space thus existed alongside regulation and discipline in the Cistercian monastery—the realities of material space in an earthly world. Heavenly landscapes and the geography of eternity were simultaneously facilitated and tempered by terrestrial monastic concerns.

CHAPTER 4

Community, Discipline and the Body: The Cistercian Chapter House

The Chapter House and its Purpose

According to Hélinand of Froidmont, the Cistercian chapter house was the holiest and most sacred part of the monastery, although an early reference to the buildings of an abbey does not mention the chapter house at all.¹ The *Ecclesiastica Officia*, however, indicates that the chapter house was a site of diverse and significant functions, ranging from the liturgical to the administrative.² Functional similarities between the Cistercian church and the chapter house have also been noted at Rievaulx abbey chapter house, where the connection between church and chapter house was strong.³ This may also be seen in the inclusion of an altar within the chapter house that was used occasionally for the celebration of mass. More specifically, the sacred nature of the chapter house may be seen in the liturgical use of the site.

After the morning mass, the prior rang the bell to summon the brothers to the

---

¹ Hélinand of Froidmont, *Epistola ad Galterum*, PL 212, col. 758: 'Nullus locus sit sanctior capitulo, nullus reverentia dignior, nullus diabolo remotior, nullus Deo proximor'; *Statuta* 1134, 12, t. 1: 'Duodecim monachi cum abbate terciodecimo ad coenobia nova transmittantur: nec tamen illuc destinentur donec locus [...] domibus et necessariis aptetur [...] domibusque oratorio, refectorio, dormitorio, cella hospitum et portarii, necessariis etiam temporalibus'.

² Griesser, 'Ecclesiastica Officia', pp. 234–77. For a comparison with Augustinian chapter houses, see S. Bonde, E. Boyden, and C. Maines, 'Centrality and Community: Liturgy and Gothic Chapter Room Design at the Augustinian Abbey of Saint Jean-des-Vignes, Soissons', *Gesta,* 29:2 (1990), 189–213.

³ Fergusson and Harrison, 'Rievaulx Abbey Chapter House'.

chapter house, to which they processed in the same order as they had gone into the church. Coming into the chapter house, the community bowed to the abbot. The *Ecclesiastica Officia* tells us that the reader then stood to begin the reading of the martyrology, bowing at the benediction. After the first reading, the monks rose and faced the east reciting the verse in the collect—the *Pretiosa in Conspectu Domini*. This verse was begun by one of the community, who said the *Gloria Patri* and prostrated himself at the seats of his superiors to rise to his knees again at the start of the prior's recitation of the *Kyrie*. The rest of the community were then either to prostrate themselves or bow, depending on where they were seated, and when the priest declared, 'And lead us not into temptation', the monks again rose, bowed, and knelt once more. At the conclusion of the *Pater Noster* the same routine was followed. Reading of the *Regula S. Benedicti* and the announcement of general and administrative matters then ensued, the *lector* reading from the *tabula* usually kept in the cloister. An office for the commemoration of the dead then preceded the hearing of transgressions committed by members of the community, together with their correction and punishment. I shall return to this portion of the chapter house liturgy shortly.[4]

The *Ecclesiastica Officia* also reveals that the chapter house was the place in which important visitors were received, including bishops, abbots of other orders or even kings.[5] When these dignitaries arrived in the chapter house, the community was to rise and bow, as was ordered if the visitor addressed the community. These visitors, it seems, did not generally stay for the duration of the chapter meeting, although if the abbot wished them to be present it was not forbidden that outsiders could remain. The prior or another presumably high-ranking member of the order was in charge of looking after these guests, but if a visitor was of a lower status, such as a monk, cleric, or lay person, a brother was to lead them out of the chapter house, rather than the prior.[6] Otherwise, the reception was similar to that ordered for visitors of a higher status. Other English monastic chapter houses were also used for the reception of visitors and for the carrying out of business. Perhaps the most notable example was at Westminster abbey, where the Benedictine polygonal chapter house was used for the King's council of feudal magnates from the thirteenth century, the first recorded secular meeting taking place here in 1257.[7]

At the conclusion of chapter, the *Ecclesiastica Officia* indicates that everyone stood and faced eastwards for the concluding prayers. After these had been uttered, the community then collectively bowed to the east and left the building.[8]

---

[4] For the preceding, see Griesser, 'Ecclesiastica Officia', pp. 234–35.

[5] Griesser, 'Ecclesiastica Officia', p. 237. That kings did indeed visit Cistercian houses may be seen in the *Calendar of Close Rolls* 1244, pp. 222–23, where it is recorded that Henry III sent a gift of wine to Fountains abbey after he had made a visit there. This is but one of numerous examples of royal visits.

[6] Griesser, 'Ecclesiastica Officia', p. 234.

[7] S. E. Rigold, *The Chapter House and the Pyx Chamber, Westminster Abbey* (London: HMSO, 1988), p. 5.

[8] Griesser, 'Ecclesiastica Officia', p. 237.

As outlined in the previous chapter, the emphasis on eastern orientation was a particularly meaningful one not only for Cistercians, but for all Christians. In the chapter house, facing the east also meant facing the abbot, whose pulpit stood at the eastern end of the building—as the altar stood at the eastern end of the church. In non-Cistercian houses, the eastern wall was often a very decorated site. At Westminster, for example, a painting of the Last Judgment covered the wall behind the abbot's seat.

The nature of the chapter house building itself varied during the thirteenth century, even in Cistercian abbeys. Usually a rectangular structure in monasteries, more secular sites introduced polygonal chapter houses especially over the course of the thirteenth century (although a much earlier case is to be found at Worcester c. 1120).[9] There are some monastic examples of the polygonal type—Westminster, as mentioned, Cockersand, and two Cistercian examples at Abbey Dore[10] and Margam in Wales. In Yorkshire, Cistercian chapter houses remained a more conservative rectangular shape, although the Rievaulx site shows a hemicyclic apse (Fig. 4.1), and all were mainly completed by the end of the twelfth century. The chapter houses are situated on the eastern range of the cloister, a position that has been described by Roberta Gilchrist as the most inaccessible part of the monastery for outside visitors.[11] Entrance to the chapter houses was via the cloister door, while the interior floor was often below cloister level, as at Jervaulx, Byland, and Rievaulx. The height of the twelfth-century buildings with their vaulted ceilings meant that the chapter houses would have been disproportionately higher than the other east range buildings if the floors were not lowered. The vaulting of the ceilings has been noted by Glyn Coppack, who identifies this design feature as often the earliest example of Gothic influence in northern Cistercian buildings, an indication, perhaps, that in the second round of building programmes undertaken by almost all the Yorkshire Cistercian houses in the later twelfth century, the remodelling of the chapter house was a priority. One exception to this is the chapter house at Kirkstall, which remained a rather gloomy space until a fourteenth-century eastern addition opened up the space to more light.[12]

---

[9] For this general information on the chapter house and its architectural development, see George H. Cook, *English Monasteries in the Middle Ages* (London: Phoenix House, 1961); M.-A. Dimier, *Les moines bâtisseurs: architecture et vie monastique* (Paris, 1964); Kinder, *L'Europe cistercienne*. For an account of comparable French buildings, see B. Beck, 'Les salles capitulaires des abbayes de Normandie, et notamment dans les diocèses d'Avranches, Bayeux et Coutances', *L'Information d'Histoire de l'Art*, 18 (1973), 204–15.

[10] See Joe Hillaby, '"The House of Houses": The Cistercians of Dore and the Origins of the Polygonal Chapter House', *Transactions of the Woolhope Naturalists' Field Club*, 46:2 (1989), 209–45.

[11] Gilchrist, *Gender and Material Culture*, p. 166; Gilchrist, 'Community and Self', 55–64.

[12] This building may provide some hint as to the nature of the first chapter house at Fountains, which Coppack thinks must have been a 'low, dark room for its ceiling was

The interior layout of the chapter house buildings followed similar patterns in all the Cistercian houses. Seating in the form of tiered steps—amphitheatre style—was provided around the perimeter of the building (Fig. 4.2), while, as mentioned previously, the abbot's place was at the eastern end. The lectern used for the recitation of the martyrology and other readings was also at the eastern end of the chapter house.[13] Other crucial features of these buildings included the tombs of abbots, and in some Cistercian houses, shrines to abbots. An example of the latter is the shrine to abbot William at Rievaulx abbey, which occupies the arch of the north window to the side of the main doorway into the chapter house, and was completed in 1250 (Fig. 4.3).[14] The whole chapter house was generally aisled—Fountains, Jervaulx, and Byland contained three aisles while the chapter house at Roche abbey had two aisles. The piers along the aisles supported the vaulted roof in these buildings, and at Fountains, for example, were almost decorative in the use of mouldings and leaf capitals.

The physical layout of the Cisterian chapter house was thus designed to facilitate a myriad of liturgical, administrative, domestic, and devotional practices. The site was consequently both a public one, in the sense that the entire community of choir monks contributed to and witnessed the rituals and events that occurred in the building. Yet the chapter house could also be interpreted as one of the most 'private' of monastic sites. Although, as previously mentioned, visitors were received in the building, the spatial functions of the chapter house were also extremely intimate. It was in the chapter house that Cistercian monks were reminded daily of their reforming agenda to uphold the simplicity and austerity of monastic life as expressed in the Regula S. Benedicti, and it was in the chapter house that issues relating to the sanctity of the community—the monastic 'body corporate'—were iterated and reiterated via bureaucratic procedures, rituals of remembrance, and rituals of accusation, confession, and punishment. In this way, the chapter house might be understood to function in part as a disciplinary space—one of the sites where the monastic community (as well as the individual monastic body) was constructed.

Ideas of disciplinary space within the chapter house should be understood in the broadest sense. Although actual and physical punishment in the form of beating during the chapter house liturgy was described by Cistercians and by other monks as *accipians disciplina*, we must look at more than instances of bodily punishment in conceptualizing the nature of space of the chapter house. To a limited degree, Michel Foucault's understanding of discipline as a productive force is useful here.[15] In *Discipline and Punish* Foucault understood discipline in the context of power relations between carefully delineated groups

---

the wooden floor of the dormitory above that was only 2.14m (7ft) above ground level'. See Coppack, *Fountains Abbey*, p. 30.

[13] The base of one of these lecterns survives at Byland abbey, now in the site museum.

[14] For the importance of commemoration in the building programme at Rievaulx, see Fergusson and Harrison, *Rievaulx Abbey: Community, Architecture, Memory*, pp. 83–101.

[15] Foucault, *Discipline and Punish*.

Fig. 4.1: Rievaulx abbey chapter house, showing hemicyclic apse (Photo, author)

Fig. 4.2: Rievaulx abbey, tiered seating in the chapter house (Photo, author)

Fig. 4.3: Rievaulx abbey, shrine to abbot William in the chapter house (Photo, author)

of people. Whereas we may assume that discipline—whether it be in the form of law, sets of rules, or in the medieval monastic sense of physical subordination—is in some way repressive or confining, Foucault argued that discipline is productive. That is, discipline and disciplinary practices are the means by which subjects are created to fit their institutional environment.

Extending Foucault's paradigm in the context of the Cistercian monastery, disciplinary practices might therefore include any ritual which stresses, enables or enforces institutional solidarity and conformity. Like the space of eternity imagined and enacted during processions around the monastic cloister, the site of the chapter house functioned as a site in which community was imagined and articulated through the disciplinary practices of word, prayer, and deed. Thus, the notion of discipline as a productive tool in the creation of meaningful spaces is useful in relating chapter house practices to wider Cistercian cultural principles.

Community

Creating Community

In 1132, Archbishop Thurstan of York arrived at St Mary's abbey, where he and his entourage attempted to enter the chapter house in response to a request from Prior Richard. The abbot of St Mary's, however, met the archbishop at the door of the building and refused Thurstan entry. It was not lawful, argued the abbot, that Thurstan should visit them with so great a retinue, or that a secular like the archbishop and members of his party should be present 'at the secret meetings of the chapter'. The archbishop himself could be permitted to enter alone, offered the abbot, 'so that the discipline of the order might not be disturbed by the insolence of clerics'.[16] It was from the ensuing riot, in which the archbishop's party tried to force its way ito the chapter house while the monks inside attempted to block their entrance, that the thirteen monks who eventually founded Fountains abbey were persuaded to follow Thurstan and leave St Mary's forever. Thus, the chapter house provided the catalytic site for the dispersal of one community, at least in part, and the creation of another. That the articulation of community in this space was a perpetual process can be seen in the direction of chapter house ritual outlined in the *Ecclesiastica Officia*.

The *Ecclesiastica Officia* states that part of the *Regula S. Benedicti* should be read after the preliminary prayers signalling the beginning of the chapter house meeting, while Michael Casey has pointed out that the primary theme running throughout many other early Cistercian documents such as the *Carta Caritatis* is 'fidelity to the Rule of St Benedict'.[17] It is a commonplace that Cistercian mo-

---

[16] *Mem. F.*, pp. 8–9.

[17] Michael Casey, 'The Dialectic of Solitude and Communion in Cistercian Communities', *Cistercian Studies Quarterly*, 23 (1988), 273–309 (p. 296).

nasticism was formulated with the desire to return to the simplicity of cenobitic life expressed in the *Regula S. Benedicti,* and this desire is recorded in a number of the later foundation histories of Cistercian houses, including that of Fountains abbey.[18] The cenobitic premise on which the *Regula* itself is based was of course a fundamental part of Cistercian life. In terms of the space of chapter house ritual in particular, we may see that the reading of the *Regula* was one way in which the obligations of each individual monk, as well as notions of community were publicly reinforced.

The reading of the *Regula* was primarily important in memorial terms, as a reminder to each monk that they were part of a cenobitic community devoted to upholding the tenets of the text. The *Regula* itself, as Mary Carruthers argues, was not meant to be read in the same way as other texts that made up the *lectio divina*. Rather, the *Regula*'s own textual references to ears not eyes, indicate that the practice of hearing the *Regula* aloud was an important part of remembering not only the words themselves, but also the connotations that they held. Thus, mnemonic tropes like that of Jacob's Ladder—architectonic in both form and signification—helped the monk to recall the steps of humility, argues Carruthers.[19] The *Ecclesiastica Officia* provides no explicit instruction as to how the *Regula* is to be interpreted or remembered in this way; yet the very position of the *Regula* in the daily ritual reveals that it is this reading that connects the overall memorial and commemorative narrative of the chapter house liturgy.

A common Cistercian purpose is stressed in the reading of the *Regula*, a purpose that is activated by shared listening and shared mnemonic practice. Like the practice of *meditatio* advocated by the author of the *Speculum Novitii*, the routine of hearing and remembering the *Regula S. Benedicti* functioned as a communally shared reiteration of the Cistercian body corporate. Remembrance of a shared tradition of cenobitic monasticism was symbolized by the public reading of the *Regula* in the most private space of the monastery—the chapter house. In this sense, we might perceive the reading of the *Regula* as possessing both sociological and spiritual functions. The activation of collective memorial practices served as one way in which the Cistercian community could position itself in the past, present, and future: as the repository of cenobitic tradition, as the immediate audience to these monastic values and as the community in which the eternal future could be anticipated. Again, parallels can be made with the *Speculum Novitii*, which stresses the eschatological dimension of monastic life in its formulation of collective memory. Likewise, other rituals such as the *Benedictus aquae* procession around the cloister used the material site of the monastery to express the endlessness of eschatological time and space. This introductory part of the daily chapter house rite immediately positioned each monk within an affirming, memorial, and commemorative space.

---

[18] *Mem. F.*, p. 12, for instance, where Thurstan's letter talks of the desire of the thirteen monks who left St Mary's to return to strict obedience to the *Regula*.

[19] Carruthers, *The Book of Memory*, p. 27. See also E. Manning, 'La Règle de S. Benoît', p. 227, 'De humilitate'.

Forms of Commemoration

In terms of discipline, as mentioned previously, the reading of the *Regula* was one of many rituals or practices in Cistercian life that marked out the boundaries of those within the community and those without it. The shared experience of listening to the *Regula* in the chapter house each day not only declared community; but it created or produced community, too. The communal agenda of this first part of the chapter house ritual was then extended by further articulations of commemoration, beginning with the reading of the necrology—an office for the commemoration of the dead. The *Ecclesiastica Officia* tells us that after the reading of the *Regula S. Benedicti*, whoever was the lector for the day would then read out commemorations for 'our brothers, fathers, benefactors and all the faithful departed', while this was followed by the response 'May they rest in peace'. Readings were then delivered and the commemoration of the dead ended this element of the liturgy.[20] At this point, the liturgy still focused on members of the Cistercian community, underscoring, perhaps, the ease with which the dead found themselves still part of the landscape of the living. The written and spoken word was not the only way in which the chapter house symbolized a particularly Cistercian commemorative space. The building itself was important in this way, too.

Studies of Cistercian chapter houses have consistently emphasized the very visible nature of abbatial authority in these buildings. Joe Hillaby, for example, in his work on the polygonal chapter houses at Abbey Dore and Margam, found that 'the personality of their respective abbots [...] provided the context within which they could exercise their ambitions'.[21] Hillaby decided that the forceful personality of Abbot Adam (c. 1186–1215) was the main factor contributing to the use of the non-Cistercian polygonal design at Dore, and that the growth of shrines to abbots in the thirteenth century reflected a developing interest in 'cults of personality' within Cistercian houses. Peter Fergusson and Stuart Harrison, too, stressed the influence of Aelred in the chapter house renovations of the mid-twelfth century at Rievaulx,[22] arguing that Aelred's interest in the cult of saints and the veneration of the founder abbot of Rievaulx—William—led to the remodelling of the Rievaulx chapter house. Other expressions of abbatial influence in the chapter houses of Yorkshire may be found, however. Perhaps the most evident of these is the use of the chapter house as a site for abbatial burial.

Abbatial tombs in Cistercian chapter houses are common. At Fountains abbey, for example, the chronicle of the abbots records that a great proportion of the thirty-eight heads of this house were interred in the chapter house.[23] Roy

---

[20] Griesser, 'Ecclesiastica Officia', p. 235. See also J. Laurent, 'La prière pour les défunts dans l'ordre de Cîteaux', *Mélanges St Bernard*, 1953, pp. 391–92.

[21] Hillaby, 'House of Houses', p. 227.

[22] Fergusson and Harrison, *Rievaulx Abbey*, p. 99.

[23] *Chronica Abbatum Fontanensium, Mem. F.*, pp. 132–40. For the iconography of abbatial tombs, see L. Butler, 'Cistercian Abbots' Tombs and Abbey Seals', in *Studies in Cistercian Art and Architecture* 4, ed. by M. Parsons Lillich (Kalamazoo: Cistercian

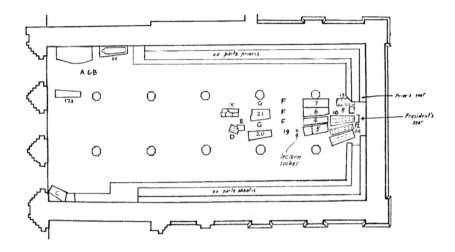

Fig. 4.4: Fountains abbey, abbatial tombs in the chapter house (Plan from R. Gilyard-Beer, 'The Graves of the Abbots of Fountains', Yorkshire Archaeological Journal, 59 (1987), 45–50 at 50 (Courtesy, The Yorkshire Archaeological Society)

Gilyard-Beer has listed the ten abbots not buried at the abbey, including Stephen of Sawley, who died at Vaudey while on visitation and was buried in the chapter house there.[24] Nineteen abbots in total were buried in the chapter house at Fountains, while the remainder were interred in the church. The position of the graves in the chapter house is given in Gilyard-Beer's diagram (Fig. 4.4). That tombs were a very prominent part of the chapter house design may be seen in the positioning of these burial plots, which were generally at the eastern end of the building, in close proximity to the area that the abbots occupied in life. The same arrangement of graves is to be found at other sites. At Rievaulx, there are three identifiable abbots' grave slabs out of the ten in the chapter house—those of the sixth abbot, William, the twenty-second abbot, Peter, and the twenty-

---

Publications, 1993), pp. 78–88. As the thirteenth century wore on, lay burials were introduced into chapter houses. I will discuss these in the final chapter of this book.

[24] R. Gilyard-Beer, 'The Graves of the Abbots of Fountains', *Yorkshire Archaeological Journal*, 59 (1987), 45–50; *Mem. F.*, p. 138: 'Sepultus est coram sede presidentis, in capitulo monasterii Vallis Dei; ubi miraculis choruscat'. Abbot Robert of Pipewell (ob. 1180) also died while visiting another house; however, his body was brought back to Fountains from Woburn and buried in the chapter house.

Community, Discipline, and the Body 115

Fig. 4.5: Byland abbey, tomb in the chapter house (Photo, author)

seventh abbot, John.[25] At Byland, only one of the original tomb covers remains, although in very good condition (Fig. 4.5).[26] This is also at the eastern end of the chapter house building.

As visual memorials, graves in the chapter house provided a significant link between the authority of the abbot in life and in death. The abbots' continued

---

[25] Glyn Coppack and Peter Fergusson, *Rievaulx Abbey* (London: English Heritage, 1994), p. 19.

[26] The others having been stolen. See Harrison, 'The Architecture of Byland Abbey', p. 40.

occupation of the eastern end of the chapter house building even after death, underscored the pervasive and eternal nature of their authority. This was further extended by the use of the chapter house in some Cistercian houses, such as Rievaulx, as sites for shrines to abbots. The shrine to abbot William at Rievaulx served to perpetuate the memory of William himself, as well as functioning as a physical reminder of the importance of the office of abbot more generally. Matthew of Rievaulx's poem *De Willelmo Primo Abbate Rievallis* demonstrates the veneration that the office of abbot could encourage: even in death, William continued to bless the Cistercians of Rievaulx abbey with spiritual guidance.[27]

The link between the material landscape of the chapter house and the commemorative liturgies performed within it were made visible by the presence of abbatial tombs within the building itself. Ideas of community premised on memorialization which were articulated in the liturgical functions of the chapter house could be visibly linked with the sites of authority and the community of the Cistercian dead who lurked underfoot and who were named and made present in word. Such rituals of commemoration within the chapter house's landscape of remembrance were immediately followed in the daily liturgy by a sequence of practices directed at the living community. This was expressed primarily in terms of the practice of discipline.

Discipline

Accusation

The ritual of accusation, confession and judgment in the chapter house followed the commemoration of the dead, according to the *Ecclesiastica Officia*.[28] The procedure began with the naming of the accused and a short and direct statement of his transgression by the accuser: 'He did this'. On hearing his name, the accused was not to respond to the charges in his seat, but was directed to come forth to be interrogated by the person in charge of the chapter that day. Like the statement of accusation, the interrogation was simple: 'Quid dicis?', said the interrogator, while the accused responded by prostrating himself in the centre of the chapter room floor, saying, 'Mea culpa'. None of the community was allowed to speak to the monk prostrated on the ground. The accused monk was then to confess his sins, receive his penalty, and finally return to his seat. The *Ecclesiastica Officia* provides other guidelines for the process and nature of accusation in the chapter house. It was ruled that an accused monk could not then call his accuser to answer a charge himself that day. It was also ruled that all accusations should be dealt with on the spot, that if someone did not comprehend the charge levelled at him then it would not be repeated, and that no-one

---

[27] Wilmart, 'Mélanges', p. 55: 'Anglia te genuit; vallis clarissima mores/ Sacros instituit; tua nos doctrina beavit'.

[28] For the following, see Griesser, 'Ecclesiastica Officia', pp. 235–36.

should accuse another on the basis of mere suspicion—all accusations had to be supported by visual or aural evidence.

Two preliminary questions arise from this portion of the ritual. What sort of offences were considered in the chapter house? And how did the monks themselves both understand accusation and respond to it? Thirteenth-century legislation, especially the codifications of that period, provides some insight into the sorts of offences considered to be serious by the Cistercian General Chapter. The legislation distinguished between what was considered to be enough to establish the state of 'light guilt' for less serious transgressions ('levi culpa') and 'gravi culpa' or serious sins. An example of the former is speaking during the long periods of silence; an example of the latter is homicide.[29] It is not always clear, however, as to why some crimes should be considered more grievous than others. For example, the codifications of 1237 and 1257 give the penalty for sorcery ('sortilegia')—which we may expect to be treated extremely seriously in a monastic environment—as being only six days in 'levi culpa' for choir monks and *conversi*, although abbots, priors, and subpriors were to be deposed.[30] There also seems to have been a distinction between transgression that harmed other members of the community and transgression that harmed outsiders, even though the actual crime may have been the same. One example of this is homicide, where exile was the penalty for a monk or lay brother who murdered a secular, but confinement within the monastery seems to have been the penalty for killing another monk. Violence toward other monks and clerics, breaking the silence, allowing secular people into the monastery, harbouring fugitives, and theft were all concerns of the General Chapter during the thirteenth century and it was during the daily chapter that individual houses were to deal with these problems. It was also important that new legislation was publicized, and the chapter house provided a communal forum in which this could be done. That this publicization was expected may be inferred from the General Chapter's view that ignorance of a regulation was not a sufficiently convincing defence for a monk accused of a crime.[31]

In the *Speculum Novitii*, the author talks about responses to accusation in the chapter house, particularly from the point of view of the accused. The text begins by urging the novice to be prepared for harshness in chapter. The novice ought to 'arm' himself—not with rebuttals or denials, but with 'the armour of God', 'the helmet of providence, the breastplate of patience and the shield of mercy'. The novice should use these virtues as defence against both just and unjust accusations.[32] From these preliminary observations, the novice is

---

[29] For these crimes, see B. Lucet, *Les codifications cisterciennes de 1237 et de 1257* (Paris: Editions du Centre National de la Recherche Scientifique, 1977), pp. 120–30.

[30] Lucet, *Codifications*, p. 277. The General Chapter of 1291 reaffirmed the punishment and extended it to include that whoever was convicted of sorcery was to be last in all things. See *Statuta* 1291, 3, t. 3.

[31] *Statuta* 1204, 8, t. 1: 'De libello definitiorum praecipitum est a generale capitulo ut ab omnibus quam citius potuerunt, habeatur, ut nullus abbatum de cetero ignorantia se excuset'. This ruling was reiterated in *Statuta* 1240, 13, t. 2.

[32] Mikkers, 'Speculum', pp. 54–55.

immediately aware not only that many elements of the chapter of faults are likely to be severe, but they may also be seemingly unfair. That unfair accusations were not considered to be permanently damaging may be seen in a tale told by Jocelin of Furness. A monk who had been wrongly called a fool in life was brighter than others in heaven, according to Jocelin, as compensation for the 'false and murky judgment' he had endured on earth.[33] More strikingly, the process of accusation and confession is evoked by Stephen's use of military imagery; the underlying analogy is between chapter and battle.[34]

'When you step forward for judgment', instructs the *Speculum*, 'think of Christ's trial before the Praetor Pilate when the Pharisees acted as his accusers and the soldiers flogged him all without legitimate cause'. And further, the novice is to 'think of the trials of the holy martyrs as they stood before kings and judges presiding with their staffs. Think of your own death and harsh trial before God when thousands of devils will shout accusations against you and a million of them bear false witness against you'.[35] Like the practices of creating mental pictures to reshape the memory or to see beyond the immediate confines of the church, the visualizing or imagining of a related scriptural scene encourages the novice to think outside his immediate physical circumstances. Yet the site in which the novice is advised to think on these images remains integral to the nature of the novice's experience. In the case of the church or even in the novice's integration into the monastery, the formulation of mental pictures is related to the intensely private and devotional practice of meditation. In the chapter house, however, the situation in which the novice is encouraged to construct images of judgement, trial and accusation is public and penitential. And although the imagery of battle is used in this tract, the novice is also instructed to withhold any defence, and to remain almost entirely silent throughout the process. The only words that should be said, are, 'I will make amends', or 'I do not remember. I will make amends'.[36] There are also words of advice for the monk who believes that he is being dealt with in a harsh way. The novice should direct his thoughts to his accuser, thinking of him as 'the razor of God who wishes to remove your unsightly hair so that you will appear fairer in beauty'. Furthermore, the novice must remember that the intent of the accuser is to free him 'from the deformity of sin'; for this, the accuser should be recompensed 'on that same day with some favour'.[37]

---

[33] G. J. McFadden, 'An Edition and Translation of the Life of Waldef, Abbot of Melrose by Jocelin of Furness' (unpublished PhD thesis, Columbia University, 1952), ch. 19, 70.

[34] For some discussion of the use of military imagery in Cistercian literature, see Newman, *Boundaries of Charity*, pp. 29–37.

[35] Mikkers, 'Speculum', p. 55.

[36] Mikkers, 'Speculum', p. 54.

[37] Mikkers, 'Speculum', p. 55. For another example of Cistercian interest in the themes of accusation, see Alan of Meaux's poem on Susanna and the Elders. This poem appears in London, British Library, MS Harley 285, fols 2–9^v and in London, British Library, MS Egerton 832, fols 1–4. It has been transcribed and edited by J. H. Mozely, 'Susanna and the Elders: Three Medieval Poems', *Studi Medievali* n.s. 3 (1930), 27–52. There is some

In the chapter house, therefore, the process of justice was related primarily to correction. This was sometimes achieved through accusation, as Stephen of Sawley's text indicates. After the Lateran Council of 1215, however, the sacrament of confession became the main instrument with which Cistercians were subjected to correction and forms of discipline.

Confession

The well-known *Omnis utriusque sexus* decree of the Fourth Lateran Council in 1215 declared that all Christians had to confess their sins to their priest at least once a year.[38] This decree has often been cited as the means by which a 'culture of guilt' was established in western Europe, although it has been shown that rituals for the restoration of baptised Christians who had seriously sinned had long been present in the Christian world.[39] From the earliest penitential rites, sinners were required to acknowledge or confess the sin, as well as to show repentance or sorrow at their lapse. Historians have tended to stress the public nature of penitential systems as one important reason for their success in the discipline of Christians, and it has been demonstrated that public penance survived well into the scholastic era. The actual process of confession, too, has been understood as a public declaration of guilt, although the *Omnis utriusque sexus* decreed that confession should be made only to a priest.[40]

Cistercian confessions were highly regulated and, not surprisingly, took place more frequently than the once a year prescribed by Pope Innocent III for the laity. In 1232, the General Chapter decreed that Cistercians should confess once every week. In 1233, the Chapter legislated that lay brothers were to confess once each year at their local abbey and in 1241, Cistercian monks were

---

discussion of the poem in Rigg, *A History of Anglo-Latin Literature,* p. 160 et seq. I would like to thank Professor Rigg for his advice on this and other poems.

[38] The decree *Omnis Utriusque Sexus* is in *Decrees of the Ecumenical Councils*, ed. by N. P. Tanner, 2 vols (London: Sheed & Ward; Washington, DC: Georgetown University Press, 1990).

[39] Jean Delumeau, *Sin and Fear: The Emergence of a Western Guilt Culture 13th–18th Centuries,* trans. by E. Nicholson (New York: St Martin's Press, 1990); Michel Foucault, *The History of Sexuality. Volume One: An Introduction,* trans. by R. Hurley (Harmondsworth: Penguin Books, 1990); and most famously, Sigmund Freud, *Civilization and its Discontents,* trans. J. Riviere and ed. by J. Strachey (London: Hogarth Press and the Institute of Psycho-Analysis, 1973). See also Thomas T. Tentler, *Sin and Confession on the Eve of the Reformation* (Princeton: Princeton University Press, 1977).

[40] Mary C. Mansfield, *The Humiliation of Sinners: Public Penance in Thirteenth-Century France* (Ithaca and London: Cornell University Press, 1995), p. 79: 'Until Carlo Borromeo invented the confessional, Lenten confession was a surprisingly public affair'. See also John Bossy, 'A Social History of Confession in the Age of the Reformation', *Transactions of the Royal Historical Society*, 5, ser. 25 (1975), 21–38. For the state of confessional practice prior to the Fourth Lateran Council, see A. Murray, 'Confession Before 1215', *Transactions of the Royal Historical Society*, ser. 6:3 (1993), 51–81.

forbidden to confess to anyone outside the order.[41] This ruling was maintained in the codifications of 1257 and by Pope Alexander IV in 1255, who said that absolutions obtained by people outside the order were illegal and invalid.[42] The text of the mid-thirteenth century rulings on confession declares that:

> Abbots and monks should confess at least weekly if they have a reason to confess. Lay brothers who are in the abbey should do the same. Whoever is in a grange should confess whenever there is a chapter. No-one should omit to confess at least once a year, unless there is a strong reason.[43]

The 1257 addition to this text decreed that 'monks and lay brothers should not presume to confess to any person outside our order'.[44] Within Cistercian houses, however, it was agreed that abbots could hear confessions from visitors. Thus, Cistercian confessions were to be kept within the order, a ruling that was extended in 1257 by the General Chapter, which indicated that monks who confessed outside the order would be excommunicated.[45]

In the Cistercian chapter house, confession was a public event. The presence of an audience in the chapter house transformed the space into a public one, while the monk in the centre of the room became the object of the audience's attention. During the ritual of confession, his words were heard by the rest of the community, his contrition was the subject of public spectacle and his punishment was witnessed by all. However, it should be recognized that there was also a place for more private confession in the Cistercian monastery. In the *Ecclesiastica Officia*, we find that this also occurred in the chapter house. After the daily chapter meeting, the community left the building except for those who were kept back for confession.[46] Monks were also able to confess at other specified times during the day, although it is not clear from the information given in the *Ecclesiastica Officia* whether these individual confessions also took place in the chapter house. At these confessions, the prior and monk exchanged greetings. The monk then briefly listed the faults for which he asked forgiveness. The prior then gave him absolution, ordered the requisite penance and counselled him (or reproached him) briefly if needed.

The difference between the two modes of confession in the Cistercian monastery could be explained in terms of public and private spaces for confession. However, it may also be useful to distinguish between the types of

---

[41] *Statuta* 1232, 8, t. 2. This was reiterated in the codifications of 1237 and 1257. See Lucet, *Codifications*, *De Confessione* 1237, 6, 4; 1257, 6, 4. For lay brothers, see *Statuta* 1233, 5, t. 2.

[42] Lucet, *Codifications*, p. 122.

[43] Lucet, *Codifications*, p. 274.

[44] Lucet, *Codifications*, p. 274.

[45] *Statuta* 1221, 5, t. 2 for confessions of visitors; *Statuta* 1257, 5, t. 2.

[46] For the following, see Griesser, 'Ecclesiastica Officia', p. 237. Certain abbots gained a reputation for their skill at hearing confessions. One example is abbot Richard of Fountains, who is said to have been a wonderful observer of consciences and a healer of spiritual wounds. See *Mem. F.*, p. 74: 'Conscientiarum scrutator mirabilis, et spiritualium vulnerum curator'.

transgression that were considered to be suitable for private confession and the types of transgression that were dealt with publicly, as the result of accusation rather than self-monitoring. I have mentioned previously that accusations made in the chapter house were more likely to concern the breaking of institutionally defined rules, or wrongful action. In private confessions, the subjects for correction are more often wrongful thoughts. The individual was thus subject to two varieties of discipline within the overall discourse of confession. First, public confession was the disciplinary measure of the wider Cistercian institution. Second, private or individual confession was the means by which the individual was encouraged to discipline himself.

Some examples of the subjects for private confession are given by the author of the *Speculum Novitii*. Apart from the sin of remembering the non-monastic past, which I have discussed in the first chapter of this book are listed suitable subjects for confession: 'idle thoughts', discontent, pride, anger, disobedience, unkindness, laziness and setting a bad example.[47] As noted previously, these transgressions are intensely individual and personal. One of the most significant implications of this is that the monk is—to a large extent—responsible for surveying his own behaviour. The *Speculum* notes that the novice should confess these thoughts 'adding or subtracting to what I am offering here in the measure you feel you have transgressed in these matters'. Thus, the Cistercian monk is directed to observe his own thoughts even before they become actions. Other examples of subjects for private confession are lack of proper devotion or reverence. The novice is urged to say 'I delay going to prayer, to Mass, to the other exercises [...] I have no fear of expending precious time on trifles, on idleness and listlessness because both my fear of God and my good resolution are defective'. Again, he should say 'I run more slowly to the Divine Office than to the table. I bow absent-mindedly, I obey unwillingly and later 'I do not keep my profession in the measure necessary or desirable. I do not recite the *debitum* [prayers prescribed for the dead], nor do I say the prescribed psalms and prayers with the proper intent for their designated benefactors. Also I have given things away or accepted things, and done this and that without permission'.

An early fourteenth-century manuscript from Kirkstall abbey also addresses private confessions, this time in a lengthy poem.[48] The act of confession in this poem is first related to the judgement day, then examples of sins or transgressions are given. The subjects for confession are similar in nature to Stephen of Sawley's guide in the *Speculum Novitii*. These relate to the proper living out of the monastic life. Disobedience is again mentioned, as are subjects such as leaving the monastery, 'sinning with one's feet' in a wavering commitment to the monastic profession.[49] Anger should also be confessed, especially in relation

---

[47] Mikkers, 'Speculum', pp. 45–46 for the following relating to confessions.

[48] London, British Library, MS Arundel 248, fols 133–34^v.

[49] 'Pertinax et contumax inobediendo/ hiis et multis aliis modis superbiendo'; 'Pedibus illicite peccavi intendo/ viam patis deferens ac schelera currendo/ ad lusus et spectacula varia gradiendo/ inconstans et uistabit ac vagus existendo'.

to disruption and delinquency in profession.[50] Using the senses in improper ways is noted, particularly the eyes, but also the ears, which hear willingly frivolities.[51] Other standard transgressions such as the breaking of the commandments are included in this treatise.[52] Private confession therefore served to underscore public discipline in producing truthful accounts of otherwise inaccessible individual thoughts of crime and sin.[53]

The legislative controls of the General Chapter sought to maintain communal order, and the public nature of the implementation of these rules was one strategy for doing this. On the other hand, there was certainly a place for confession without an audience, and other regulations show that Cistercians were keen to keep individual cases of transgression to the forum in which they were corrected. It was, for instance, forbidden that faults and secrets exposed in the chapter house should be discussed anywhere else.[54] Lucet has noted that the thirteenth-century compiler of the codifications was most concerned with issues of penitential jurisdiction. Lucet also detected a perceptible evolution in the canon privileges accorded to Cistercians regarding confessions.[55] This indicates that Cistercians themselves believed—as did the Church—that confession was a critical part of their world. Equally important, however, in the process of discipline were correction and punishment.

Punishment

Dealing with Disorder

The thirteenth century was a time when the Cistercian General Chapter showed

---

[50] 'Per iram et per odium peccavi tumescendo/ rixas et contumelias clamorem preferendo/ indignans in animo malicia servendo/ sic et multipliciter per iram deliquendo'.

[51] 'Peccavi visu nimium illicita videndo/ in carnis volupteribus occulem figendo/ spectaculis ac lusibus me vanis in miscendo/ superbum nequam lividum occulsum habendo'; 'Per auditum aurum patavi deliquendo/ vana stulta noxia libenter audiendo/ aurem dectractoribus ac frivolis prebendo/ ascultans illicita utilique spernendo'.

[52] 'Nomen domini in vanum assumendo'.

[53] For a similar view, see Foucault, *History of Sexuality. Volume One,* esp. p. 58 et seq.

[54] Griesser, 'Ecclesiastica Officia', p. 237: 'Hoc etiam caveatur, ne aliquis extra capitulum loquatur alicui vel significet de culpis seu de secretis causis, que in capitulo pertractantur'; *Statuta* 1217, 10, t. 1: 'Item prohibemus omnino ne quis monachus vel conversus praesumat proclamatori signo aut verbo indignationem ostendere, vel sententiam abbatis aut prioris seu eius qui praeest capitulo audeat iudicare vel de eo loqui aut significare extra capitulum, nisi ei qui potest corrigere. Qui vero aliter praesumpserit, sex diebus continuis in capitulo verberetur'.

[55] Lucet, *Codifications*, p. 128: 'Une évolution est perceptible dans l'application du privilège du canon. Malgré les concessions accordées, l'ordre cistercien sollicite de nouvelles bulles pour lever les excommunications encourues en cas de violence, mais décide que les voies de fait sur un abbé constitueront un délit spécial entraînant excommunication qui ne pourra, désormais, être levée que par l'autorité pontificale'.

great interest in the development and definition of forms of punishment. In 1206, it was authorized that prisons could be constructed in Cistercian monasteries. These were built for the purpose of housing fugitives and criminals, such as the monk of Jouy who was imprisoned for life in 1226 for wanting to kill the abbot with a razor.[56] By 1229, it was decreed that prisons must be built in all Cistercian monastic abbeys to house criminals, while throughout the century there is a plethora of examples of monks being punished for various misdeeds by being put in chains.[57] The mid-century codifications delineate quite specifically the types of crimes that should be punished with imprisonment,[58] while graffiti discovered on the wall of a prison cell at Fountains abbey indicates the hopelessness that one monk felt while incarcerated. The graffito reads 'Vale libertas'.[59]

Not all crimes or transgressions, however, were punished with incarceration, whether in actual prisons or by other forms of bodily confinement, and this remained the case throughout the thirteenth century.[60] The General Chapter distinguished between punishments appropriate for grave transgression and those appropriate for less serious crimes. In cases of 'gravi culpa', punishment could be expulsion from the order. This was the penalty for sodomy.[61] Periods of exile from the monastery were often prescribed for monks who were perceived to be disruptive. This was the case with a monk from Foigny, who was sent to Villarium in Brabantia in 1218 for disputing the abbot's sermon, and for a lay brother of Bonacumbae who was sent to a faraway house in 1232 for excesses in his own abbey.[62] Exile was also deemed to be an appropriate punishment for some homicides, in particular those committed by a Cistercian against someone outside the order.[63] Removing criminal or sinful monks from the abbey does not seem to have been replaced by incarceration as a form of punishment in the later thirteenth century, even after the General Chapter's ruling of 1240. Rather, as

---

[56] *Statuta* 1206, 4, t. 1: 'Qui voluerint carceres facere, faciant, ad fugitivos suos et maleficios, qui talia meruerint'. For the monk of Jouy, see, *Statuta* 1226, 25, t. 2.

[57] *Statuta* 1229, 6, t. 2. For an instance of chained monks, see, *Statuta* 1226, 41, t. 2, where a monk of Foucarmont abbey is said to be held 'in vinculis'.

[58] Lucet, *Codifications*, pp. 277–78: 'In singula abbatiis ordinis nostri in quibus fieri poterit, fortes ac firmi carceres construantur, ubi ad abbatis arbitrium retrudantur et detineantur, secundum quod sua exegerint crimina, criminosi. Criminosos autem hic vocamus indicibili vitio laborantes, fures, incendarios, falsarios, homicidas. Abbas vero si falsarius fuerit deponatur.'

[59] Coppack, *Fountains Abbey*, p. 75.

[60] For contextual material on the development of prison systems during the medieval period, see Ralph Pugh, *Imprisonment in Medieval England* (London: Cambridge University Press, 1968); *The Oxford History of the Prison: The Practice of Punishment in Western Society*, ed. by N. Morris and D. J. Rothman (New York and Oxford: Oxford University Press, 1995), pp. 27–30 for monastic prisons.

[61] *Statuta* 1221, 9, t. 2. This ruling was modified in *Statuta* 1224, 21, t. 2.

[62] *Statuta*, 1218, 47, t. 1; *Statuta* 1232, 24, t. 2.

[63] Lucet, *Codifications*, p. 276.

Jane Sayers has pointed out, abbots now possessed 'twin weapons of exile and prison'.[64] Abbots themselves, not surprisingly, appear frequently in the *Statuta* of the General Chapter, and we find in some of these cases that abbots were deposed from office for serious offences. The abbots of Jerpoint and Mellifont abbeys in Ireland, for example, were deposed in 1217, both for various manifestations of inhospitable and contumacious behaviour.[65]

Excommunication was the ultimate penalty for transgression, although it appears that Cistercians tried to avoid using this punishment where possible. The early statutes show that a formal ritual of expulsion occurred on Palm Sunday in which those convicted of certain crimes would be excommunicated in the chapter house.[66] This legislation was reiterated verbatim in the codices of the mid-thirteenth century,[67] although later some modification occurred. Whereas the initial reasons for excommunication were related to the committing of criminal acts, a statute of 1282 reveals that disruption and disobedience became sufficient reasons for excommunication.[68] More frequent than excommunication, however, were less terminal means of correcting disorderly monks and lay brothers, such as exile, mentioned above.

The confession and punishment of grave sin was an extremely formal and public ritual, which took place in the church, cloister and chapter house. The guilty monk was first beaten in chapter, and then—with his cowl over his head—was led out of the building, accompanied by an older monk, who was to counsel him and say a special prayer for him in chapter.[69] At the church, the monk was to prostrate himself on the ground outside the building after having removed his cowl from his head. The monk was ultimately to return to the chapter house, in which he was to prostrate himself again. Similarly, for the rest of the day, the monk was to prostrate himself during the *opus dei*, whether in the church or while other work was taking place.

A monk who was decreed to be in 'levi culpa' was also subject to bodily mortifications. The monk was to prostrate himself at the presbytery steps at the *Kyrie* and remain there until after the *Deo gratias*. He was also to eat outside the refectory in a place decided by the abbot, and was not to go with the rest of the community to the refectory for a drink either. If the prior or subprior were in 'levi culpa', they were deprived of some of their liturgical duties, such as reading or leading the singing of the Divine Office. Abbots, too, were subject to similar discipline meted out by the General Chapter. In 1225, for example, the abbot of Pilis was punished for having a bath on Sunday and for shaving while in the bath. He had to give up the abbot's *stallum* for forty days, was held to be in 'levi culpa' for six of those days, was to spend two days fasting on bread and

---

[64] Sayers, 'Violence in the Medieval Cloister', p. 536.

[65] *Statuta* 1217, 78, t. 1; *Statuta* 1217, 79, t. 1.

[66] *Statuta* 1183, 10, t. 1.

[67] Lucet, *Codifications*, p. 275.

[68] *Statuta* 1282, 15, t. 3.

[69] For the following, see Lucet, *Codifications,* pp. 275–76.

water, and was also to fast on bread and water every Friday until Easter.[70] Fasting on bread and water was a common part of punishments for various transgressions, such as breaking the silence,[71] while an abbot who failed to visit his daughter house(s) was also declared to be in 'levi culpa'. This was the case with the abbot of Fountains, who was to fast on bread and water for a day, and held to be in 'levi culpa' for three days after failing to visit his daughter house of Lysa in Norway in 1210.[72]

Thus, punishment tended to focus on the correction of the disorderly body, whether through deprivation of food or by removing that body from the monastic environment altogether. One consistent feature of punishments in Cistercian houses was their public nature. The prostrated monk before the presbytery steps, and the abbot absent from his usual seat in the church, were visible manifestations of discipline in action. Humiliation was also an important part of these punishments—the guilty monk was physically humbled by prostration and spiritually humbled by being excluded from certain day-to-day devotional privileges. Similar principles may be discerned in one of the most common forms of punishment in Cistercian monasteries—beating. As with other punitive regimes such as prostration, the target of beating was the body.

Accepting (the) Discipline

The *Regula S. Benedicti* stated that beating with birch rods was appropriate chastisement for monks who continually transgressed and that the Bible advocated this disciplinary measure.[73] Forms of corporal punishment had therefore long been present in monastic houses, and Cistercian monasteries were no exception.[74] It is important to remember at this point that beating may also have taken place more routinely in the chapter house, and was not necessarily related to the commission of a particular crime. In other words, beating functioned in punitive ways and as a manifestation of more general mortification.[75] The *Eccle-*

---

[70] *Statuta* 1225, 22, t. 2.

[71] *Statuta* 1221, 6, t. 2.

[72] *Statuta* 1210, 35, t. 1. The relationship between Lysa and Fountains remained tense. In 1212 the General Chapter reported that 'monachi de Fontanis [...] qui multa enormia occasione visitationis in domo de Lisa commiserunt, emittantur de domibus propriis'. (*Statuta* 1212, 32, t. 1). Finally, it was decided that Fountains was no longer responsible for the Norwegian house (*Statuta* 1213, 11, t. 1.).

[73] Manning, 'La Règle de S. Benoît', p. 222. The biblical references are Proverbs 19. 25 and 29.

[74] See Louis Gougaud, *Devotional and Ascetic Practices in the Middle Ages* (London: Burns, Oates, and Washbourne, 1927) pp. 179–204 for a résumé of the practice. For an iconographic relationship between beating and chapter house sculptural programmes in France, see Leon Pressouyre, 'St Bernard to St Francis: Monastic Ideals and Iconographic Programs in the Cloister', *Gesta*, 12 (1973), 71–92 (pp. 78–81).

[75] These voluntary beatings—often self-inflicted—are most famously associated with the millennial cults described by (*inter alia*) Norman Cohn, *The Pursuit of the Millennium*, rev. edn (Oxford and New York: Oxford University Press, 1970). Cohn relates the heretical self-flagellant sects of the thirteenth century and after to the

siastica Officia provided the regulations on the practice of beating, saying that once it has been ruled that a monk should be beaten certain caveats must first be taken into account. First, the person who has accused the guilty monk of the particular transgression for which he is being punished was forbidden to do the beating himself. The same was true for whoever led the chapter meeting. The person to be beaten should, at the order of the abbot, get up from his seat at once and placing his cowl before him, kneel and stretch his arms over his head. He should say nothing but 'I am guilty. I will make amends'. No-one else in the room should speak or intercede as the beating then took place, while the member of the community who was designated the task of carrying out the beating should not stop, unless instructed to do so by the abbot.[76]

The participation of the whole community in the ritual of beating may be seen to have had a deterrent effect. However, the performing of corporal punishment by a designated member of the community also had the effect of making the entire audience in the chapter house complicit in the administering of punishment. Other cases occasionally reported to the General Chapter reveal that violent punishment was not unknown. In 1226, for example, the abbot of the Hungarian house of Sanctae Crucis was to be deposed for beating a monk to death.[77] More usually, however, the beatings in chapter were regulated, and formed a routine part of certain punishments. Apostates, for example, who had been returned to the monastery were to be beaten in chapter every Friday for a year and on these days were to fast on bread and water, while monks and lay brothers who had broken the silence were to be beaten in chapter.[78] Punishment in the form of beating was therefore, like confession, a public ritual, involving the members of the Cistercian community as either perpetrators, witnesses, or as victims.

The Space of the Chapter House

Themes of memory, community, and discipline were all important elements in the Cistercian chapter house during the thirteenth century. These themes did not exist independently, but rather intersected in the use and function of the chapter house itself. There are a number of ways in which we can read the space of the chapter house through the intersection of its principal functions.

Trial in Life, Trial in Afterlife

First of all, the Cistercian chapter house was evidently a site where theological assumptions and disciplinary strategies met. The use of the building for practices

---

'crusading *pauperes* before them' (p. 128).

[76] Griesser, 'Ecclesiastica Officia', pp. 236–37.

[77] *Statuta* 1226, 27, t. 2. For a more detailed discussion of violence, see chapter 6.

[78] *Statuta* 1221, 11, t. 2; *Statuta* 1221, 6, t. 2.

such as confession and commemoration could be easily related to Christian doctrinal principles. This is especially clear in the relationship between the chapter house as a site of trial and Cistercian understandings of purgatory. The *Speculum Novitii*, for example, suggests that the Cistercian monk has already been 'predestined' for heaven. Already saved from the pains of hell by Mary, Cistercians are equated with the elect and therefore have the privilege of hope.[79] But they also live in a place of penance and a place of suffering and trial.[80] The consolation is that the Cistercian will be cleansed at the moment of death or shortly thereafter. This implies that the process of purgation is at least started in this world, and possibly also completed.[81] The space of the chapter house may be identified with this purgatorial period of trial that follows judgement, in that both are sites where the body and soul are cleansed of the stain of sin within the protective and corrective sphere of the Cistercian world. Whereas the *interim* space of purgatory in the church is tempered by the possibilities of intercessionary prayer, the relationship between the chapter house and purgatory is defined in terms of suffering, bodily pain, and the possibility of communally directed redemption.

The relationship between times of trial on earth and in the afterlife was of particular concern to Cistercians during the medieval period, while interest in the space of purgatory on earth was popularized by the Cistercian *Tractatus de Purgatorio Sancti Patricii,* written by a monk of Saltrey in Huntingdonshire in the late-twelfth/early-thirteenth century.[82] This version of purgatory was certainly known by the Yorkshire Cistercians, and by other Cistercian houses.[83] The *Tractatus* located purgatory in Ireland, and described it as a dark hole into which one descends for a time in order to avoid purgation after death. St Patrick's purgatory was strongly Cistercio-centric and, as Carol Zaleski has shown, was

---

[79] Mikkers, 'Speculum', p. 53.

[80] Mikkers, 'Speculum', p. 65.

[81] See, however, William of Auvergne, who says that Purgatory is where penance is completed and therefore occupies the entire period between death and the resurrection— A. C. Bernstein, 'Esoteric Theology: William of Auvergne on the Fires of Hell and Purgatory', *Speculum,* 57 (1982), 517–21.

[82] An edition by E. Mall is in *Romanische Forschungen,* 6 (1891), 147–96. A later date has been proposed by F. M. Locke, 'A New Date for the Composition of the *Tractatus de Purgatorio Sancti Patricii*', *Speculum,* 40 (1965), 641–44.

[83] The *Tractatus* is contained in an early-thirteenth-century MS from Jervaulx, now Dublin, Trinity College, MS 171. It appears in a collection of saints' lives, at fols 58–70. Incipit: 'Narratio cuisdam de purgatorio Hibernie et de gloria paradisi terrestris'. The *Tractatus* was also in the library of Meaux. See David Bell, 'The Books of Meaux Abbey', *Analecta Cisterciensia,* 40 (1984), 25–83. Jocelin of Furness refers to St Patrick's Purgatory. But see Locke, 'A New Date', who disputes that Jocelin knew the exact *Tractatus* story. For the versions and manuscript history of the *Tractatus,* see H. D. L. Ward, *Catalogue of Romances in the Department of Manuscripts in the British Museum,* 3 vols (London: Printed by order of the Trustees, 1883–1910), 2, p. 435 et seq. And also Carol Zaleski, 'St Patrick's Purgatory: Pilgrimage Motifs in a Medieval Otherworld Vision', *Journal of the History of Ideas,* 46:4 (1985), 467–85.

particularly attractive to Cistercians in terms of both its penitential character and its transforming power.[84] The trend toward identifying purgatory as an earthly space, in which one may choose to undergo penance, has also been related to new topographical interest in finding other significant devotional or spiritual sites on earth.[85]

Connections may be found between ideas of purgatory and trial on earth in the rhetoric of Cistercian narratives describing the suffering of saints. As with forms of tribulation encountered in the chapter house (unjust accusation and bodily mortification) and in the darkness of St Patrick's purgatory, suffering on earth is continually linked to the possibility—or even inevitability—of rising to glory in heaven. This is evident in the writing of Matthew of Rievaulx, whose short poems on the suffering of St Andrew and St Lawrence reiterate this common theme. St Lawrence, for example, is described as victorious in his martyrdom,[86] while the burning of this saint is a particularly relevant metaphor for the fires of purgatory. The experience of St Andrew is also presented in terms of heavenly reward for atrocious suffering in life, but more importantly, the saint assumes an intercessionary quality as a result of his torture and death— 'You have been sent, to purge our sins'.[87] St Andrew is said to have experienced arrest, incarceration, torture, and ultimately death, yet, according to Matthew of Rievaulx, did not cease to listen and instruct the people in Christian ways until he died. Veneration of the saints was, of course, not peculiarly Cistercian.[88] However, Matthew's admiration for Lawrence and Andrew may be understood to reflect the wide range of meanings Cistercians found in the experiences of trial and suffering undergone by martyrs. Not only were saints like Lawrence venerated because, like Christ, they triumphed over pain for their faith. But tribulation could also be instructive, and stories of devotion in the face of physical agony provided the Cistercian monk with 'mental pictures' (as the *Speculum Novitii* puts it) which could be utilized when the monk found himself in a similar situation in the chapter house.

Theological ideas, therefore, could be used to interpret the chapter house and the practices that took place within it. The relationship between trial and

---

[84] See Carol Zaleski, *Accounts of Near-Death Experience in Medieval and Modern Times* (New York: Oxford University Press, 1987), p. 38 and p. 78. See also D. R. French, 'Ritual, Gender and Power Strategies: Male Pilgrimage to St Patrick's Purgatory', *Religion*, 24:2 (1994), 103–15, who argues that Cistercian popularization of St Patrick's purgatory reinforced the exclusively male nature of the site, thus restricting access to women.

[85] See Robert Easting, 'Purgatory and the Earthly Paradise in the *Tractatus de Purgatorio Sancti Patricii*', *Cîteaux*, 37 (1986), 23–48.

[86] Wilmart, 'Mélanges', pp. 62–63.

[87] Wilmart, 'Mélanges', p. 63.

[88] For another account of popular representations of the suffering of saints, see Caroline Walker Bynum, *Fragmentation and Redemption: Essays on Gender and the Human Body in Medieval Religion* (New York: Zone Books, 1992), especially pp. 285 et seq. on Jacobus de Voragine's interest in the torture of saints and his 'obsession with martyrdom' (p. 286).

punishment that occurred daily in the Cistercian monastery could be equated with both purgatorial trials to be faced on earth or in the hereafter, or with the devotional precedent of the trials of Christian martyrs or saints, as the *Speculum Novitii* advises. Yet the trials faced by those accused of transgression in the chapter house also took place in daily ritual which also emphasized communal solidarity and friendship.[89] Thus, when the novice is advised to think on chapter as battle, it is with the understanding that the novice will realize that this correction 'is an extra pittance sent to you from heaven'.[90] The articulation of community in the liturgy preceding the rituals of accusation, confession, and punishment are extremely important, therefore, in revealing that the disciplinary practices played out in the chapter house are part of the community of Cistercians—living and dead—whose task it was to maintain and protect that community and its members.

The Body Social, the Body Corporate

Sociologists of institutions have long identified certain disciplinary strategies in the organization of institutional life, and the medieval monastery may be seen to encompass similar principles. Scholarship on 'total institutions', pioneered by Erving Goffman,[91] has consistently used the monastery as an example of a disciplinary organization that may be likened to hospitals, prisons, army barracks, and others. Public confessions are cited by Goffman as an institutional strategy to eradicate a private self, for example, while the relationship between members of the hierarchy is defined by both difference (the abbot is more autonomous than the novice, for instance) and by processes of 'levelling', where all members of the monastery are held to be the same.[92] Sociologically, the chapter house could be interpreted as the site within the monastery where such strategies are played out.

Historians of monasticism have been less than enthusiastic in accepting sociological models such as Goffman's 'total institution' as comparable to the monastic life. Michael Casey considers that complete uniformity and control is impossible in a monastery, where 'potential for variety' exists in the harmonious relationship between solitude and community. 'It is only when the monastery degenerates into a total institution', argues Casey, 'demanding uniform observance in a way that the Rule of St. Benedict would never have envisaged, that solitude and community become alternatives'. For Casey, 'the attraction of con-

---

[89] See McGuire, *Friendship and Community*.

[90] Mikkers, 'Speculum', p. 55.

[91] Erving Goffman, *Asylums. Essays on the Social Situation of Mental Patients and Other Inmates*, repr. edn (Harmondsworth: Penguin Books, 1991).

[92] Goffman cites the *Regula S. Benedicti*, chapter 2, which decrees that there should be no difference between people in the monastery, that everyone should be loved equally and that class distinctions that existed outside the institutional environment should no longer apply inside the monastery. See *Asylums*, p. 111.

templation is a gift of grace; it cannot be institutionalized'.[93] Jean Leclercq, too, disagrees that a monastery can be equated with institutions in which people are incarcerated. Leclercq traces the development of the equation of monastery and prison from Tertullian to St Bernard, agreeing that ideas of imprisonment in early Christian and medieval literature were significant in expressing separation from the world, yet refuting the notion that monasteries may more generally be understood in the same way as 'des casernes, des hôpitaux psychiatriques, des prisons et des camps de concentration'.[94]

Disciplinary aspects of the medieval Cistercian abbey are not extensively discussed by either Jean Leclercq or Michael Casey in their respective discussions of total institutions. Yet, these aspects are precisely those elements which led Goffman to establish the relationship between different types of institutional environment. Returning to the themes of accusation, confession, and punishment which I have discussed above, we find that Goffman isolated similar themes in other institutional environments. In the prison, for example, the putative guilt of all the inmates may be equated with the daily mortification of accused monks in the chapter house.[95] In the asylum, Goffman likens the group therapy sessions where patients are obliged to discuss intensely personal matters to the practice of confession in the public chapters of religious houses.[96] And in the concentration camp, inmates forced to stand by as others are tortured to death are—in Goffman's view—witnessing the same 'experiential mortification' as the monk who watches his brethren 'accepting the discipline'.[97]

Both Michael Casey and Jean Leclercq insist that the voluntary nature of membership in a monastery is the fundamental difference between life in an abbey and life in other 'total institutions'. That no-one is forced to become or remain a Cistercian monk is, for both these scholars, the point where sociological theory fails in an attempt to describe all institutions within the same paradigm. Michel Foucault's more recent work on the nature of institutional life and the social theories that it describes, takes issue with the idea of voluntary participation in institutional regimes.[98] Foucault argues that all disciplinary institutions work by excluding, segregating, and normalizing. In the monastic chapter house, this would be manifested in the exclusion of non-Cistercians from certain parts of the meeting, the segregation of those witnessing confession and punishment and the monk in the centre of the room who is the object of correction, and the normalization of that guilty monk by means of beating or other corrective strategies. Likewise the rituals of reading the *Regula* and commemorating the dead are part and parcel of naming the community in which

---

[93] Michael Casey, 'In Communi Vita Fratrum: St Bernard's Teaching on Cenobitic Solitude', *Analecta Cisterciensia*, 46 (1990), 243–61 (pp. 245–46).

[94] Jean Leclercq, 'Le cloître est-il une prison?', *Revue Ascétique et de Mystique*, 47 (1971), 407–20.

[95] Goffman, *Asylums*, p. 49.

[96] Goffman, *Asylums*, p. 39.

[97] Goffman, *Asylums*, p. 40.

[98] Foucault, *Discipline and Punish*.

each monk finds himself present. More broadly, Foucault contends that individuals in these environments—by consenting to fit into the normalizing regime—are in fact products of the discipline that the regime seeks to exert. For Foucault, in other words, institutions fashion docile subjects and produce members of the community. The success of this disciplinary economy may be seen when members of these institutions monitor and correct themselves. When the author of the *Speculum Novitii* advises his novice that: 'My dear friend, you must choose to be flogged with Christ and remain in subjection rather than deny him with Peter, by aspiring to any position',[99] Foucault would argue that the Cistercian institution has succeeded in producing a conformist subject. The issue of freedom in Foucault's institutional regimes is, therefore, an illusion that the institution uses to maintain compliance.[100] For Michael Casey and Jean Leclercq, freedom is the absolute condition and result of monastic life and monastic discipline.

The medieval Cistercian chapter house was a space where these ostensible dialectics of freedom and regulation, individual and community, belonging and alienation were drawn together in the performance of rituals designed to carve out a workable and distinctive Cistercian community. The liturgy of the chapter house described in the *Ecclesiastica Officia* was formulated so that community, or institution, was privileged. The commemoration of the dead and the careful articulation of cenobitic monasticism in the reading of the *Regula*, for instance, reminded those present in the chapter house that the site of the chapter house was a specifically communal Cistercian space, as did the visible evidence of abbatial tombs. The following aspect of the ritual—involving confession, accusation, and punishment—was designed to fortify the community against the individual trangressor, just as it was designed to fortify the individual monk against transgression itself. The body corporate and the individual body were thus both created and disciplined in the chapter house.

Returning to Michel Foucault's understanding of the productive and complex nature of discipline in institutional milieux, some useful conclusions may be drawn. It is clear from the various rituals played out in Cistercian chapter houses of the thirteenth century that community, or the monastic body, was in a daily process of not only enunciation but also creation. The existence, therefore, of regulations such as the *Regula S. Benedicti* or the *Statuta* of the General Chapter was only the preliminary articulation of uniformity of observance and institutional discipline. We might also see that the chapter house, although ostensibly concerned mainly with the quotidian administration of business and the enforcement of those regulations, was in reality a crucial site in which the idea of cenobitic life was constructed. As the cloister encouraged Cistercian monks to imagine the eternity of paradise, and as the church led monks towards the *lux*

---

[99] Mikkers, 'Psalmodie', p. 283: 'Tu autem, dilectissime, magis elige cum Christo flagellari in subiectione quam negare cum Petro in cuiuslibet dignitatis ambitione'.

[100] See J. Mohanty, 'Foucault as a Philosopher' in *Foucault and the Critique of Institutions*, ed. by J. Caputo and M. Yount (University Park: Pennsylvania State University Press, 1993), pp. 27–40 says that 'Foucault's power succeeds by hiding its own mechanism' (p. 33).

mundi, the chapter house also played its part in expressing the transcendent qualities of daily life. For the community envisaged in practices of commemoration, correction, and confession was not simply the community that existed in the thirteenth-century present. Rather, it was the community of eternity that could be imagined each day in the chapter house liturgies. As each monk looked east towards the abbot's pulpit and as he lay on the floor with his face close to the graves of previous abbots, the past, present, and future were inextricably combined into a Cistercian world in which his body and that of the monastery itself were one and the same.

CHAPTER 5

Blood, Body, and Cosmos: The Infirmary

Recent research into Cistercian infirmaries reveals that although the infirmary complexes of the Yorkshire houses differ in various ways, the dates of the buildings fall within the same one-hundred-year period.[1] There were already fairly detailed twelfth-century Cistercian regulations regarding the use of the infirmary space, but it is not until the thirteenth century in most of the Yorkshire houses that we see any significant building there. These thirteenth-century dates of the northern English Cistercian infirmaries have been frequently cited, but rarely analysed.[2] Even David Bell's careful study of the English Cistercians' medical interest notes only that the thirteenth-century infirmaries do not preclude the likelihood of infirmaries existing prior to that time, and that 'the size and splendour of some thirteenth-century infirmaries might perhaps imply an interest in medicine which Bernard would have found excessive'.[3] In archaeological reports on the various sites, too, the possible

---

[1] The exceptions to this generality are the infirmaries of Byland and Rievaulx. According to Stuart Harrison, the infirmary at Byland was complete—with the rest of the monastery—by the end of the twelfth century. See Harrison, 'The Architecture of Byland Abbey'.

[2] For example, M.-A. Dimier, 'Infirmeries cisterciennes', in *Mélanges à la mémoire du Père Anselme Dimier: 1/2*, ed. by B. Chauvin (Arbois: Benoît Chavin, 1982), pp. 804–25; Nicola Coldstream, 'Cistercian Architecture from Beaulieu to the Dissolution', in Norton and Park, eds, *Cistercian Art and Architecture in the British Isles*, pp. 139–59 (p. 155). The exception is Rievaulx abbey, which was built in the mid-1150s. See Fergusson and Harrison, *Rievaulx Abbey: Community, Architecture, Memory*, p. 123.

[3] David N. Bell, 'The English Cistercians and the Practice of Medicine', *Cîteaux*, 40 (1989), 139–74 (p. 162 and p. 171). Also Nancy Siraisi, *Medieval and Early Renaissance Medicine: An Introduction to Knowledge and Practice* (Chicago: University of Chicago Press, 1990), p. 14 for a similar understanding of Bernard's attitude toward medicine and treatment of the body.

reasons for the proliferation of this complex in the thirteenth century have not been forwarded.[4]

Concepts of material and imagined space may provide some insight into Cistercian interest in the infirmary during this period. The cultural practices associated with the site tell us that the infirmary was devoted almost exclusively to the regulation, care and classification of the monastic body. Practices such as periodic bloodletting served to describe not only the tumultuous and subjective body itself, but also the spaces the body represented. The material place of the infirmary implied a range of meanings or imagined spaces, spaces that were signified by cultural practices that positioned and defined the monastic body in relation to landscape, to institution, to spiritual discourse, and to cosmos. In this way, the infirmary was a site where a number of spaces were described, a fixed topography, but with boundless vistas.

The Infirmary

The topographical features of the thirteenth-century Yorkshire infirmaries may be described primarily in terms of separation. The complexes themselves were generally at some distance away from the main cloister, east of the east range. At Kirkstall abbey, the thirteenth-century choir monks' infirmary complex was accessible from the main part of the monastery by a long passage that continues from the south cloister arcade, south of the parlour to the infirmary hall (Fig. 5.1). At Fountains abbey, the choir monks' infirmary was entirely rebuilt under the leadership of Abbot John of Kent, and like the Kirkstall site was accessible via a similarly sited passageway. Prior to the thirteenth-century alterations, the Fountains infirmary had been closer to the church, but, as Glyn Coppack has indicated, this earlier structure had to be demolished in order to make way for extensions to the south transept. The new infirmary was constructed over the River Skell. At Rievaulx abbey, the infirmary site was separated from the east range by a separate cloister (Fig. 5.2). The Rievaulx site is particularly interesting in that the major work to the infirmary was—like Byland—already complete by the beginning of the thirteenth century, although some alterations took place during the period to 1500 east of the infirmary cloister.[5]

At Roche, the infirmary hall is no longer evident, but the lay brothers' infirmary may still be seen, and to the southeast of the main claustral buildings, the infirmarer's lodgings survive in part. The infirmary complex at Byland is also one of the two (with Meaux and Roche) where there is now no trace of the

---

[4] For example, Coppack, *Fountains Abbey*; Stephen Moorhouse and Stuart Wrathmell, *Kirkstall Abbey Vol. 1. The 1950–64 Excavations: A Reassessment* (Leeds: West Yorkshire Archaeology Service, 1987), pp. 51–56; W. St John Hope and H. Brakspear, 'Jervaulx Abbey', *Yorkshire Archaeological Journal*, 21 (1911), 303–44.

[5] For Fountains, see Coppack, *Fountains Abbey*, p. 59. For Rievaulx, see Fergusson, *Architecture of Solitude*; Nikolaus Pevsner, *The Buildings of England, Yorkshire North Riding* (Harmondsworth: Penguin Books, 1966), pp. 299–303; Fergusson and Harrison, *Rievaulx Abbey: Community, Architecture, Memory*, pp. 111–135.

Blood, Body, and Cosmos: The Infirmary

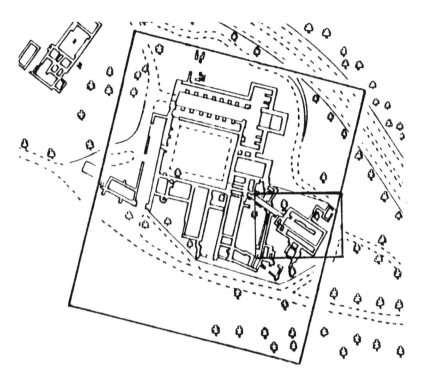

Fig. 5.1: Plan of Kirkstall abbey showing the infirmary to the south-east of the cloister (Courtesy, West Yorkshire Archaeology Service)

Fig. 5.2: Rievaulx abbey, the infirmary cloister; Rievaulx abbey, detail of the infirmary cloister arcade (Photos, author)

infirmary hall, chapel, kitchen, or cloister. This has engendered some controversy over exactly where the complex originally stood, with some suggesting that the infirmary site was southeast of the east range while others identify this space as the abbot's lodging.[6] Harrison's alternative site for the Byland infirmary (Fig. 5.3) nevertheless shows a large complex, which occupied a fairly standard position east of the east range.

David Bell has argued that the eastern site of Cistercian infirmaries was 'dictated more by the practical consideration of water supply', than by any other reason and that easy access to the monastery's supply of water, which was generally to the south of the site, must be seen as the main determining factor in monastic selection of an infirmary site.[7] More generally, it can be contended that sites with a distinguishable infirmary cloister, such as that of Rievaulx, reinforce the idea that the infirmary was an entirely different type of space from other significant monastic areas; not only was the complex at a distance from the symbolic heart of the monastery—the cloister proper—but once an alternative cloister was established, the complex assumed its own topographically meaningful identity. I shall return to the question of siting later.

It is worth noting at this point that there were often two infirmaries within the one monastery, one for choir monks and one for *conversi*. These were invariably distant from each other. An example may be found at Kirkstall abbey. The lay brothers' infirmary at Roche Abbey was a thirteenth-century building, while the choir monks' complex was an earlier structure (Fig. 5.4).[8] And at Fountains abbey, there were also two infirmaries, both of which were extremely generously proportioned; the lay brothers' infirmary hall measured 37 x 22 m, while Glyn Coppack describes the choir monks' infirmary at Fountains as 'one of the largest aisled halls ever built in medieval England' (Fig. 5.5).[9] The practice of including a separate infirmary for the *conversi* seems to have been an arbitrary one. Dimier suggested that it was only in places where there was a lay brother community of sufficient size to warrant the building of separate infirmary that this was done, but at Rievaulx, for example, where it is claimed that there were five hundred lay brothers by 1200, there is no separate infirmary provision for such a population.[10]

---

[6] For the former argument, see Peers, *Byland Abbey*, For the rebuttal see Harrison, 'The Architecture of Byland Abbey', p. 49.

[7] David N. Bell, 'The Siting and Size of Cistercian Infirmaries in England and Wales', in *Studies in Cistercian Art and Architecture*, vol. 5, ed. by M. Parsons Lillich (Kalamazoo: Cistercian Publications, 1998), pp. 211–38. I would like to thank Professor Bell for very generously sending me a copy of this paper prior to its publication.

[8] A. Hamilton-Thompson, *Roche Abbey Yorkshire* (London: HMSO, 1957).

[9] Coppack, *Fountains Abbey*, p. 59.

[10] Dimier, 'Infirmeries cisterciennes', p. 805; *Life of Ailred*, p. 38: 'Hinc est quod post se Rievalli reliquit monachos bis sepcies decem et decies quinquaginta laicos fratres pater recedans ad Christum'. Also note that the same criticism is made of the inadequate lay brothers' living quarters at Rievaulx. The site is on a hill at this point, so it was virtually impossible to keep building west; this is certainly a reason for the absence of building in this direction.

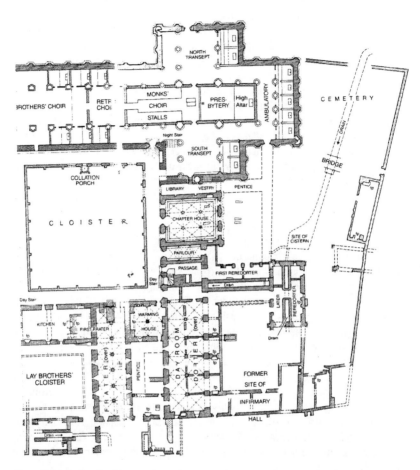

Fig. 5.3: Byland abbey, plan showing the site of the infirmary (Courtesy, English Heritage)

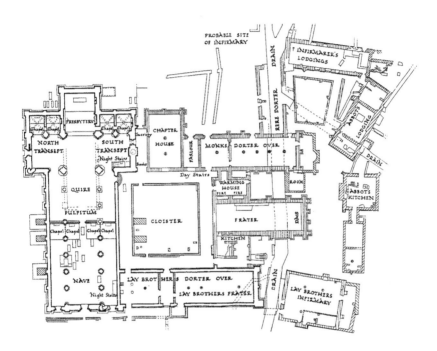

Fig. 5.4: Roche abbey, plan of the precinct showing the choir monks and the lay brothers' infirmaries (Courtesy, English Heritage)

Fig. 5.5: Fountains abbey, the infirmary (Photo, author)

It should also be noted that from the thirteenth century, the statutes of the Cistercian General Chapter begin to report requests to establish infirmaries on granges. These 'alternative' infirmaries are, in all cases, requested by continental houses. In 1257, the abbey of Casanova was given permission to build an infirmary on one of the granges, because of its distance from the abbey proper.[11] And two years later, an infirmary was built on a grange of Fossanova in Italy, in which monks and lay brothers were allowed to eat meat if necessary.[12] These are the only two cases I have found for the thirteenth century, although for the English houses, and especially for the Yorkshire ones with their large tracts of land and their many granges, it is unusual that there would be no similar requests.

There were various buildings within the infirmary complex. The main structure was generally an aisled hall, while the rest of the complex included a separate kitchen, a cellar, a chapel, and infirmarer's lodgings. The hall, however, was the nucleus of the infirmary area and as such was generally the largest of the infirmary buildings, as in the case of Fountains abbey which has already been mentioned.[13] The practice of dividing the infirmary hall into bays was a standard one; at Fountains, there were eight bays in the single storey building for the choir monks, and at Rievaulx the infirmary had ten bays. Aisles ran around the bays and there was at least one fireplace in the hall—at Fountains there were two fireplaces, one on each of the north and south walls. A number of the bays, which demarcated the interior of the infirmary hall, were themselves divided into 'compartments', some of which were recessed, and it was within these recesses that the monks' beds were put. This was a fairly standard ward arrangement, adopted not only by religious houses, but by lay hospitals as well.[14]

Ideas of separation also mark the organization of the internal life of the infirmary. The inhabitants of the infirmary—who were not always the sick[15]—

---

[11] *Statuta* 1253, 37, t. 2.

[12] *Statuta*, 1255, 36, t. 2.

[13] For a similar early design, see Kenneth Conant, 'Medieval Academy Excavations at Cluny', *Speculum*, 38 (1963), 1–45 (p. 13). For some discussion of other monastic infirmaries, see C. H. Talbot, 'Monastic Infirmaries', *St Mary's Hospital Gazette*, 67:1 (1961), 14–18 and M. Candille, 'Un hôpital conventuel du XIIIe siècle. L'Infirmerie ou Salle des Morts de l'abbaye d'Ourscamp', *Hôp. Aide. Soc. Paris*, 1 (1960), 83–96.

[14] See P. Flemming, 'The Medical Aspects of the Medieval Monastery in England', in *Proceedings of the Royal Society of Medicine* (1928), 25–36; R. M. Clay, *Medieval Hospitals of England* (London: Frank Cass, 1966); E. Prescott, *The English Medieval Hospital 1050–1640* (London: Seaby, 1992); B. Durham, 'The Infirmary and Hall of the Medieval Hospital of St John the Baptist at Oxford', *Oxoniensia*, 56 (1991), 17–75; N. Orme and M. Webster, *The English Hospital 1050–1570* (New Haven and London: Yale University Press, 1995).

[15] St John Hope, 'Fountains Abbey', mentioned the presence of elderly monks—*sempectae*—in the infirmary, although Bell, 'The English Cistercians and the Practice of Medicine,' (p. 160, note 108) is suspicious of this claim. As regards the suggestion that the thirteenth century saw an ageing monastic population (and that this could be one reason for an increase in monastic infirmary space), see G. Minois, *History of Old Age*, trans. by S. Hanbury Tenison (Chicago and Cambridge: Polity Press, 1989). This claim

were separated from the life of the cloister in various ways. This is evident even from early examples of Cistercian usages and legislation. The *Ecclesiastica Officia* contains a number of chapters pertaining to the use of the infirmary, to the duties of the monk in charge of the infirmary and to the treatment of the monks confined within.[16] Monks who had been admitted into the infirmary were not to go from the infirmary into the church unless an Office was being celebrated there, while there was to be no lingering in the cloister on the way. It was also forbidden to communicate with the other monks. Monks who were too ill to make the journey from infirmary to church were instructed to celebrate the Divine Office in the infirmary (or in the infirmary chapel if there was one), and the *Ecclesiastica Officia* specifies that the Hours were to be chanted in the same way as in the church. It was also legislated in 1262 that there should be a separate chalice for the sick, for the greater praise of Christ, as well as for the greater good of the community.[17] Monks who had been able to attend the Divine Office in the church were not permitted to serve at Mass, unless it was harvest time, when, presumably, the able-bodied were helping with work in the fields. The behaviour of an ill monk within the infirmary is also specifically regulated. The sick monk could only speak to the infirmarer and to the cellarer, who served the food. The only exception to this was when the bell had sounded the canonical Hour at the end of a meal, and a monk who needed some assistance to rise from the table was allowed to request this.

Themes of separation in the infirmary were manifested not only in the demarcation of the site and its inmates, but also in the separation of this space from outside visitors and guests of the monastery. This is evident from the reports to the Cistercian General Chapter of incidents where abbots of various houses were punished for allowing seculars into the infirmary space. Instances of misuse of the infirmary crop up as early as 1189, when it was generally decreed that any abbot, prior, or cellarer of any Cistercian house who allowed seculars into the kitchen of the monks' or lay brothers' infirmary was to spend every Friday on bread and water.[18] When, in 1192, a Spanish abbot and others disobeyed the rule, they were required to spend three days in 'levi culpa', as well as the usual bread and water punishment.[19] The thirteenth-century examples of breaches of this rule appear, in the main, in the early part of the century. In 1205 there are three examples of misuse of the infirmary space. The abbot of Carciloci who allowed bishops, archbishops and other abbots into the monks' in-

---

seems to be based on Conrad of Mainz's 1261 decree that every monastery should have an infirmary for old people, rather than on a determinable demographic shift—which for this period is enormously difficult to establish, particularly for populations such as the old, or the poor.

[16] Griesser, 'Ecclesiastica Officia', pp. 255–56 for the following.

[17] *Statuta* 1262, 8, t. 3. The separate chalice may support David Bell's contention that understandings of infection may have been important in the siting and uses of Cistercian infimaries. See Bell, 'Siting and Size'.

[18] *Statuta* 1189, 27, t. 1.

[19] *Statuta* 1192, 13, t. 1.

firmary spent three days in 'levi culpa', one of those on bread and water, and was obliged to give up his position as abbot for forty days. The abbot of Rigniaci was subjected to similar punishment the same year. But by far the worst infringement of the rule was committed by the abbot of Pontigny, who let the Queen of France and her large retinue—at least some of whom were women—into the monastery, where they heard the sermon in chapter, took part in the procession in the cloister, ate in the infirmary and stayed in the infirmary for two nights. For this, the abbot received the penalty of giving up his post until Easter, being in 'levi culpa' for six days, and fasting on bread and water for a day. While the General Chapter was dealing with his misdemeanours, the abbot was told to get rid of the decorative tiled pavements he had installed in the monastery, as they were a scandal to everyone.[20] More generally, secular people—and this presumably means hired servants—were not permitted to serve in the infirmary until 1257.[21]

Discipline continued to be enforced within the infirmary if a monk was disruptive, or had either complained about or transgressed the rules of the infirmary. If the abbot saw fit, and if the monk's physical condition allowed it, a continually disruptive monk would be brought into the chapter house and judged for his perversity there. Again, depending on the severity of the illness and the abbot's decision, the monk could be beaten. In the same spirit, a monk who suffered what was perceived to be a condition that did not weaken him too much—a cut or an abscess, for example—was forbidden to lie down for more than a short time, was not allowed to alter his diet in the refectory, and was not to shirk any of his duties.[22]

Ideas of separation in the positioning and administration have been interpreted in a number of ways. David Bell postulates that the separation of the infirmary mirrored something of a monastic preoccupation with health—as evidence that monks were faintly 'modern' in their insistence that the disease-ridden and infectious should keep well away from the robust.[23] Likewise, the expansion of Cistercian infirmaries during the thirteenth century has been cited as evidence that Cistercians were becoming more interested in medicine in general. The hiding away or enclosure of the sick in the monastic milieu has been contrasted with the display of the sick in other medieval medical situations.

---

[20] *Statuta* 1205, 41, t. 1; *Statuta* 1205, 25, t. 1; *Statuta* 1205, 10, t. 1.

[21] Lucet, *Les codifications cisterciennes,* p. 130. This is reiterated in *Statuta* 1274, 12, t. 3.

[22] Griesser, 'Ecclesiastica Officia', p. 256. The reference to the food refers to the eating of meat, allowed in the infirmary but nowhere else. This was a ruling that was repeated time and time again during the course of the thirteenth century. See Lucet, *Codification 1237,* 'De non comendendis carnibus extra infirmitorium'.

[23] See for instance Bell, 'Siting and Size'. Bell suggests that despite the notorious vagueness of medieval understandings of infection, the separation of the sick from the healthy may indicate that infirmaries were positioned at a distance from the monastery proper as a preventive measure against the spread of disease. Bell also suggests that the infirm may have been seen to have attracted the 'stigma of sickness' as a result of divine retribution, and that these 'suffering sinners' were best 'kept at a distance'.

According to Roberta Gilchrist, for instance, the siting of leper houses and borough hospitals on the fringes or edges of medieval urban centres can be seen as a way of exposing the diseased body at the same time as segregating it from the community.[24] Visitors to an urban centre would encounter the hospital at the periphery of the town. In religious houses, Gilchrist argues, the religious infirmary operated within principles of enclosure more than display.

The Body

Custodians of the Body

According to the *Ecclesiastica Officia*, the relationship between the inhabitants of the infirmary and the rest of the monastery was mediated by the infirmarer, or the *servitor infirmorum*. Although the *servitor* was required to be present every day in chapter, it was forbidden that he reveal any of the business and events that took place there to any of the monks in the infirmary. Again, the silence was not to be broken by the infirmarer and he was to communicate to the cellarer only by sign. He was also required to act as a porter for the incoming monks, carrying the *scyphum* and the daily allowance of ale (*iustitiam*), to the infirmary, and was responsible for lighting the candle at Matins, bringing the necessary books to the infirmary chapel for the office and returning them to the book cupboard before Compline. The *servitor infirmorum* was only allowed to read or work in the infirmary if he was not required to tend the sick. If there were enough people in the infirmary to serve themselves at mealtimes, then the servitor was allowed to eat in the refectory with the other monks. If, however, there was only one inmate, or there was a monk who required constant supervision in the infirmary, then the *servitor* was obliged to remain with him. Other duties required of the *servitor infirmorum* included performing the *mandatum* rite on Saturdays, reporting to the refectory and chapter house if a monk was to return to the cloister, lighting the fire in the infirmary and cleaning the bowls used to receive the blood from periodic bloodletting. When a monk died, it was the *servitor infirmorum* who was responsible for washing the corpse and placing it on a grey cloth on the ground, and it was also the *servitor infirmorum* who was to report the death to the community by ringing a bell, or by writing the news up on the *tabula* in the cloister.[25]

The duties of the monk in charge of the infirmary were clearly not, for the main part, what we might describe as medical ones. The function of the *servitor* was essentially administrative, rather than what we would understand to be medical. So who were the medical practitioners in the Cistercian houses if the *servitor infirmorum* was more of an orderly? Several sources indicate that there

---

[24] Roberta Gilchrist, 'Medieval Bodies in the Material World: Gender, Stigma and the Body', in *Framing Medieval Bodies*, ed. by S. Kay and M. Rubin (Manchester: Manchester University Press, 1994), pp. 43–61.

[25] Griesser, 'Ecclesiastica Officia', pp. 276–77 for the preceding.

was not necessarily a single *medicus* or *physicus* always responsible for the infirmary; rather, medical practitioners from outside the abbey were summoned. This is not to say that there was no medical knowledge within the monasteries themselves, or that the Cistercian houses were entirely reliant on the expertise of non-Cistercians in medical matters.[26] However, who is to practise medicine is not delineated in any great detail in the *Ecclesiastica Officia* and the strong presence of *medicus* and *physicus* in charters as witnesses testifies to the receptiveness of the Cistercians to 'outside' intelligences.

Talbot and Hammond's biographical register of medical practitioners in medieval England (MPME) turns up the names of several practitioners associated with the Yorkshire Cistercian houses, mainly as witnesses to charters.[27] During the twelfth and thirteenth centuries, there are sixteen names variously described as *medicus* and *physicus*. Adam, *physicus* of Fountains Abbey, for example, was witness to a number of charters, and Talbot describes him as 'presumably a leech of the West Riding of Yorkshire [...] summoned to the monastery for medical practice and attesting legal documents'.[28] So too was Benedict, a *physicus* of York in the early thirteenth century, whose brother William (identified as *medicus*) often appears in the same charter as witness.[29] Paulinus of Leeds, something of a local identity, witnessed at least three charters for Rievaulx in the late twelfth century,[30] as did Richard, *physicus* of York, and Stephen, *physicus*.[31] At Jervaulx, the foundation history mentions Peter de Quincy, a Savignac monk who was 'expert in the art of medicine.'[32]

---

[26] Which is the implicit opinion of E. A. Hammond, 'Physicians in Medieval Religious Houses', *Bulletin of the History of Medicine*, 32 (1958), 105–20 (p. 113, note 50). Hammond says that all large Benedictine houses had a trained physician, but that 'I venture no such conjecture [...] for the Cluniacs, Cistercians and Carthusians, whose service in England was generally without distinction'.

[27] C. H. Talbot and E. A. Hammond, *The Medical Practitioners of Medieval England: A Biographical Register* (London: Wellcome Historical Medical Library, 1965) (hereafter MPME).

[28] MPME, p. 2. Also *Abstracts of the Charters and Other Documents Contained in the Chartulary of the Cistercian Abbey of Fountains*, ed. by W. T. Lancaster, 2 vols (Leeds, 1915), 1, p. 43, no. 3; 1, p. 385, no. 16a; 1, p. 405, no. 108; 2, pp. 754–55, no. 38. The MPME dates Adam's involvement with Fountains as being thirteenth century, but see Faye Getz, 'Medical Practitioners in Medieval England', *Social History of Medicine*, 3:2 (1990), 245–83 (p. 255), who indicates that Adam appears as witness to charters under Bernard, Prior of Newburgh and Ralph, Abbot of Fountains and should thus be given a late-twelfth century date.

[29] MPME, p. 23. And Lancaster, *Abstracts of Charters*, 2, pp. 651–52, no. 22 and 2, p. 652, no. 23.

[30] See R. Holmes, 'Paulinus of Leeds', *Miscellanea* (Leeds: Publications of the Thoresby Society 4, 1895), pp. 209–25; *Cartularium Abbathie de Rievalle*, ed. by J. C. Atkinson (London: Publications of the Surtees Society 83, 1889), nos 86, 167, and 168.

[31] MPME, p. 272 for Richard and p. 326 for Stephen. Also Farrer and Clay, eds, *Early Yorkshire Charters*, 2, 292.

[32] See *Monasticon Anglicanum*, 5, p. 568. Also mentioned in MPME, pp. 252–53.

Recently, the MPME has been amended and added to by Faye Getz, and several new names associated with the Yorkshire Cistercian houses may now be included. Getz notes that the MPME was not an exhaustive list of all medieval English 'doctors', and that the diversity of medical practitioners 'that is a hallmark of medieval English medicine' was well-represented in its pages, but incomplete.[33] The addenda pertinent to the Yorkshire houses begins with the brief chronological correction to Adam of Fountains. Moving down the list, we find Aelred of Rievaulx himself and later on, Walter Daniel. The former is included on the basis of his unspecified 'healing' and the latter on the basis that he refers to himself as *medicus*.[34] Aelred's healing refers to a number of stories in Walter Daniel's *Vita Ailredi*. One of these involves the sub-prior of Revesby, a daughter house of Rievaulx, who:

> had long been the victim of very sharp attacks of fever [...] his veins were so dried up that he could scarce retain the panting breath in his body. His frame was so wretchedly wasted that it looked like the hollowed woodwork of a lute; eyes, face, hands, arms, feet, shins, blotched and misshapen.[35]

Aelred came to visit the sub-prior in the infirmary and told him to return to the choir the next day, which—miraculously—he was able to do. Aelred, himself a sick man, was ultimately confined to the separate quarters built for him when he became ill, where two of the brethren looked after him until he died.[36]

This story shows that the term 'medical practitioner' may have a number of implications and meanings. Treating, tending to, and curing the sick took many forms and could be expressed in different ways. Getz's inclusion of Aelred in her list of medical practitioners is based on the epistemological assumption that what constitutes a 'doctor' is an extremely broad category in this period. This assumption, with which I agree, is sympathetic to medieval understandings of medicine and the role of the *medicus*, *physicus*, *phlebotomarius*, *infirmarius* and others. Not only was healing a technical practice, but it was also, in a Christian context, an extremely powerful practice. This power could, of course, be exerted in beneficial ways, as Aelred's story demonstrates. The miraculous element in healing is the focus in this tale, and was not an uncommon theme in other Cistercian writings of the same period. An example of this is found in an early

---

[33] Getz, 'Medical Practitioners', p. 245.

[34] Getz, 'Medical Practitioners', p. 255. See also E. J. Kealey, *Medieval Medicus: A Social History of Anglo-Norman Medicine* (Baltimore: Johns Hopkins University Press, 1981), p. 123 for Aelred and p. 149 for Walter Daniel.

[35] *Life of Ailred*, p. 29: 'Supprior itaque eiusdem domus vir religiosus et timens acutissimis febribus tenebatur longo iam tempore, quarum immoderato ardore sanguis vitalis propre modum exhaustus et vene corales exsiccate vix modice palpitanntem spiritum circa precordia permiserant residere. Ipsa precordiorum exterior fabrica miserabiliter artefacta concavum cithare lignum formare visebatur [...]'.

[36] *Life of Ailred*, p. 39: '[I]ussit sibi fieri mausoleum iuxta communem cellam infirmorum et ibi consistens duorum solacio fratrum curam tocius infirmitatis sue subiecit [...]'. See Fergusson and Harrison, *Rievaulx Abbey: Community, Architecture, Memory*, p. 131 for the plan of the infirmary complex at Rievaulx.

thirteenth-century manuscript from Rievaulx, which contains a collection of miracles of the Virgin. It recounts the story of a man who suffered so grievously from the *mal des ardents*, that he amputated his own foot. Mary appeared to him after he cried to her for help, miraculously restoring the foot and healing him.[37]

The practice of medicine could also operate in dangerous ways, and could be abused. This is apparent from a thirteenth-century case brought to the Yorkshire Assize of 1218–19. Simon, a monk and *physicus* from an unspecified house, was outlawed for homicide and arson after the jurors heard that:

> Simon the monk was a doctor and went many times to William of Tanton's house to cure his [William de Tanton's] wife who was sick. At last he [Simon] had intercourse with her, and made a confidant of William de Vacher, so that this William often went to the lady as Simon's messenger. And William [de Vacher] heard that Simon and his associates planned to kill William of Tanton his lord, and he immediately went to him and told him the whole story and how Simon had had intercourse with his wife. And Simon heard how he had disclosed his plan and came to the house of William de Vacher and killed him.[38]

Of more importance among those who could be construed as medical practitioners are the abbots of the Cistercian houses. This is particularly the case within the context of bloodletting. It is through this routine practice that we can discern the efficacy of regulatory controls being exercised over the body of each monk. It is clear from the legislative controls implemented by the Cistercian institution that the body was a complex site. The body was separated in infirmity from the rest of the abbey, regulated by both medical practitioners and the abbot and normalized by bloodletting.

Blood and the Body

Bloodletting was a practice undertaken in both monastic infirmaries and non-monastic hospitals. As something of a lens through which to look at the meanings of the infirmary site more generally, the practice of bloodletting shows the many ways in which the body was understood and defined in the monastic world of the thirteenth century. The practice of bloodletting was entrenched in the monastic world, and in medieval medicine in general. Cistercian legislation, as outlined previously, did not provide detailed medical advice to particular practitioners. But what does become clear from the instructions for bloodletting is just how much influence the abbot wielded over decisions concerning the

---

[37] London, British Library, MS Arundel 346. The disease is described as 'in ardoris languore'. *A Catalogue of Romances in the Department of Manuscripts in the British Museum*, ed. by H. D. L. Ward and J. A. Herbert (London, 1883–1910), p. 620 notes that *mal des ardents* was an epidemic of erisypelas (OED defines this as a streptococcal infection producing inflammation and a deep red colour on the skin) that 'ravaged' the north of France in 1128 and 1129, and was recorded by Anselm of Gembloux, *PL* 160, col. 251. For another example see Herbert of Clairvaux's *Liber Miraculorum*, *PL* 185, col. 1141, where a monk is cured of his tendency to harm himself by an angelic surgeon.

[38] Clay, ed., *Three Yorkshire Assize Rolls*, pp. 377–78, no. 1045.

monastic body. This would suggest that there was a distinction between medical cases and the practice of phlebotomy in the monastery. Although medical practitioners were summoned from outside the abbey on a frequent basis, it was the abbot, representing the authority of the Cistercian order, who was almost solely responsible for the bloodletting.

The *Ecclesiastica Officia* rules that bloodletting, or *minutione*, should take place four times per year.[39] It was stated that monks ought to be bled for as much and for as long a time as the abbot decreed. The practice assumes a quasi-liturgical aspect in this rule, with extremely precise directions for preparation and recovery. A fire was to be lit in the warming house in preparation for the monk to be bled, who should eat in the refectory beforehand if he wished. Lay brothers, it seems, were to be bled in the abbey proper with the choir monks.[40] The *minutor*, the one to do the bleeding, came to the warming house by order of the prior as did those to be bled.[41] After the bleeding, the monk was to rest in the cloister or the chapter house, or sit down on his bed until he was ready to join the choir. He was not to sing or read in church or cloister, or fulfil any of his usual duties until instruction came from the abbot, the cellarer, the sacristan, or the prior. If the monk had a particular duty for the week, he was required to find someone to stand in for him until he had recovered from the bleeding.[42]

There were also rules governing the performance of the Divine Office by monks who had recently been bled. Those recovering from bleeding should sit down as soon as they arrived at the first entrance of the church, separate from the other monks, and after prayer they were to indicate by sign that they were to remain seated. Recuperating monks were not allowed to prostrate themselves, nor incline their heads in a low bow, apart from at the *Magnificat*, the *Te Deum Laudamus*, and at the benediction and reading of the Gospel. After the canonical Hour, a monk who had been bled had to leave the church before the rest of the community. In the refectory, he was excused from reserving a portion of his food for the poor, and was allowed to receive assistance from the monk sitting next to him if he was not able to cut up his own food. After the meal, when the community returned to church, those who had been bled were allowed to say the thanksgiving prayer—the *gratiarum actio*—while seated, and were allowed to leave the church after the benediction for rest if they wished.[43]

At Vigils, however, a monk who had been bled could either go to the church or the infirmary to chant the Hour, the latter site if that is where the other sick monks were. If the infirmary community were in the chapter house, then the bled monk was to join them there. The readings for this hour were significantly shorter than those taking place in the church. At the usual time for reading and working, the monk who had been bled should rest, and on the fourth day after

---

[39] Griesser, 'Ecclesiastica Officia', p. 253.

[40] Noschitzka, 'Codex Manuscriptus 31', p. 101; *Statuta* 1134, 43, t. 1; *Statuta* 1180, 11, t. 1: 'Conversi non in grangiis, sed in abbatia minuantur'.

[41] Noschitzka, 'Codex Manuscriptus 31', p. 101.

[42] Griesser, 'Ecclesiastica Officia', p. 253.

[43] Griesser, 'Ecclesiastica Officia', p. 253.

the bleeding, when he was allowed to return to work, he was given a lighter workload or was excused from certain duties altogether.[44]

Body and Institution

There are a number of ways of looking at the practice of bloodletting in the monastic milieu. One view is to read bloodletting as functioning as an intimate strategy of institutional control. Given the monastic debt to Galenic physiology, to which I will return shortly, the draining of the blood could be seen to represent the restoration or rebalancing of the body's fluids. This was based on an understanding that the body, made up of unstable and disorderly humours, was subject to intense and inevitable fluctuation in heat and moisture. Blood itself, an inherently unstable fluid, was to be drained at particular times of the year. This practice was not, of course, peculiar to the Cistercian order.[45] Bede, for example, had indicated extremely precise times for the draining of the blood, noting that it was between the eighth Kalends of April and the seventh Kalends of June that the blood was most active, and that bloodletting was ideally conducted during this period.[46] At other times of the year, the weather and the moon served as indicators of the practicability of bloodletting. Cistercian practice operated under the same sorts of assumptions.

In the Cistercian context, the regulation of bodily temperature by the elimination of 'bad blood' can also be read as being indicative of the normalizing strategies of Cistercian authority. Bodies, as anthropologists and cultural historians have long argued, are subject to processes of acculturation.[47] In the monastery, these processes might include the penitent and symbolic acts of tonsure or the wearing of a distinctive habit—external ways in which the body was used to express the adoption of the culture of monasticism. Routine bloodletting could be seen to be a similar process. Pervasive monastic agendas of normalization and regulation could be seen to be present in this act—the divesting of the unruly and unstable elements in a body to ensure the order and

---

[44] Griesser, 'Ecclesiastica Officia', pp. 254–55.

[45] For an historical overview of the practice of bloodletting by various religious orders, see L. Gougaud, 'La pratique de la phlébotomie dans les cloîtres', *Revue Mabillon*, 2:13 (1924), 1–13 and more recently L. Moulin, *La vie quotidienne des religieux au moyen âge* (Paris: Hachette, 1978), pp. 158–61. A short overview of bloodletting as a medical technique may be found in I. Kerridge, 'Bloodletting: The Story of a Therapeutic Technique', *Medical Journal of Australia*, 11–12 (1995), 631–33. For continued medieval enthusiasm for this practice, see Bell, 'The English Cistercians and the Practice of Medicine', at p. 164, who cites the discovery of blood-soaked earth at the site of an Augustinian hospital at Soutra in Scotland. Bell says that reports from the archaeologists estimate the quantity of blood in the ground to be as much as 300,000 pints, deposited at the (poorly drained) site over a period of 400 years.

[46] *De Minutione Sanguinis*, *PL* 90, col. 959.

[47] Gail Kern Paster, *The Body Embarrassed: Drama and the Discipline of Shame in Early Modern England* (Ithaca and New York: Cornell University Press, 1993), p. 4.

uniformity envisaged by the *Regula S. Benedicti*: an emphasis on maintaining an internal status quo and therefore external institutional stability.

So it is possible to argue that the practice of bloodletting could be seen as what Michel Foucault might describe as a strategy for making the body docile. This might further be supported by the fact that the director of the routine bloodletting was, according to Cistercian legislation, neither the infirmarer nor an outside medical practitioner, even if these were the people most likely to perform the actual opening of the veins. Rather, it was the abbot who determined the quantity of the blood to be drained from the body, and it was the abbot who decided how frequently this was to take place. By the end of the twelfth century, the General Chapter had reinforced the abbot's role over the routine bloodletting, ruling that the abbot was the only one who could decide when lay brothers were to be bled, and that anyone found in contempt of this practice would lose the opportunity to be bled altogether.[48]

The segregating in church of monks who had been bled could also be seen as significant in terms of control: these monks, as unstable bodies, were distanced from the most sacred areas of the church, and relocated to the periphery until order within their bodies had been regained. The fact that all monks were subject to bloodletting and that the practice was routine supports the notion that the practice was not entirely medical, but rather more institutional. In this way, the bodies recuperating from bloodletting in the infirmary were the newly stable bodies constructed by the monastic institution, a corporeal reminder that monastic regulation involved interior just as much as exterior discipline.

This sort of view could be seen to be replicated in non-monastic medical discourse, especially from the thirteenth century onwards. In medical manuscripts from the later Middle Ages, we find more interest in illustrative recording of the precise areas of instability in the body, areas from which blood is best drained. An English manuscript of the early fifteenth century— 'bloodletting man'—shows the exact points on the body where blood is to be drained (Fig. 5.6). The bloodletting bowl is on the ground, ready to receive the blood, while there are mnemonic notes to the medical practitioner around the central figure indicating to which veins the diagram refers. The body as a text on which instability could be read and then eliminated appears in medical compendia in relation to disease in general as well. A German manuscript in the later fourteenth century incorporates the labelling into the body itself (Fig. 5.7). 'Disease man' is a text for physicians, who would relate particular ailments to particular bodily sites. And slightly later, the same idea is found in 'wound-man' (Fig. 5.8), who served as a mnemonic to surgeons that external instability can cause internal instability, too.

To return to the process of bloodletting, we could argue that in terms of medical discourses which constructed the body as an inherently disorderly site, that draining tempestuous fluids was a process of order. In the context of the monastic world, we might also see bloodletting as a medical and as an institutional practice. In this milieu, the custodians of the body—abbot and *minutor*—

---

[48] *Statuta* 1184, 6, t. 1.

are also the agents of normalization. The body itself, then, is the subject of discipline.

Body and Cosmos

However, there are more ways of looking at the practice of bloodletting than as a purely institutional procedure. In Cistercian and non-Cistercian anatomical discourses, the body was potentially unstable because it was thought to be composed of humours—parts of the internal body that were intimately related to the natural world.[49] One representation of this relationship between body and cosmos appears in a typical twelfth-century manuscript illustration in which we see the body itself positioned in an orderly universe (Fig. 5.9). East is at the top of the folio, west at the bottom, north to the left, and south to the right. The body is standing in a square, which distinguishes it from the rest of the natural world without. On the edges of the pages are the winds, and short descriptions of the temperatures and so on associated with each direction. Within the square, the top part of text reads: 'the macrocosmos is composed of four elements—fire, air, water and earth. Likewise, the microcosmos, that is man, is composed of four elements, fire, air, water and earth'.

This common discourse of body and cosmos derives, of course, from the medical theories of Galen, whose influence in the Middle Ages has been well documented.[50] In Galenic discourse, the connection between these four humours and the four elements produced cosmological theories of the body in which air and springtime were associated with blood and a sanguine temperament; earth and autumn with black bile and melancholy; water and winter with phlegm and a phlegmatic temperament and fire and summer with yellow bile and choler.[51] Jacquart and Thomasset have suggested that this basic doctrine faced modifica

---

[49] See Siraisi, *Medieval and Early Renaissance Medicine* for an account of this type of illustration. For its other iconographies in relation to the body, see Michael Camille, 'The image and the self: unwriting late medieval bodies' in Kay and Rubin, eds, *Framing Medieval Bodies*, pp. 62–99.

[50] For some of the main contributions to this vast field, see Rudolph E. Siegel, *Galen's System of Physiology and Medicine* (Basel and New York: Karger, 1968); Owsei Temkin, *Galenism: Rise and Decline of a Medical Philosophy* (Ithaca and New York: Cornell University Press, 1973); Danielle Jacquart and Claude Thomasset, *Sexuality and Medicine in the Middle Ages*, trans. by M. Adamson (Princeton: Princeton University Press, 1988), esp. pp. 48–52; Thomas Laqueur, *Making Sex: Body and Gender from the Greeks to Freud* (Cambridge, MA: Harvard University Press, 1990); Siraisi, *Medieval and Early Renaissance Medicine*.

[51] Jacquart and Thomasset, *Sexuality and Medicine*, p. 51, for a diagrammatic representation of this principle. Also Marie-Christine Poucelle, *The Body and Surgery in the Middle Ages*, trans. by R. Morris (Cambridge and Oxford: Polity Press in association with Basil Blackwell, 1990), pp. 160–78, 'The Body Under Attack from Nature', for a fourteenth-century expression and expansion of the same idea by John de Mondeville.

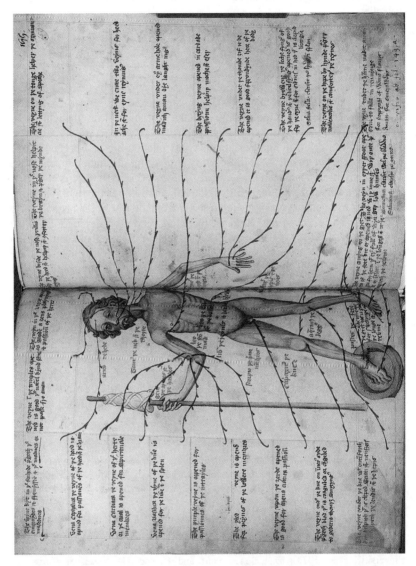

Fig. 5.6: 'Bloodletting man', London BL MS Harley 3719, fols 158ʳ–159ʳ (Courtesy, British Library, London)

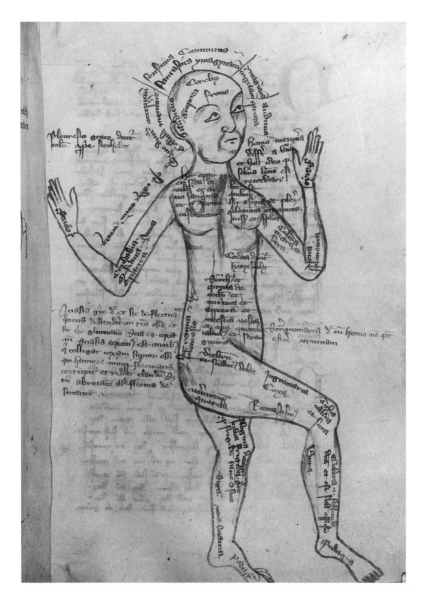

Fig. 5.7: 'Disease man', London BL MS Arundel 251, fol. 37 (Courtesy, British Library, London)

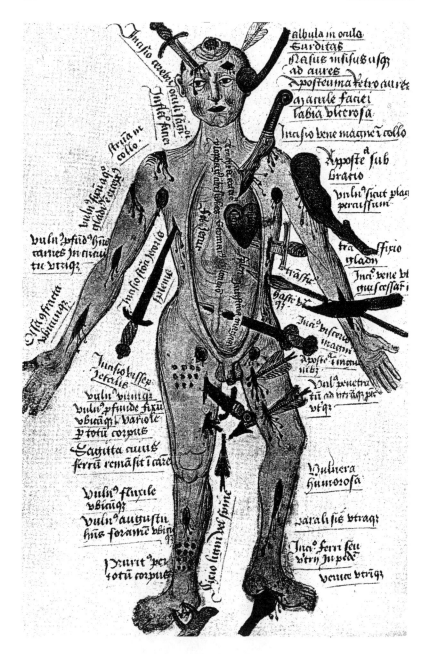

Fig. 5.8: 'Wound man', London Wellcome MS 290, fol. 53 (Courtesy, The Wellcome Library, London)

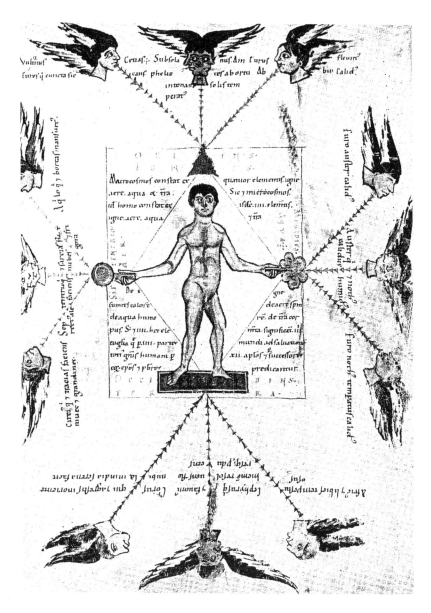

Fig. 5.9: Man positioned in relation to the cosmos, Wien, Österreichische Nationalbibliothek, MS 12,000, fol. 29ʳ (Courtesy, Bildarchiv der ÖNB, Wien)

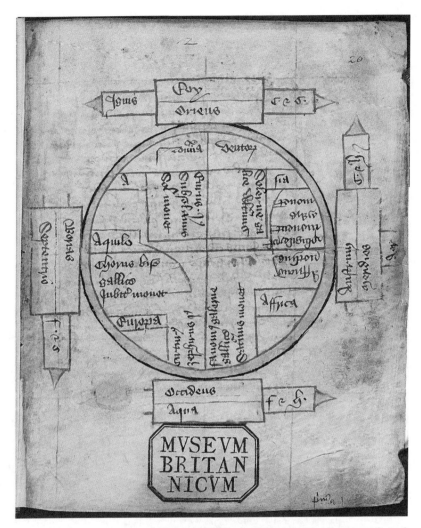

Fig. 5.10: The cosmos, London BL MS Sloane 795, fol. 20 (Courtesy, British Library, London)

tion 'as early as antiquity', when tripartite structures were found to underlie the flow of blood around the body.[52]

Nonetheless, the Cistercian debt to the traditional Galenic model is apparent from even the most cursory of medical notes in the extant Yorkshire manuscripts dealing with medicine. Humourology provides the basis for understanding anatomy and the relationship between body and cosmos. One example is found in a late-twelfth/ early-thirteenth century manuscript from Byland, which contains among other items the beginnings of what promises to be an alphabetized and exhaustive list of herbs. The list is abandoned soon after the letter 'B', however, and a brief summary of physiology follows:

> There are four winds. There are four ranks of angels in heaven. There are four times of the year: spring, summer, autumn, winter. There are four humours in the human body: red bile, black bile, blood, and phlegm.[53]

This simple résumé goes on to describe (briefly) the characteristics of these humours; the black bile exists in summer under the spleen, and it is cold and dry. The red bile, on the other hand, is hot and dry, and should be drained in the summer.[54] The information here is also concerned with the correct times of year to adjust the temperature of the body, particularly in relation to bloodletting. After the Ides of July, bloodletting should be suspended, according to this manuscript, as it was prior to this time that the blood was at its most active, turbulent and disruptive.[55] The summary ends abruptly with a quadrapartite summmary of the body's compostition.[56]

Similar information appears in a manuscript from Fountains Abbey, which like the Byland manuscript contains primarily non-medical tracts.[57] The material relevant to medicine, however, namely an anonymous commentary on Aristotle's *De Physiognomia* and *Meteorum* (Book IV), relates anatomy to cosmology. The appearance of anatomy and cosmology in the same compendium continues in a thirteenth-century manuscript from Kirkstall, a reproduction of a well-known *Medulla Philosophorum*, which includes everything from descriptions of the sea to explanations of the properties of gold and silver, from

---

[52] Jacquart and Thomasset, *Sexuality and Medicine*, p. 50.

[53] Cambridge, Trinity College, MS 1214, fol. 100: 'Quatuor sunt venti. Quatuor angeli celi. Quatuor tempora anni; vernum, estas, autumnas, hyems [sic], Quatuor humores sunt in humano corpore; colera rubea, colera nigra, sanguis et flegma'.

[54] Cambridge, Trinity College, MS 1214, fol. 100: 'Colera nigra habitat in estate sinistra sub splene [et est] frigida et sicca ... Colera rubea calida et sicca qui estatis tempora corpora exusta reddit'.

[55] Cambridge, Trinity College, MS 1214, fol. 100: 'Post idus iulii suspendetur medice flebotomii'.

[56] Cambridge, Trinity College, MS 1214, fol. 100: 'In quatuor partes divimus corporis humanum: caput, stomachum, ventrem atque vesicam'.

[57] Oxford, Bodleian Library, MS Laud. Misc. 527. Dated 12th–14th century by H. O. Coxe in *Bodleian Library Quarto Catalogues II: Laudian Manuscripts*, repr. edn (Oxford, 1973).

tracts on digestion and indigestion to explanations of plants and herbs, from discourses on blood to discourses on water.[58]

Other medical manuscripts turn up equally standard material. Oxford Bodley MS 514 is a thirteenth-century manuscript from Jervaulx abbey.[59] The manuscript contains five commentaries or summaries of various well-known medical tracts sandwiched between Geoffrey of Monmouth's *Historia Regum Anglorum*[60] and the *Liber Translationibus S. Cuthberti*.[61] The medical texts themselves are a gloss on Hippocrates' *Prognostica*, a commentary on Theophilus, a commentary on the *De Pulsu de Philaretus*, a commentary on the *Isagoge* of Johannitius (with some medical notes following) and a summary of an incomplete and unidentified medical text. The medical portion of this manuscript which has attracted some attention has been the commentary on Johannitius.[62] This commentary was, according to Mark Jordan, fairly typical of a series of manuscripts including Oxford, Bodleian MS Digby 108 and Chartres MS 171,[63] although Oxford, Bodleian Library, Bodley 514 includes a long digression on Galen's *Tegni* before the real commentary begins.

Primarily interesting in this commentary is the information given on the theory and practice of medicine. Here we discover the same sort of assumptions about the nature of the human body that appears in the material outlined above.

---

[58] Cambridge, Jesus College MS Q.G.28, fols 69b–198. David N. Bell, *An Index of Authors and Works in Cistercian Libraries in Great Britain* (Kalamazoo: Cistercian Publications, 1992), p. 208 notes an edition of this *Medulla Philosophorum* from Paris, Bibliothèque Nationale, MS Lat. 15879 by M. de Boüard, *Une Nouvelle Encyclopédie Médiévale: le Compendium Philosophie* (Paris: E. de Boccard, 1936). This edition does not take into account the Kirkstall manuscript. A curious medical formula that seems to be based on quasi-magical ideas rather than the cosmological/anatomical relationship appears in Lancaster and Paley-Baildon, eds, *Coucher Book of the Cistercian Abbey of Kirkstall*, p. 40. The recipe is against the falling sickness, and includes the charm words 'Dealbagneth, Debagneth, Degluthun', which were to be chanted whilst making the sign of the cross. This is the only example of its type that I have seen for the Cistercian houses of England.

[59] See *Summary Catalogue of Western MSS in the Bodleian Library*, ed. by F. Madan, 7 vols (Oxford: Clarendon Pres, 1922), 2, Madan's manuscript number 2098.

[60] Oxford, Bodleian Library, Bodley MS 514, fols 1–36b. I would like to thank the Bodleian Library, Oxford, and especially the staff of the Duke Humfries library for their assistance with this manuscript.

[61] Oxford, Bodleian Library, Bodley MS 514, fol. 80 et seq. A late-thirteenth-century table of contents on fol. 1 shows a *Glosae Prisciani de Constructione*, now missing.

[62] Mark D. Jordan, 'Medicine as Science in the Early Commentaries on Johannitius', *Traditio*, 43 (1987), 121–45; P. O. Kristeller, 'Bartholomaeus, Musandinus and Maurus of Salerno and Other Early Commentators of the Avicella with a Tentative List of Texts and Manuscripts', *Italia Medioevale e Umanistica*, 19 (1976), 57–87. For Johannitius more generally, see D. Jacquart, 'A l'aube de la renaissance médicale des XIe–XIIe siècles: "L'Isagoge Johannitii" et son traducteur', *Bibliothèque de l'Ecole des Chartes*, 144 (1986), 209–40. The commentary is on fols 57b–63 of the Bodley manuscript.

[63] The Chartres manuscript was destroyed during the Second World War, but the Digby manuscript is intact.

The commentator views the *Isagoge* as an addition to material already elucidated by Galen, and claims that Johannitius' purpose was to explain and add to Galenic principles. More generally, the commentator reiterates the Galenic idea that medicine is divided into theory and practice and that the practice of medicine is subject to the theory.[64] Medical theory is also divided into components, such as the contemplation of natural things and the study of the human body.[65]

Computation of the best time of year for bloodletting was therefore extrapolated from cosmological understandings just as much as from institutional and disciplinary requirements. This is borne out by the presence in medical manuscripts of simple diagrams illustrating the cosmos, with notes on winds, directions, temperature and so on (Fig. 5.10). One thirteenth-century English example shows a very basic representation of the cosmos, with notes on the main features of its order.[66] The west is associated with water, for instance, and there are divisions between *Europa* to the left of the circle and *Africa* to the right. This sort of diagram is extremely common in medieval medical manuscripts, including Cistercian ones, and accompanies tracts on the care of the body, including bloodletting.[67] In Oxford, Bodleian Library, MS Laud Misc. 527 for example, a wheel diagram of the four winds appears followed by a short explanation (with the preliminary caveat 'sed ista figura est mala facta!').[68] This type of simple diagram is positioned in the manuscript in clear relation to the medical material.

Looking at the tempestuous body as part of a cosmos that could be symmetrically and diagrammatically ordered, we can see that the idea of bloodletting

---

[64] Oxford, Bodleian Library, Bodley MS 514, fol. 57^v: 'Medica divisit in duo partes, in theorecticam et in practicam [...] Practica est subiecta theorie [...]'.

[65] Oxford, Bodleian Library, Bodley MS 514, fol. 57^v: '[i]n contemplacionis naturalis rerum'; 'componencium humanum corpus'. See also D. Furniss, 'The Monastic Contribution to Medieval Medical Care: Aspects of an Earlier Welfare State', *Journal of the Royal College of General Practitioners*, 15 (1968), 244–50 for the view that the monastic contribution to medicine should be seen in terms of its social welfare agenda, particularly in the establishment of monastic hospitals. For the state of medical education in England during the medieval period, see Faye Getz, 'Charity, Translation and the Language of Medical Learning in Medieval England', *Bulletin of the History of Medicine*, 64 (1990), 1–17. For an overview of the development of medieval medical theory, see J. M. Riddle, 'Theory and Practice in Medieval Medicine', *Viator*, 5 (1974), 157–84.

[66] London, British Library, MS Sloane 795, fol. 20.

[67] See Peter Murray-Jones, *Medieval Medical Miniatures* (London: British Library in association with the Wellcome Institute for the History of Medicine, 1984), pp. 20–21 for a very similar diagram included in his discussion of Galenism. Also F. Wallis, 'Medicine in Medieval Calendar MSS', in *Manuscript Sources of Medieval Medicine*, ed. by M. R. Schleissner (New York and London: Garland Publishing, 1995), pp. 105–43 concerning the strong presence of medical material in manuscripts otherwise devoted to astrological information and other forms of 'computing'.

[68] Oxford, Bodleian Library, MS Laud Misc. 527, fol. 115.

might be read as part of the natural order of things. The idea of draining blood at particular times of the year, based on cosmological principles, may simply be the restoration of natural order to a body that in itself is a micro-cosmos, a mini-world. As the climates fluctuate—but according to patterns—so the body responds and replicates these changes. Bloodletting is thus a form of cosmological balancing. In the monastic world, we might also return to the site of the infirmary as support for the idea that Galenic understandings of body and cosmos were omnipresent. David Bell has pointed out that the east was seen in Galenic and Hippocratic thought to be the best geographical location for the maintenance of health, on the grounds that east was the most temperate and stable of climates.[69] The infirmary, to the east of the cloister, thus fulfilled a certain 'cosmological requirement' in its siting. The bodies within the infirmary—the microcosmos within the macrocosmos—signified the natural world and its ordained behaviours.

However, east was also an important direction in liturgy and spirituality generally. In Cistercian, as in other monastic theologies, east was associated with the location of the earthly paradise, and it was associated with light. In the natural world, the east was the source of light from the sun, while in the spiritual world, the light from the east could also symbolize Christ as the lux mundi. East was thus a direction and location in which the natural and the spiritual worlds met. We have seen that Cistercian building programmes in the thirteenth century certainly bear out the significance of the eastern end of the church as an important site of intercessionary prayer, a site where the division between this world and the next was mediated and negotiated through supplication and private masses. In the context of the infirmary site, the building's eastern location was also redolent with spiritual meaning.

The Subjective Body

The tempestuous body associated with the infirmary did not only signify the tempestuous cosmos. It also signified the difficult path toward union with God that every Cistercian had chosen to take. This may be traced in narratives which represent the idea of the body. Bodily discipline is partly represented in Cistercian narratives as a necessary element of a monk's spiritual growth. The *Speculum Novitii* advocates care of the body in forbidding total abstinence, for example,[70] but also uses the site of the body to practise and demonstrate proper behaviour. According to the *Speculum*, if a novice is not hungry at mealtimes, he should eat a little, thinking of Christ 'standing in the cold at the door, waiting for what you have left over', or 'make a cross from five breadcrumbs and say to yourself "Here are the feet, there the hands nailed to the cross; here the head rested, there through his side flowed his mercy"'.[71] The author says: 'should you

---

[69] Bell, 'Siting and Size of Medieval Infirmaries', p. 8.

[70] Mikkers, 'Speculum', p. 57.

[71] Mikkers, 'Speculum', p. 56.

happen to have a nocturnal emission while sleeping, do not be too upset about it. But if you should be invited to serve at a private mass, be sure to make a sign in the presence of the brethren and humbly excuse yourself, as is our custom, so that everybody will know about your defilement during the night'.[72] The body of the novice is made 'docile', as Foucault would call it, by this constant reflection of its actions, while like Matthew of Rievaulx, the novice is reminded that his 'corruptible body [...] is destined to become a heap of dirt and food for worms'.[73]

The ideal Christian body is also occasionally described in terms of medical knowledge. One example appears in the *Narratio* of Hugh of Kirkstall, who includes some common medical knowledge within the spurious letter of Thurstan, archbishop of York. In the letter, Thurstan is said to have written to the archbishop of Canterbury, describing the secession of a group of Benedictine monks from St Mary's, York, to found a new Cistercian abbey at Fountains. Thurstan tells the archbishop that the leader of the group, the prior Richard, believed that St Benedict provided 'medicine for their [his monks'] bodies', and 'health for not only their souls, but also for bodies'.[74] Benedict himself is said to have been moderate and regular in his own diet, which, as Thurstan notes, is medically healthy: 'Moreover, everyone knows that physicians very decidedly uphold the theory that there is no better way of avoiding unhealthy conditions than for every man to aim at a moderate and regular frugality in his diet'.[75] The inclusion of this 'common-sense' information is, of course, primarily related to Christian monastic emphasis on frugality and moderation. However, this information is also important in indicating that what was construed as medical knowledge could be related to the monastic life, and to the body, in very specific ways.

Other descriptions in Cistercian literature provide a more complex picture of monastic understandings of the body. In the first quarter of the thirteenth century, Matthew, the precentor of Rievaulx abbey, wrote to his dear friend William, the prior of Byland. Matthew was in pain when he wrote his letter— 'from the bottom of my feet to the very top of my head', he cried, 'there is no peace in my bones', and he spoke of the great pressure under which he lived and worked.[76] Matthew's letter to William was not the only time he mentioned the stress of his office, and it was also not the only time Matthew talked about his own body in terms of pain. In another letter (*Ad Richardum*), the precentor described in great detail an incident in which his body seemed to fall apart,[77]

---

[72] Mikkers, 'Speculum', p. 57.

[73] Mikkers, 'Speculum', p. 65: 'Id est, corpori corruptibili dic, quod vas stercorum est et esca vermium futurum'; Wilmart, 'Mélanges', p. 73: 'vas stercorum, esca vermium'. cf. Job 25. 6.

[74] *Mem. F.*, p. 14.

[75] *Mem. F.*, pp. 14–15.

[76] Paris, Bibliothèque Nationale, MS Lat. 15157, fol. 88ᵛ: 'Pressura magna circumdedit me'; fol. 89ʳ: 'Officia [...] precentoris plenum est angoris et magne solicitudinis'.

[77] Transcribed in Wilmart, 'Mélanges', pp. 74–79.

while in a shorter poem (*Cuidam Amico*), he tells us that the burden of his office exceeded his strength, which was ebbing away.[78]

Matthew saw his own body as a site where strength and weakness struggled for dominance, and where the demands of monastic life were manifested in discomfort, sadness, and confusion. Matthew of Rievaulx's narrative representations of his own body, for example, reveal the precentor's concern for his personal health and the meaning of bodily pain. Pain is described in a number of ways in Matthew's letters and poems, but is related most often to ideas of weakness. In his letter to Richard, Matthew begins by advising that he will give an account of his 'infirmitatis'—'ne dicam infelicitatis'—and then goes on to recount a great constriction of his stomach, which seemed to rupture and burst his skin. The assault on his body moved to his head, as if a sharp razor had struck his crown and all the skin had been torn out, and he felt as though he should be carried from the singing of the choir to the tomb. The agony in his stomach is described as labour pains—'sicut dolor parientis'.[79]

Matthew says that he began to mourn, sighing long and loud and constantly beating his breast, while calling out to God 'My lord where are you now?' He curses his ill-fortune as precentor, saying that he is falling to pieces under the strain of his office, and that there is more burden than honour in it. His office is injurious and almost impossible to bear; it exceeds his strength and never ends.[80] His responsibilities suffocate his body—'corpusculum destruit'—and preclude the possibility of lasting good health. Yet it is not with lurid retellings of his suffering, or his physical weakness, that Matthew is essentially concerned. His own response to this attack is what has made the precentor anxious, as have, even more significantly, the meanings of this response.

When Matthew realizes that his physical symptoms are related to the stress of his job as precentor, the language of his narrative changes from an account of his dialogue with himself to the reasons for his tribulations. He understands that his condition should be seen as a test, and that the answer to his malaise ought not to be found in his body, but in the example of Christ—'Here is our doctor, among us here'.[81] Confidence in eternal grace will remedy weakness, which is not, as the first part of his letter would indicate, a corporeal matter. Rather, the end of pain and weakness is to be found in the spiritual life. Matthew has moved from talking about the body as essentially weak to the possibilities of strength in Christ. The experience of infirmity is primarily spiritual and psychological.[82]

Another of Matthew's poems raises similar issues in relation to his monastic

---

[78] Wilmart, 'Mélanges', pp. 67–68.

[79] Wilmart, 'Mélanges', p. 75.

[80] Wilmart, 'Mélanges', p. 75: '[s]emper incipit, numquam explicit, vires excedit'.

[81] Wilmart, 'Mélanges', p. 76.

[82] For descriptions of pain and illness in the mystical tradition, see Bynum, *Fragmentation and Redemption*, pp. 188–90. For the relationship between medicine and the otherworlds, see R. Seiler, 'Mittelalterliche Medizin und Probleme der Jenseitsvorsorge', in *Himmel, Hölle und Fegefeuer. Das Jenseits im Mittelalter*, ed. by P. Jezler et al (Zurich: Schweizerisches Landesmuseum, 1994), pp. 117–24.

duties, and again, Matthew uses his own physical weakness as a narrative tool. This time, Matthew appeals directly to the abbot to relieve him of the burden of his office. He describes his body as prostrated at the abbot's feet, and cries for the abbot to pay attention to his cries for relief; he begs for the abbot's rebuke. He is exhausted, having led the night office for a month, and now suffers from insomnia.[83] It is not right, Matthew argues, that the office of the precentor should rest on one set of shoulders. The stress of it means that the job is badly done and Matthew, in particular, is becoming too old to cope with the demands on him.[84] This poem is intensely personal, and unlike the letter to William of Byland, there is no recourse to theology in it. Matthew's own body is very much the focus.

Yet Matthew also describes the body in more abstract terms: to his friend William, he writes intimately, 'we are two from the flesh of one';[85] to the prior of Beverley he writes that the body is vile froth, a vessel for dung and food for worms.[86] In the narrative context of the latter comment, Matthew's reference is specifically related to notions of the body as a site for worldly sin and temptation. This letter to the prior of Beverley is primarily concerned with converting the prior to the Cistercian way of life, thus the dangers of the non-Cistercian world provide a foil for the eulogized narrative.[87] One example is Matthew's interest in poverty. Another is his comment on food and greed, where the body desirous of carnal pleasures is bestialized. And later in this same letter, Matthew says that one feature of man is vile matter.[88]

The complexity of Cistercian discourse on the body has been discussed mainly in relation to Bernard of Clairvaux.[89] It is worth noting here that St Bernard's vision of the body is not particularly hostile. Only infrequently does he refer to it as a vessel of sin.[90] This is generally because the body is seen as requiring maintenance—feeding and so forth, not because it is inherently flawed. It should also be remembered that cosseting the body, or paying undue

---

[83] Wilmart, 'Mélanges', p. 67: 'Uno mense novem noctes duxi vigilando/ Si michi non credis, tu peccas in dubitando/ Egit eas gemitus insompnes absque sopore/ Inde notare potes in quanto vivo dolore.'

[84] Wilmart, 'Mélanges', p. 68: 'Precentoris unus humeris imponere nostris/ Non decet. Inquiris causa? Obstat valitudo/ Que? Dolor interior me cogit prorsus obire/ Et genus hoc mortis ad tempus nescit abire/ Regula iuris habet: eis pausatio detur/ Quos iugis morbus aut deprimit egra senectus.'

[85] Paris, Bibliothèque Nationale, MS Lat. 15157, fol. 89ᵛ. Matthew often writes in an extremely intimate manner; another example is the opening lines of his letter to Henry of Beverley: 'Amantissimo fratri H. De Beverlaco [...] in illum suspirare cuius pulchritudinem sol et luna mirantur.' See Wilmart, 'Mélanges', p. 72.

[86] Wilmart, 'Mélanges', p. 73.

[87] Wilmart, 'Mélanges', pp. 69–70, where a poem describing the delights of Fountains Abbey refers to that house as 'egros medicamine sanas' and 'fons celestis medicine.'

[88] Wilmart, 'Mélanges', p. 73.

[89] For example, Bynum, *The Resurrection of the Body*.

[90] For example, *De Consideratione*, 2, 4, 7.

attention to its care, is represented as sinful. One instance is described in a collection of exempla tales from Clairvaux abbey. A brother of Clairvaux, who took great care of his body and procured strange medicines, experienced a vision in which the Virgin Mary appeared to the community and gave each monk a spoonful of liquid from a precious vessel which she carried. When the brother approached the Virgin for his share, she sent him away, saying that he was a wise doctor and that he had no need of her medicine, as he took such care of himself. The brother, mortified, begged her pardon for his selfishness, which was granted.[91]

The Cistercian monastic body was thus represented in a number of ways. Personal narratives such as that of Matthew of Rievaulx reveal that the body could be a source of concern for Cistercian monks, not so much because the body was the means by which a monk could transgress, but because it was the site in which abstract ideas of spiritual strength and weakness were made immediately present. Other descriptions reveal that regulation and order of the body were of paramount concern to Cistercians. Representations of the body as requiring discipline to promote spiritual growth are based on perceptions of the body as inherently disorderly. Institutional regulation of the body, through such means as bloodletting, was supported by individual understandings of the body as a complicated and tumultuous space, in relation to not only the Cistercian institution, but in relation to the cosmos, spirituality and the quest for union with God.[92]

Body and Landscape

In this way, the body was also intimately connected with the landscape of the Cistercian monastery, as it is only within this specific environment that the monk may find the way to God and the tempestuous body may become stable. The Cistercian monastic landscape was—as one thirteenth-century poem described—the source of eternal medicine.[93] It was also heaven on earth, the earthly manifestation of heavenly space, a site that was suffused with celestial longing. The infirmary, as part of this metaphorical economy, represents the blurring of boundaries between the material world and the world of eternity. The monastic infirmary, which was the subject of so much development during the

---

[91] Paris, Bibliothèque Nationale, MS Lat. 15912, fol. 133. This story also appears in Herbert of Clairvaux, *Liber Miraculorum*, *PL* 185, col. 1366.

[92] For some critical analysis of the construction of the medieval body, see Jacques Le Goff, 'Corps et idéologie dans l'Occident médiévale: la révolution corporelle', in *L'Imaginaire médiéval: essais*, ed. by J. Le Goff (Paris: Editions Gallimard, 1985), pp. 123–27; Kay and Rubin, *Framing Medieval Bodies*. For the ways in which the body has been described historically, see for instance Roy Porter, 'History of the Body', in *New Perspectives on Historical Writing*, ed. by P. Burke (Cambridge: Polity Press, 1991), pp. 206–32.

[93] Matthew of Rievaulx, *Epistulare Carmen de Fontibus*, in Wilmart, 'Mélanges', p. 69: 'fons celestis medicine.'

thirteenth century, should be understood as more than a space of medicine. Cistercians were able to incorporate medical discourse and particularly Galenic physiology to express the agendas of enclosure and freedom that underpinned monastic life. But at the same time, specific spiritual agendas were also at work in constructing the monastic body. The body was described—through cultural practices such as bloodletting—as a space where institution, cosmos, spirituality, and landscape were all present, while the medieval infirmary was a site where these many meanings were signified. Renewal of Cistercian interest in the infirmary during the thirteenth century was part and parcel of Cistercian anxiety about the body and the spatial worlds it was destined to encounter.

CHAPTER 6

Status, Space, and Representation: The Cistercian Lay Brother

Serlo of Fountains remembers that, at a time of great peace and prosperity in the early days of the abbey's foundation,

> the enemy of peace, our ancient foe [...] threw their [i.e., the monks'] camp into confusion, casting contention among the citizens, discord among the brethren [...] dissension among the holy brotherhood. And lo, suddenly the thorns of scandal began to spring up, the shoots of bitterness to come forth, the poison of discord to be diffused, the darts of calumny to fly hither and thither. They attacked each other [...] All their indignation was kindled against the abbot, a mutiny broke out, and the sons rose against their father.[1]

But the guidance of abbot Richard was enough to confound the stratagem of the devil, and the community 'returned to their right minds, were amazed at their reckless presumption, [and] repented'. From that day in the mid-twelfth century, says Serlo, 'no such thing was ventured on at Fountains, nothing was rashly attempted by the holy brotherhood against the rule of the order'.[2]

From the later twelfth century and throughout the thirteenth, however, other Cistercian houses did not escape dissent and disorder so easily. From 1200–1300 there are many reports to the Cistercian General Chapter of violent episodes and insurrection in various monasteries. Almost one hundred of these cite lay brothers (*conversi*) as either taking part in the trouble or as being the instigators.[3] A complex picture of the lay brother emerges from these reports of

---

[1] *Mem. F.*, pp. 112–13.

[2] *Mem. F.*, p. 113: '[e]t ab illa die et deinceps apud Fontes, nichil tale praesumitur, nichil a conventu sacro, contra formam ordinis temere attemptatur'.

[3] From the statutes 1200–1300, there are ninety-three incidents reported to the General Chapter which specifically cite lay brothers as having been involved, or having possibly been involved. How to measure what constitutes a disturbance or an incident of some

violence during the thirteenth century, a picture that constructs him as a potential 'threat from within', a danger in the midst of Cistercian stability, and a figure on the fringes of monastic life. Narrative representations of the lay brother replicate this view of the *conversus*, but, paradoxically, are also able to represent the lay brother as the epitome of monastic humility. Some examples are to be found in Herbert of Clairvaux's *Liber Miraculorum*, Caesarius of Heisterbach's *Dialogus Miraculorum* and other exempla collections such as the *Exordium Magnum*. An English example is Hugh of Kirkstall's *Narratio*.

For continental Cistercian houses, it is clear that from the early thirteenth century, the Cistercian lay brother came to occupy an ambiguous social and spiritual space within the order, serving as a measure of both probity and deviance. This contrasts vividly with the less complex situation during the early years of Cistercian expansion, when it seems that literate nobles and others who would normally become choir monks were demonstrating their desire for absolute humility by choosing to become lay brethren instead. In 1188 the General Chapter was forced to legislate against nobles entering the lay brotherhood—despite its clear and attractive association with simplicity, humility, and piety—presumably on the grounds that the popularity of the lay brotherhood could lead to depletion in the ranks of choir monks.[4] Such unequivocal virtue attached to the idea of the *conversi* was not to last, as this chapter will explore.

In England during the same period, we find a similar ambiguity in the prescribed spaces allocated to the *conversi*, both within the monastery and without. Although instances of disruption are few, the lay brother seems to have

---

seriousness is, of course, not always easy to determine, as I will discuss later on in this paper. The incidents which I accept as constituting or referring to a disturbance are *Statuta* 1200, 28, t. 1; 1201, 48, t. 1; 1202, 14, t. 1; 1202, 33, t. 1; 1202, 46, t. 1; 1203, 50, t. 1; 1204, 34, t. 1; 1205, 14, t. 1; 1205, 60, t. 1; 1206, 19, t. 1; 1206, 23, t. 1; 1206, 50, t. 1; 1206, 59, t. 1; 1207, 61, t. 1; 1208, 33, t. 1; 1209, 11, t. 1; 1209, 19, t. 1; 1212, 46, t. 1; 1212, 52, t. 1; 1213, 25, t. 1; 1213, 46, t. 1; 1217, 32, t. 1; 1217, 78, t. 1; 1218, 42, t. 1; 1219, 31, t. 1; 1219, 40, t. 1; 1220, 14, t. 1; 1220, 44, t. 1; 1223, 25, t. 2; 1225, 26, t. 2; 1226, 23, t. 2; 1230, 12, t. 2; 1232, 24, t. 2; 1236, 40, t. 2; 1236, 57, t. 2; 1237, 66, t. 2; 1237, 69, t. 2; 1238, 51, t. 2 incidents at Preuilly and at Pontigny; 1238, 52, t. 2 incidents at Eberbach and at Grandselve; 1240, 41, t. 2; 1241, 19, t. 2 an incident at Eberbach again; 1241, 26, t. 2; 1242, 58, t. 2; 1243, 31, t. 2; 1243, 43, t. 2; 1243, 70, t. 2; 1246, 31, t. 2; 1247, 20, t. 2; 1247, 61, t. 2 incidents at Blandecques and Heiligenkreuz; 1251, 74, t. 2; 1252, 51, t. 2; 1253, 35, t. 2; 1255, 27, t. 2; 1255, 28, t. 2; 1256, 7, t. 2; 1258, 4, t. 2 another incident at Berela; 1259, 34, t. 2; 1260, 25, t. 2 incidents at S. Vincenzo, S. Martino del Monte, S. Sebastiano, and Casanova; 1260, 27, t. 2; 1261, 32, t. 2; 1262, 43, t. 3; 1266, 27, t. 3; 1266, 53, t. 3; 1267, 48, t. 3; 1268, 1, t. 3; 1268, 37, t. 3; 1268, 54, t. 3; 1269, 19, t. 3; 1269, 71, t. 3; 1270, 32, t. 3; 1271, 17, t. 3; 1272, 5, t. 3; 1273, 16, t. 3; 1275, 22, t. 3; 1275, 24, t. 3; 1276, 14, t. 3; 1277, 28, t. 3; 1278, 13, t. 3; 1279, 67, t. 3; 1280, 16, t. 3; 1280, 19, t. 3; 1280, 25, t. 3; 1281, 33, t. 3; 1289, 14, t. 3; 1291, 36, t. 3; 1292, 8, t. 3; 1293, 9, t. 3. This list is a more conservative estimate than that produced by James Donnelly, *The Decline of the Medieval Cistercian Laybrotherhood* (New York: Fordham University Press, 1949), who appends a list of 123 'revolts' to his study. I will discuss the problems in Donnelly's analysis later.

[4] *Statuta* 1188, 8, t. 1.

been increasingly marginalized from the mid-thirteenth century, both in status and in practical terms. The physical and spiritual spaces occupied by Cistercian *conversi* had always been carefully demarcated, yet it is from the thirteenth century that we find these spatial demarcations more and more carefully iterated in topography and in the written word. As the General Chapter legislated against what appears at first sight to be increasing outbreaks of lay brother violence in France especially, we find in England that the lay brother's function within the monastery becomes more carefully delineated. Attractive as it would be to postulate a direct connection between the two situations, such a link would be erroneous in itself. Yet what we do see throughout the thirteenth century in England and in the continental houses, is the formulation of the lay brother as a figure of ambiguity, whose presence and function within the monastery and its sites needed to be very precisely and deliberately defined. In this way, the lay brother was quasi-secular—even liminal—not unlike the Cistercian novice who, by virtue of his institutional status, tested the imagined boundaries of monastic space.

The Status of the Cistercian Lay Brother

Lay brothers were a constituent part of the Cistercian order almost from its legislative inception,[5] although the *Regula S. Benedicti* was not the body of regulation that lay brothers were to follow. Rather, their vocation and conduct was established by the *Usus Conversorum*, a set of rules apparently written by Stephen Harding in 1119. This was the basis for the *Regula Conversorum*, written at Clairvaux around 1174.[6] The purpose of lay brothers in the Cistercian abbey was to some extent auxiliary, their main task being manual work (providing the choir monks with food and attending to any farming), leaving the choir monks free to concentrate on prayer.[7] Often, lay brothers' work was described as the work of Martha, while the choir monks' more contemplative life was seen as the work of Mary.[8] The lay brother was also the go-between, acting as monastic agent for any transactions with the 'outside world', again, allowing the choir monks to remain within the cloister, as the *Regula S. Bene-*

---

[5] For general accounts, see Jean de la Croix Bouton, *Histoire de l'Ordre de Cîteaux* (Westmalle: k, 1959); M. LaPorte, 'Origine de l'Institution des Frères', in *Dictionnaire de spiritualité ascétique et mystique, doctrine et histoire*, published under the direction of Marcel Viller, SJ (Paris: G. Beauchesne et ses fils, 1937–1995), vol. 5, cols 1194–1207; Conrad Greenia, 'The Laybrother Vocation in the Eleventh and Twelfth Centuries', *Cistercian Studies*, 16 (1981), 38–45; Lekai, *Cistercians: Ideals and Reality*, pp. 334–36.

[6] For a résumé of the *Usus Conversorum*, see Donnelly, *The Decline of the Medieval Cistercian Laybrotherhood*, p. 64. To Donnelly's summary ought to be added the edition of the *Usus Conversorum* by J. Lefebvre, *Collectanea Ordinis Cisterciensis Reformatorum*, 17 (1955), 84–97.

[7] *Summa Carta Caritatis*, ch. XIX and ch. XX.

[8] For example, Nicholas of Clairvaux, *Ep*. 36, *PL* 196, col. 1632.

dicti had stipulated.[9] The lay brother vocation was to be a life of humility, work, patience, and prayer.

Important to remember is that the spiritual status—as well as the vocational status—of Cistercian lay brothers was represented in the early documents as the same as that of choir monks. That is, both groups were to expect the same spiritual redemption and salvation. Other aspects of the lay brotherhood were also similar to the ways in which choir monks lived. *Conversi*, like choir monks, were obliged to undergo a year's novitiate or period of probation before being accepted into the monastery.[10] The *Usus Conversorum* states that after the year had elapsed, the lay brother should come to chapter, where he should divest himself of all property and formally make his profession. Prostrating himself, the lay brother was to ask for mercy and then, kneeling before the abbot, he should promise obedience *usque ad mortem*. The abbot must then pray for the new *conversus* to have perseverance and the rest of the chapter should respond, 'Amen'. The ceremony ended with the lay brother kissing the abbot.[11] From the point of initiation, however, a *conversus* could never then become a choir monk,[12] and after 1188, as mentioned, it was decreed that nobles who joined the Cistercian order were not to be lay brothers. Jean Leclercq has observed that 'les convers vivent dans un état humble, c'est-à-dire l'humilité, mais aussi de l'humiliation'.[13] By the thirteenth century, this was manifested in various different ways.

Lay brothers were physically distinguished from choir monks by virtue of a different habit and a beard.[14] The beard was particularly distinctive and gave rise to both the lay brothers' nickname—*barbati*—and, in some cases, derision.[15] Lay brothers' clothing was to be black, rather than the white undyed wool worn by choir monks,[16] and even more significantly, lay brothers did not take part in

---

[9] Manning, 'La Règle de S. Benoît', p. 263—ch. 66 of the *Regula*: 'Monasterium autem si possit fieri ita debet construi, ut omnia necessaria, id est aqua, molendinum, hortus, pistrinum, vel atres diverse intra monasterium exerceantur, ut non sit necessitas monachis vagandi foras, quia omnino non expedit animabus eorum'. Also *Statuta* 1134, 6, t. 1 headed 'Quod non debeant monachis extra claustrum habitare'.

[10] De la Croix Bouton and van Damme, eds, *Les plus anciens textes de Cîteaux* for the early legislation in the document known as the *Summa Carta Caritatis* ch. XXI.

[11] Lefebvre, 'Usus Conversorum', p. 94, *De Professione*. Another MS of the *Usus*, Paris, Bibliothèque Nationale, MS Lat. 430, adds some lines on apostasy from the *Summa Carta Caritatis* after the standard description of the initiation, saying that a *conversus* who leaves his vocation and assumes the habit of a monk or regular canon will never be received into the order again (fols 97–98ʳ).

[12] *Summa Carta Caritatis*, ch. XXII.

[13] Jean Leclercq, 'Comment vivaient les frères convers?', *Analecta Cisterciensia*, 21 (1965), 239–58 (p. 241).

[14] De la Croix Bouton and van Damme, *Les plus anciens textes de Cîteaux*, *Exordium Parvum*, ch. XVII.

[15] Leclercq, 'Comment vivaient les frères convers?', p. 244.

[16] Lefebvre, 'Usus Conversorum', p. 96: 'Solis tamen fabris conceditur habere camisias nec tamen nisi nigras'.

the recitation of the Divine Office with choir monks. Their liturgical duties ended with the ability to recite the *Pater Noster*, the *Credo*, the *Ave Maria*, and the *Miserere*. And there was no *lectio divina* for the *conversi*; rather, the *Usus Conversorum* indicates that lay brothers are known for their simplicity and illiteracy.[17] Lay brothers were also forbidden to participate in the election of the abbot, or to be present in the chapter house (where the business of the monastery was carried out every Friday) except on certain feast days. On special occasions, the lay brothers were allowed to listen to the sermon read to the monks in the chapter house, but they were to stand outside the building in the cloister arcades and listen through the windows when there was insufficient space for them inside.[18]

Space within the Monastery

Demarcation of the monastic site reiterated the separateness of *conversi* and *monachi*. The lay brothers' living quarters were either the west wing of the cloister, or on the granges. Not unlike the eastern claustral range which incorporated a first-floor dormitory, the western range of the Yorkshire houses served as both accommodation and as a place of work. The ground floor was partitioned into bays, which served as storage units, the cellarer's office, and a reception area (Fig. 6.1), while the second floor (now missing in all the Yorkshire ruins) was the lay brothers' dormitory. The lay brothers' refectory tended to be part of the southern end of the west range, and also on the ground floor. The separation of the lay brothers' quarters from those of the choir monks was clearly effected from the earliest days of the Yorkshire houses, although the inclusion of separate refectories and infirmaries was often a later development. At Fountains abbey, for example, the west range had been part of the abbey even before the more permanent stone buildings were erected from 1144, but the lay brothers' infirmary and refectory were part of the remodelling of the west range from 1160.[19] At Byland abbey, the west range was finished in 1165, and was, according to Stuart Harrison, the earliest surviving building of the

---

[17] Lefebvre, 'Usus Conversorum', p. 96: 'Nullus habeat librum nec disceat aliquid nisi tantum pater noster et credo in deum miserere mei deus et hoc non littera sed corde tenus'. And *conversi* were also called *illiterati*. But see Brian Stock, *The Implications of Literacy: Written Language and Models of Interpretation in the Eleventh and Twelfth Centuries* (Princeton: Princeton University Press, 1983), who stresses that being described as *illiterati* was not necessarily a derogatory term, but rather a statement of fact.

[18] *Statuta* 1181, 2, t. 1 on elections. On space, see Chrysogonous Waddell, 'The Place and Meaning of Work in Twelfth-Century Cistercian Life', *Cistercian Studies*, 23:1 (1988), 25–44 (p. 35).

[19] Coppack, *Fountains Abbey*, notes that the although the entire west range was demolished to make way for the new and much larger structure in the 1160s, the east wall of the range remained intact in order 'to ensure the enclosure of the cloister' (p. 44). This remodelling was not completed until almost two decades later.

Fig. 6.1: Fountains abbey, two photographs showing the west range, interior (Photos, author)

monastery to be completed (Fig. 6.2).[20] The likely site of the lay brothers' infirmary at Byland was southwest of the range, and was probably built after the completion of the west range itself.[21]

At Cîteaux, Clairvaux, Pontigny, and other houses in France, and at Byland, Kirkstall, and Beaulieu in England, the typical Cistercian cloister plan, based on the Bernardine prototype, was modified slightly to establish distance between the west range and the cloister proper. Access from the west range to the cloister was via a lane, which served as a cloister walk for the *conversi*. At Byland, the thirty-five niches that held seats along the eastern side of the lay brothers' walk may still be seen (Fig. 6.3).[22] Without any topographical contingencies to explain the west range being made deliberately distant from the symbolic heart of the monastery, it is clear that segregation of monk and lay brother was reinforced by the use of space.[23] The existence of a separate lay brothers' cloister at Byland, too, reveals that it would have been extremely unlikely for the *conversi* community to have encountered the cloister proper during the course of their daily work, or the recitation of their meagre liturgy.

There was little reason, therefore, for the lay brother to be in the cloister at all, and with the establishment of chapels on the granges from the thirteenth century, little reason for them to be in the main monastic church either. Before the growth of the monastic grange and the building of chapels there, the lay brothers occupied the western end of the monastic church, but were unable to see the choir monks through the rood screen that divided them. On Good Friday, however, an exception was made allowing the *conversi* access to the east end of the church; when the choir monks were in the cloister chanting the Psalter, lay brothers were sent in to clean the church.[24] Otherwise, space again was manipulated to emphasize separation and difference. One significant point regarding the lay brothers' designated space within the monastic church is that the west end of the church was the initially preferred site for the burial of lay patrons and benefactors during the early to mid-thirteenth century. I would suggest that this was not only because the east end of the church was considered to be a particularly holy area, and therefore not to be tainted by the inclusion of the secular, but that the westerly end of the Cistercian church was particularly secular because of its association with the *conversi*. The peripheral nature of the lay brothers' spaces within the monastery itself signifies their association with

---

[20] Stuart Harrison, 'The Architecture of Byland Abbey', p. 26. Harrison and Fergusson, *Architecture of Solitude*, both disagree with Sir Charles Peers's 1170–77 date for the west range, on the grounds that the scallop and leaf details on the capitals and corbels indicate an earlier date.

[21] Stuart Harrison, *Byland Abbey* (London: English Heritage, 1990), p. 20.

[22] It has been suggested that private prayer was carried out here. See J. Dubois, 'Lay Brothers in the Twelfth Century', *Cistercian Studies*, 7 (1972), 161–213.

[23] Gilchrist, 'Community and Self'.

[24] Griesser, 'Ecclesiastica Officia', p. 200: '[i]nterim laici fratres barbati ecclesiam mundent' and p. 278: 'In Parasceue post primam ecclesiam et post vesperas claustrum et capitulum conversos mundare faciat.'

Fig. 6.2: Byland abbey, the west range, interior (Photo, author)

Fig. 6.3: Byland abbey, the lay brothers' lane (Photo, author)

the non-monastic world, and the mixing of lay people with lay brothers in the monastic church serves to underline the secular characteristics of the Cistercian lay brotherhood.

Within the precinct walls, the lay brothers' work was in the court of the monastery where the granary and mill house were housed, the main exception being their work in the kitchen, which occupied part of the south range of the cloister. Even here, segregation was enforced. At Fountains abbey, for example, a serving hatch from kitchen to refectory was included on the west wall, which meant that there was no physical contact between the lay brothers in the kitchen and the choir monks in the refectory. The remains of a similar hatch on the other side of the kitchen wall shows that the same principle applied to the lay brothers' refectory. The effect of both, however, was to minimize contact with the kitchen, and in the case of the choir monks, the effect was also to minimize contact with *conversi*. The same principles can be seen in the placing of the *conversi*'s labour in the court of the monastery. The separation of domestic labour from the contemplative work of the choir monks in the cloister ensured tranquillity in the cloister and at the same time reinforced the essential difference between the types of work carried out in the abbey.

Space without the Monastery

Perhaps the most striking manifestation of the difference between lay brothers and choir monks may be seen in the spaces occupied by *conversi* outside the monastery. As agents for monastic business and farmers, the Cistercian *conversi* were, at many times, outside the monastery proper altogether. These extra-monastic sites were the granges, which served as domestic centres, markets, and fairs, where lay brothers played an extremely significant role in determining the economic state of the monastery, and lastly, the less defined sites traversed by the *conversi* as intermediary or agent for the monastery—companions to nobility, or abbatial agents to London, for example. During the years of expansion from the mid-twelfth century, the granges had assumed more and more importance as centres of monastic productivity, while the lay brothers in charge of those granges were consequently important players in the burgeoning Cistercian economy.

It has recently been argued that the Cistercian order acquired property outside the monastery to 'reshape the material world and remove its secular character',[25] rather than to colonize and exploit for the purpose of fiscal profit, as some earlier historians believed. This transformation of property was almost entirely dependent on the labour of lay brothers, who, as managers of the 'mini-estates' that were the granges, maintained and enforced the expansion of Cistercian dominions. Cistercian granges were, therefore, the primary domain of *conversi*, who not only provided a ready labour force for heavy extra-claustral farming work, but who were in some position of power and authority in the practical administration of the estates. And if the view proposed by Martha Newman is

---

[25] Newman, *Boundaries of Charity*, pp. 67–88.

accepted, then lay brothers performed a vital role in the expansion of Cistercian spirituality, too.[26]

The Yorkshire Cistercian granges had an especially important role to play in the economy of the area. We know that Fountains abbey had already established at least six granges as early as 1146, which reflects that abbey's rapid expansion in both population and territory.[27] With these years of expansion in the twelfth century came more lay brethren charged with the responsibility of managing Cistercian estates. As Colin Platt has pointed out in his study of the monastic grange, Cistercians were pioneers of the monastic grange in England at least partly because 'they preferred to manage their own estates and had to devise a technique for doing so'.[28] To this end, granges were an integral part of the Cistercian economy. The buildings of the grange estates themselves appear to have been various. The remains of chapels are evident at the granges of Morcar and Beverley attached to Fountains abbey, while other domestic buildings were predominantly timber living quarters. Some of the larger granges did include more permanent stone structures. Granges continued to multiply at Fountains and other Cistercian houses during the late twelfth century, as grants of land came flooding in from some of the wealthiest and most influential northern families, the Percy and Mowbray families among them, while Glyn Coppack estimates that by 1265, the estates of Fountains abbey had reached their maximum extent. It was at this point that the utility of the *conversi* was also at its greatest extent.[29]

The wool trade, which was if not the sole source of Cistercian income in Yorkshire, at least a crucial part of it, was managed primarily from the granges, while the lay brothers themselves performed most of the commercial transactions relating to the sale of the wool.[30] The General Chapter was concerned in 1214 that the lay brothers in charge of the sale of wool were selling the fleeces

---

[26] Serlo of Fountains is interested in the great expansion of the Cistercian order during the twelfth century, often noting the establishment of granges under certain abbots. An example is Abbot Henry, who established three granges during his abbacy. See *Mem. F.*, pp. 85–86.

[27] Donkin, 'Settlement and Depopulation', pp. 142–44.

[28] Colin Platt, *The Monastic Grange in England: A Reassessment* (New York: Fordham University Press, 1969), p. 12.

[29] For the various land grants to Fountains, see Joan Wardrop, *Fountains Abbey and its Benefactors 1132–1300,* CSS 91 (Kalamazoo: Cistercian Publications, 1987). See also Coppack, *Fountains Abbey,* p. 82.

[30] For the English Cistercian wool trade, see *inter alia* F. Mullin, *A History of the Work of the Cistercians in Yorkshire* (Washington: Catholic University of America, 1932); R. A. Donkin, 'The Disposal of Cistercian Wool in England and Wales during the 12th and 13th Centuries', *Cîteaux in de Nederlanden,* 8 (1957), 109–31; 181–202; C. C. Graves, 'The Economic Activities of the English Cistercians in Medieval England (1128–1307)', *ASOC,* 13 (1957), 3–60; T. R. Eckenrode, 'The English Cistercians and their Sheep during the Middle Ages', *Cîteaux,* 24 (1973), 250–66; Coppack, *Fountains Abbey,* pp. 78–97.

at too high a price,[31] while Matthew Paris tells us that even the papacy was aware of the English Cistercians' reputation as successful wool traders. When the papal collector arrived in 1256 to demand the papal tithe for Cistercian transactions with Sicily, the assembled Cistercians were told that 'all the world knows your wealth in wool'.[32] The precise role of the lay brothers in this profitable enterprise varied from the care of the flock—which in some cases had as many as fifteen thousand sheep—to selling the wool at markets. The granges of Cistercian abbeys—traditionally the domain of the *conversi*—were the monastic outposts from which these activities were managed.[33]

The Cistercian presence at markets in Yorkshire and its environs also indicates the travelling afield of members of the monastery. It has been shown that the number of markets and fairs in the north of England had expanded rapidly from the beginning of the twelfth century, while the transformation of the thirteenth-century rural economy can be traced to the twelfth-century creation of northern boroughs and the changing demography of the north.[34] An increase in the number of markets and fairs was one result of these broader changes, and Cistercians were certainly involved in these systems of trade. Sawley abbey had a fair at Guisborough manor from September 7th–9th and a weekly market there from 1260, while Meaux had two weekly markets from the late 1270s: one at Kingston-on-Hull and one at Pocklington manor.[35] Annual fairs were also held at each of these locations. And by the early fourteenth century, Jervaulx abbey had a weekly market at East Witton Manor and two

---

[31] *Statuta* 1214, 45, t. 1.

[32] *Matthew Paris, Chronica Maiora*, ed. by H. Luard, 7 vols (London: Rolls Series, 1872–83), 5, pp. 682–88. For an example of a contractual arrangement with Italian merchants, see Lancaster and Paley-Baildon, eds, *Coucher Book of Kirkstall Abbey*, p. 226. And for a fuller contract from Fountains abbey, also with Italian wool merchants, see *CCR* 1272–79 (1276), p. 354.

[33] Graves, 'Economic Activities', p. 25. Graves estimates that during the last quarter of the thirteenth century, Fountains abbey had around twelve thousand sheep and Jervaulx abbey had as many as fifteen thousand. Another example of two *conversi* from Vaudey abbey selling fifty sacks of wool at Boston fair in 1275 is given at p. 28. For the concern of the General Chapter that profiteering may result from autonomous administration of the granges, see *Statuta* 1262, 10, t. 3.

[34] R. Britnell, 'Boroughs, Markets and Trade in Northern England, 1000–1216', in *Progress and Problems in Medieval England: Essays in Honour of E. Miller*, ed. by R. Britnell and J. Hatcher (Cambridge: Cambridge University Press, 1996), pp. 46–67; E. Miller, 'Farming in Northern England during the Twelfth and Thirteenth Centuries', *Northern History*, 11 (1975), 1–16. For the demographic changes, especially the population increase, which in Yorkshire has been estimated as a tenfold increase from the Domesday reckoning to the end of the fourteenth century, see E. Miller and J. Hatcher, *Medieval England—Rural Society and Economic Change 1086–1348* (London: Longman, 1978). For a very good general overview of the social implications of these changes, see Alan Harding, *England in the Thirteenth Century* (Cambridge: Cambridge University Press, 1993), pp. 68–105.

[35] For Sawley, see *CChR* 1257–1300 (1260), p. 32. For Meaux, see *CChR* 1257–1300 (1279), p. 214; *CPR* 1291–1301 (1301), p. 550.

annual fairs there.³⁶ Other markets and fairs were well attended by Cistercians, who were quick to recognize the profitable potential of the larger fairs, such as Boston, and in some cases, urban property was acquired to allow the monks and lay brothers better access to these events, and quasi-monastic accommodation outside their home abbey.³⁷ The Cistercian delegates to markets were both choir monks and *conversi*, although some of the earliest Cistercian statutes indicate that lay brothers were the most likely negotiators.³⁸

The third main extra-monastic duty of Cistercian lay brothers was to act as agents of the abbot and companions to nobility on various occasions. A thirteenth-century example from Fountains abbey shows that, although the abbot signed the contracts, lay brothers acted as the commercial agents in transactions for the sale of wool.³⁹ Also during the thirteenth century, the General Chapter was increasingly asked for permission to provide either benefactors or others who requested it with the service of both monks and lay brothers. In 1234, a Cistercian lay brother was sent by Henry III to Simon, Count of Ponthieu with secrets told to him by the king. Henry had not bothered to write the information, because he knew he could trust the *conversus*.⁴⁰ Examples of requests from the *Statuta* are many, and range from bishops requiring either monks or lay brothers to specific requests for lay brothers.⁴¹

Can it be proposed, therefore, that the role played by lay brothers in the economy of northern England shows that *conversi* were in some ways more autonomous and more influential than their vocation would suggest? Were *conversi*, in reality, not as marginalized as we may think? Despite the important economic role played by *conversi* which I have outlined above, by 1300, it is clear that this role was in a process of rapid and irrevocable transformation. The physical spaces occupied by lay brothers were in an unstoppable process of contraction.

By the end of the thirteenth century, the grange economy and its management was under threat and subject to extreme change. The victims of this change were *conversi*, who began to be replaced by hired labour and monastic servants from about 1208, when the General Chapter opened the way for the leasing out of granges for revenue.⁴² Alternative labour also came with donations and gifts of land from wealthy benefactors, who would often include villeins with property donated to Cistercian houses. Henry de Elland granted a carucate of land and all

---

³⁶ *CChR* 1300–26 (1307), p. 81.

³⁷ R. Donkin, 'The Urban Property of the Cistercians in Medieval England', *ASOC*, 15 (1959), 104–31. Donkin shows that in Boston alone, Fountains, Furness, Holm Cultram, Jervaulx, Kirkstead, Louth Park, Meaux, Newminster, Rievaulx, Revesby, and Sawley held property. See appendix, pp. 116–17.

³⁸ *Statuta*, 1134, 24, t. 1, where trade is associated with granges.

³⁹ *CCR* 1272–79, p. 354.

⁴⁰ *CPR* 1258–1266, p. 203.

⁴¹ For example, *Statuta* 1237, 10, t. 2; *Statuta* 1269, 46, t. 3.

⁴² *Statuta*, 1208, 5, t. 1. And see also Donnelly, *Decline of the Medieval Cistercian Laybrotherhood*, pp. 42–43 for the complex history of these statutory developments.

his villeins to Kirkstall abbey, for example, while Hervey the carpenter was given to the same abbey by Thomas Scot in the early thirteenth century.[43] Alan Harding has shown that the developing 'common culture of rights and obligations' did not mean that serfs were not treated as chattels to be bought and sold along with a lord's other property, and the evidence from the Kirkstall transaction certainly upholds this.[44]

That the *conversi* labourers were displaced as an integral part of the monastic economic system is seen in the existence of peasant settlements on some grange estates. 'It is significant', says Colin Platt, 'that [...] peasant settlements may be seen even at the immediate home farms of some abbeys: in other words, precisely where the lay brethren [...] might most readily have handled every task of the farm on their own'.[45] Even if, as Platt postulates, lay brothers continued to serve as supervisors to the other peasant workers on the granges, it was only a matter of time before it became more economically expedient to phase out *conversi*. The extra-monastic domain of lay brothers was physically shrinking from the end of the thirteenth century—and with it, their own particular form of autonomy within the Cistercian world. A bull issued in 1302 by Pope Alexander VIII allowed the full-scale renting out of Cistercian land and granges and by 1308, as James Donnelly reports, there were complaints from *conversi* in the Cistercian houses of Flanders that they were redundant.[46]

The later years of the thirteenth century also saw many of the English Cistercian houses in an advanced state of poverty, to the extent that Fountains abbey was taken into royal custody twice by the end of the century for relief of debts, while Kirkstall abbey was disbanded altogether in 1281 after having been taken into royal custody itself in 1277. Rievaulx abbey, too, was taken into royal custody in 1276, as was Jervaulx in 1275.[47] The reasons for such economic disaster have been explained by historians in various ways, ranging from seeing the taxation burdens placed on the Cistercian houses by English, French, and papal jurisdictions as the primary factor, to poor management of resources and bad business strategies, particularly the practice of advance selling of wool.[48] Despite the commercial assets and the farming resources of some of the larger houses like Fountains abbey, internal and external problems took their toll on

---

[43] Lancaster and Paley-Baildon, eds, *Coucher Book of Kirkstall Abbey*, p. 195 and p. 208.

[44] Harding, *England in the Thirteenth Century*, pp. 69–76.

[45] Platt, *The Monastic Grange*, p. 91. Platt cites the example of peasant houses at Morcar, a grange of Fountains.

[46] Donnelly, *The Decline of the Medieval Cistercian Laybrotherhood*, pp. 59–60.

[47] For Fountains, see *Calendar of Patent Rolls*, 2 Edward I, p. 59 (1274); *Calendar of Patent Rolls*, 19 Edward I, p. 431 (1291); for Kirkstall, see *Statuta* 1218, 38, t. 3; *Calendar of Patent Rolls*, 5 Edward I, p. 208; for Rievaulx, see *Calendar of Patent Rolls*, 4 Edward I, p. 152; for Jervaulx, see *Calendar of Patent Rolls*, 3 Edward I, p. 76.

[48] See for example, E. Madden, 'Business Monks, Banker Monks, Bankrupt Monks: the English Cistercians in the Thirteenth Century', *The Catholic Historical Review*, 49: 3 (1963), 341–64; Graves, 'Economic Activities'.

actual profit. It suffices to indicate here that the effect of these financial troubles on lay brothers was profound. Not only were the granges being reallocated to non-monastic labourers,[49] but the likelihood of lay brothers being useful in any way during the Cistercian 'depression' of the late thirteenth century was remote. As early as 1224, the General Chapter had prohibited the reception of monks or lay brothers by any house in debt by more than 100 marks,[50] and given the poor state of many of the Yorkshire houses by the close of the thirteenth century, it is not surprising that lay brothers and monks were dwindling in number. With the disastrous Scottish incursions into Yorkshire in the early fourteenth century, some granges were abandoned and some shrank significantly. With a period of economic turmoil followed by political instability, it was the *conversi* who found themselves on the margins of Cistercian life.

The labour originally performed by English Cistercian lay brothers did not by any means guarantee their continued influence, or even their continued presence within the Cistercian context. Such a fundamental change in the place of the lay brother must be linked to not only the economic trials of the Cistercians outlined briefly above, but also to the devaluation of *conversi* as productive agents of labour in a more general sense. Although it is clear that some of the traditional tasks, such as going to market, were still performed by lay brothers at the end of the thirteenth century, the fact that their sole position of autonomy or influence—the administration of the granges—was gradually removed negates the importance of the continuation of other tasks. Without a secure place within the monastic hierarchy, or a secure space on its estates, lay brothers were increasingly marginalized in England. These manifestations of marginalization, already present in the strict demarcation of the spiritual space of *conversi* may also be found in legislative and narrative representations of their role and place within the monastic milieu.

Representations of the Cistercian Lay Brother

Violence and the Conversi

Reading the *Statuta* of the Cistercian General Chapter, we find that lay brothers did not enjoy a felicitous reputation during the thirteenth century. We find instances of violence, outbreaks of rebellion, accounts of unspecified disruption and the weary sigh of 'Proh dolor' as Cîteaux hears of yet another instance of insurrection in the monasteries of Europe and England. For the period 1200–1300 there are some forty-seven instances of disruption and rebellion in the *c.* 200 Cistercian houses in France, thirty-seven of these involving lay brothers.[51]

---

[49] For example, in 1280, abbot Richard of Meaux leased an estate at Skerne, and three granges, See *Chronicon Monasterii de Melsa*, II, p. 175.

[50] *Statuta* 1224, 25, t. 2.

[51] *Statuta* 1200, 22, t. 1; 1200, 28, t. 1; 1200, 45, t. 1; 1202, 14, t. 1; 1202, 33, t. 1; 1204, 34, t. 1; 1205, 60, t. 1; 1207, 61, t. 1; 1209, 11, t. 1; 1209, 44, t. 1; 1212, 52, t. 1

In England for the same period, when there were sixty-six Cistercian houses,[52] we find six cases of rebellion reported to the General Chapter, only one of which involved lay brothers specifically.[53] There is also one general reference to incorrigible behaviour in 1192 and another reference to rebellious monks and lay brothers at Furness is 1193.[54]

This ostensible phenomenon of lay brother violence has been explained by historians predominantly in terms of *conversi* marginalization within the monastery. Ernst Werner, for example, emphasizes exploitation of the lay brothers in what he perceives as an inherently inequitable environment.[55] Werner relates the monastic hierarchy to the feudal structure, isolating the lay brother as the equivalent of a serf. Werner does not believe that lay brothers, who were voluntary members of the monastery, willingly reflected monastic desire to be a slave of Christ (as did choir monks). Rather, Werner argues that the imposed superstructure of cultural values present in the monastery and especially in the demarcation and enforcement of its hierarchical divisions, conspired to subjugate the *conversi*. Ultimately, the force of this argument rests on the issue of deprivation of freedom of choice. Although the lay brother has chosen to be involved in monastic life, the choice is not a free one, in Werner's view, as the alternative is starvation. And even worse, once in the monastery, the lay brother was 'chained to the cloister' in order to avoid problems with serfs.[56] Outbreaks of violence and disruption in this milieu, according to Werner's critique, are the

---

and 1213, 9, t. 1 same incident; 1217, 32, t. 1; 1219, 32, t. 1; 1220, 44, t. 1; 1220, 47, t. 1; 1225, 25, t. 2; 1225, 26, t. 2; 1228, 24, t. 2; 1230, 12, t. 2;1230, 13, t. 2; 1230, 36, t. 2; 1232, 24, t. 2; 1233, 67, t. 2; 1236, 57, t. 2; 1237, 66, t. 2; 1238, 51, t. 2; 1238, 52, t. 2; 1242, 55, t. 2; 1247, 20, t. 2; 1247, 61, t. 2; 1251, 74, t. 2; 1252, 40, t. 2; 1259, 34, t. 2; 1262, 43, t. 3; 1267, 54, t. 3; 1269, 71, t. 3; 1272, 7, t. 3; 1275, 30, t. 3; 1277, 16, t. 3; 1277, 28, t. 3; 1278, 13, t. 3; 1280, 16, t. 3; 1280, 19, t. 3; 1289, 14, t. 3; 1292, 8, t. 3; 1293, 25, t. 3; 1294, 80, t. 3. These figures are based on a capacious definition of disturbance and includes instances of violence (including assaults and homicides), conspiracy, the vague caetgory of *excessus* which I will discuss in more detail below, reports of lay brothers and monks treating guests with disdain or violence and individual reports of misdeeds.

[52] D. Knowles and R. N. Haddock, *Medieval Religious Houses England and Wales* (London: Longman, 1953), p. 104 et seq. This figure does not include the various incarnations of the same house. For example, I have counted Byland abbey only once, although it existed in 4 locations (Calder, Hood, Old Byland, and Stocking) prior to settling at its final site.

[53] *Statuta* 1207, 38, t. 1; *Statuta* 1212, 32, t. 1; *Statuta* 1218, 30, t. 1; *Statuta* 1259, 1, t. 2; *Statuta* 1279, 22, t. 3; *Statuta* 1280, 25, t. 3. There is also a report of arrogant lay brothers in Bond, ed., *Chronicon Monasterii de Melsa*, 1, pp. 432–33, which refers to an incident in the mid-thirteenth century. The only statute which refers to lay brothers specifically is *Statuta* 1280, 25, t. 3 at Kirkstall. This included choir monks as well.

[54] *Statuta* 1192, 16, t. 1; *Statuta* 1193, 58, t. 1.

[55] E. Werner, 'Bemerke zu einer neuen These über die Herkunft der Laienbrüder', *Zeitschrift fur Geschichtswissenschaft*, 6 (1958), 355–59.

[56] Werner, 'Bemerke zu einer neuen These über die Herkunft der Laienbrüder', p. 360.

inevitable consequences of inequitable conditions. Aaron Gurevich is one who has applied similar principles to medieval society, arguing that '[t]he oppressed meet violence with violence when they are no longer able to put up with being exploited',[57] while Jane Sayers, too, argues again that outbreaks of violence in the monastery are symptomatic of inbuilt inequity within the Cistercian community.[58] Sayers discusses monastic regulation and hierarchy in the context of Erving Goffman's theory of the 'total institution'.[59] In this environment, violence is the response to a number of institutional conditions designed to keep *conversi* in check and to maintain the authority of the institution. Ultimately, Sayers asserts that lay brothers were violent 'probably on account of their inability to express their grievances verbally'.[60]

The materialist view has not always found unequivocal support among historians of spirituality such as Kassius Hallinger, who emphasises what he sees as the implicit desire of the lay brother to be a slave of Christ.[61] Hallinger's view and its hostility to the materialist arguments advanced by Ernst Werner have been conveniently summarized by Bruno Lescher, who himself argues that Cistercian spirituality indicates a monastic class structure, but that prohibitions and separation within the monastery—such as the rule against reading—make sense in the context of the lay brothers' 'specialised vocation'.[62] With Jacques Dubois, Lescher defends the ruling of the Cistercian General Chapter of 1188 that all nobles who joined the order were to be made choir monks, on the grounds that the nobles were placed where they were needed most.[63] Yet, in the final analysis, Lescher leans toward a sympathetic reading of what he has described as the materialist critique, acknowledging that the 'radical equality' of Pachomian and Benedictine monasticism gave way during the twelfth and thirteenth centuries to a brand of monastic life structured around the clerical/lay divide. And, he mourns, lay brothers like himself are today still wondering 'what it means to live with "that kind of equality"'.[64]

Issues of lay brother violence become even more problematic when the *Statuta* themselves are investigated more closely. Disturbance, disruption,

---

[57] Aaron Gurevich, *Categories of Medieval Culture*, trans. by L. G. Campbell (London: Routledge Kegan Paul, 1985), p. 182.

[58] Sayers, 'Violence in the Medieval Cloister'.

[59] Goffman, *Asylums*.

[60] Sayers, 'Violence in the Medieval Cloister', p. 539.

[61] K. Hallinger, 'Woher kommen die Laienbrüder?', *ASOC*, 12 (1956), 1–105.

[62] B. Lescher, 'Laybrothers: Questions Then, Questions Now', *Cistercian Studies*, 23 (1988), 63–85.

[63] Jacques Dubois, 'The Laybrother's Life in the Twelfth Century: A Form of Lay Monasticism', *Cistercian Studies*, 7 (1972), 161–213 (p. 171). Presumably, Lescher means that nobles were more likely to be able to read Latin, which was needed for a choir monk to fulfil the *lectio divina* requirement in the *Regula S. Benedicti*. See E. Manning, 'La Règle de S. Benoît', p. 250 (ch. 48 of the *Regula*): '[o]ccupari debent fratres in labore manuum; certis iterum horis in lectione divina'.

[64] Lescher, 'Laybrothers: Questions Then, Questions Now', p. 85.

violence, revolt, and so on seem to have been very fluid and, in some cases, very vague categories. The best example of this is the common description of *excessus monachorum* or *conversorum* which denotes various offences which are not always explained. Brother William of Bonaecumbae was exiled to a monastery far away from his own house for excesses in his abbey in 1232, while the following year, the General Chapter heard of intolerable excesses by both monks and lay brothers at Pilis. The excesses on the grange of Fontenay in 1233 were instigated by two lay brothers, while the other *conversi* who were present that day were to be put on bread and water every Friday until Easter. In 1247, the lay brothers who hit the abbot and stole the abbatial seal at Boscanio were charged with excess, as were the choir monks who stood by and pretended not to see what was happening. And in 1268, the murder of the cellarer by the *conversi* of Ripalta was reported initially as *excessus*, and when the perpetrators ran away, as apostasy.[65] From this summary list of examples, it is clear that the category of excess is a broad and subjective one, that covers almost everything from unspecified disruptive behaviour to homicide.

The vagueness of such categorization by the General Chapter in the thirteenth century does problematize how modern historians have accepted the notion of lay brother violence. Strikingly different numbers used by scholars to describe revolts in the same period show these inherent difficulties in categorizing and quantifying rebellion, disturbance, and disruption. Even accounting for the difference in figures for the continental and English houses is extremely difficult. James Donnelly discovers 123 instances of 'revolt' in total for the thirteenth century; Jean Leclercq, on the other hand, finds thirty,[66] while Anselme Dimier lists only twenty-one for the years 1200–1300.[67] Donnelly decided that collusion was the main criteria for determining the seriousness of the event, while Dimier, in contrast, found that specific types of disturbance (and especially violence) were the best indicators of the weakness of human nature. Rather than concentrate on the perpetrators of crime and dissent, he chose to focus on three particular manifestations of disruption—violence against the abbot, antipathy between monks (especially homicides) and dissent among Cistercian nuns. Jean Leclercq is happy to accept the number of revolts by lay brothers given by M. Laporte (who was himself primarily concerned with Carthusians).[68] Yet Leclercq includes no account of what constitutes a revolt or a disturbance here.

The seeming contrast in *conversi* disturbance in France and England is

---

[65] For the preceding examples, see *Statuta* 1232, 24, t. 2; *Statuta* 1233, 51, t. 2; *Statuta* 1233, 67, t. 2; *Statuta* 1247, 20, t. 2; *Statuta* 1268, 1, t. 3.

[66] Leclercq, 'Comment vivaient les frères convers?', p. 250. These thirty incidents are lay brothers' revolts, 'les révoltes de convers'.

[67] M.-A. Dimier, 'Violence, rixes et homicides chez les Cisterciens', *Revue des Sciences Religieuses*, 46 (1972), 38–57. Dimier concentrates on the reports to the General Chapter, but only provides a selection of violence, riots, and homicides.

[68] Leclercq, 'Comment vivaient les frères convers?', p. 250, note 7, quoting M. Laporte, *Aux Sources de la vie Cartusienne*. III, *L'Institution des frères en Chartreuse* (Grand Chartreuse, 1960).

another difficult issue. We know from James Given's masterful study of homicide in thirteenth-century England that England was not exempt from the so-called endemic violence of medieval society.[69] Given has estimated homicide rates in both urban and rural areas from the almost complete and extant collection of Coroner's Rolls that survives for the thirteenth century, and found that rates ranged from nine per 100,000 population (in Norfolk) to twenty-three per 100,000 (Kent).[70] If we accept homicide rates as an indicator of violence within society generally, we should expect to find a higher number of reports of violence in England, rather than the small figure revealed by the General Chapter's *Statuta*.

An immediate possibility is that English cases were not necessarily reported to the General Chapter, and that they were dealt with by either the abbot of the relevant house or by the abbot of the mother house on his annual visitation. This may be particularly applicable to incidents at the less serious end of the scale of disruption or disturbance—instances where lay brothers were inhospitable to guests or visitors, for example.[71] On the other hand, it must be recognized that the General Chapter itself was—among other things—a problem-solving forum, therefore reports of breaches of discipline ought to be expected there. Given that the General Chapter functioned in this way, and given that its meetings were held in France, it is not surprising that the number of French cases reported in the *Statuta* is high. Louis Lekai rightly notes that the repetition of certain rulings year after year ought not necessarily to denote that a decision had been ignored the first time it was ruled, but rather that the General Chapter was aware of the sporadic nature of abbatial attendance at Cîteaux and repeated things a number of times to make certain that everyone knew them.[72] An example of this is the repetition, year after year, of the rule prohibiting drinking on granges.[73] In other words, all figures representing disruption themselves may seem inflated.

In England, it appears highly likely that serious transgressions may have been dealt with by the secular arm, and escaped record within the monastic milieu (and more especially the General Chapter) entirely. This is certainly evident

---

[69] James B. Given, *Society and Homicide in Thirteenth-Century England* (Stanford: Stanford University Press, 1977).

[70] Given, *Homicide*, p. 36. For Coroner's Rolls generally, see R. F. Hunnisett, 'The Medieval Coroner's Rolls', *American Journal of Legal History*, 3 (1959), 95–125; 205–21; 324–59. In contrast, T. R. Gurr has shown that for 1974, homicide rates for England and Wales were 1.97 per 100,000. See T. R. Gurr, 'Historical Trends in Violent Crime: A Critical Review of the Evidence', *Crime and Justice; An Annual Review of Research*, 3 (1981), 295–353. Gurr himself uses Given's findings as the starting point for his discussion, pp. 305–06.

[71] See, for example, *Statuta* 1251, 74 , t. 2.

[72] Lekai, *Ideals and Reality*.

[73] See, for example, *Statuta*, 1192, 16, t. 1; 1202, 10, t. 1; 1238, 5, t. 2) and others. See Donnelly, *The Decline of the Medieval Cistercian Laybrotherhood*, p. 28 et seq. for the exceptions and the development of these prohibitions.

from the surviving records of petition for the writ *De Apostata Capiendo*.[74] From this very specific set of records, which deals only with a particular offence—other information contained within the writs is, therefore, incidental to the main complaint—we find for the last fifty years of the thirteenth century several accounts of violence and criminal activity by Cistercian monks in England. The cases range from requests from religious houses to the secular arm to arrest and deliver apostates back to the monastery (with no other information), to accounts involving theft and assault. The lay brother Henry Sampson of Fountains abbey, for example, was signified for arrest on 9 October 1289, but there is no record of this in the statutes of the General Chapter that year. Another lay brother signified for arrest was John of Palfleteby, the cellarer of Louth Park, who allegedly stole money from the monastery and was spending it in apostasy. And at Rievaulx in 1279, William de Acton claimed to be a leper, stabbed the monk who was examining him, and ran off into the woods. He was pursued by two monks from the abbey, who caught him, and beat him to death.[75] This case, too, is not recorded in the *Statuta*. Out of the twenty-five Cistercian cases Logan finds for this period, seven involve lay brothers. And more significantly, these examples demonstrate that monasteries in England were perhaps more likely to utilize the secular arm in certain cases, than to go to the General Chapter in France. With the development of insular writs like *De Apostata Capiendo*, Cistercian houses (and others) were able to avail themselves of legal remedies much more quickly than the slow machinery of the Cistercian institution in Cîteaux could allow.[76]

---

[74] See Logan, *Runaway Religious*. The Public Record Office collection of the writs is C81 for the thirteenth century, although Logan utilizes other collections of writs and more general records such as the Calendar of Patent Rolls. See pp. 204–05 for the list of apostate Cistercian religious from 1250–1300, and p. 268 for the PRO sources.

[75] Logan, *Runaway Religious*, pp. 204–05.

[76] This has broader and important implications in the context of the 'Anglicizing' of the Cistercian order during this period. It has been argued that in terms of twelfth-century Cistercian historiography, for example, that foundation and other histories produced by English houses ought to be read in the light of Anglo-Norman historiography and the creation of the English past, rather than as examples of a standard 'monastic' topos that was the same in France and England. See Elizabeth Freeman, 'Aelred of Rievaulx's *De Bello Standardii* and Medieval and Modern Textual Controls', in Cassidy, Hickey, and Street, *Deviance and Textual Control*, pp. 78–102. I would argue that these sorts of insular developments can be seen in other areas as well, especially adminstrative and legal areas, and that they become particularly pronounced over the course of the thirteenth century. There is some evidence that the 'Englishness' of the Cistercians was not recognized by the crown, however. In 1295, it was decreed that all monks and religious who were French, and who lived 13 miles or less from the sea ought to be relocated further inland during the time of war with France. This included Cistercian monks and lay brothers, who were to be arrested if they were found wandering around. The abbot of Fountains, for example, was ordered to receive James and William, monks of Cîteaux and Michael the lay brother, late wardens of the church at Scardebuburgh [Scarborough]. See *CFR* 1272–1307, p. 366. And more seriously, abbots of all Cistercian

The *De Apostata Capiendo* cases are the tip of the iceberg. There are further reports in other legal sources for the period showing that the General Chapter statistics alone are not enough to provide a representative picture of the potential number of cases in England in the thirteenth century and of the activities of Cistercian lay brothers and monks there. The Yorkshire Assizes of 1218–19, for example, showed that on some occasions, the perpetrator of a criminal act would enter a Cistercian monastery after the crime. Jeremiah of Ecclesfield was appealed by Maurice of Askern for cutting off three of Maurice's fingers. The jurors said that in that same week, Jeremiah gave himself up to religion at Fountains abbey and the marginalia on the roll itself notes that Jeremiah stayed in the abbey (but in lay dress) for a long time after the deed.[77] The Coroner's Rolls of the period also turn up pertinent cases, some of which were incorporated into domestic monastic records. One such case is the proceedings concerning the death of Adam the forester of Clifford, who was beaten to death by two *conversi* of Kirkstall.[78] Other references to episodes of violence appear even in the Hundred Rolls, one example being the case of the lay brother Peter, the *grangiarum* (grange master) of Barnoldswick (a grange of Kirkstall), who cut off the ear of a serving boy for stealing two loaves of bread.[79]

The historiography of the Cistercian lay brotherhood has focused on the deviant lay brother, concentrating on manifestations of violence and reasons for discontent, as outlined above. Historians have tended to rely on the numbers of revolts reported to the General Chapter as evidence of a culture of violence among the lay brotherhood which, as I have argued, is a methodologically problematic enterprise. Accepting the *Statuta* at face value is a dangerous way of coming to conclusions regarding the 'disruptive' *conversus*, even if those conclusions are then used to articulate empathetic notions of historical class difference and institutional oppression. To flesh out these notions, more exploration of thirteenth-century Cistercian representations of the lay brother and his spiritual space within the monastery is needed. By looking beyond the numbers to read the *Statuta* in the light of exempla collections and other more 'literary' narratives, it becomes apparent that the idea of the recalcitrant Cistercian *conversus* became the subject—and from there, the product—of a number of discursive structures designed to mark out both his status within the order and his space in its spiritual world.

---

houses were forbidden to attend the General Chapter at Cîteaux in 1298 and 1299. See *CCR* 1296–1302 (1298), p. 215 et seq.

[77] Clay, ed., *Three Yorkshire Assize Rolls*, pp. 215–16.

[78] Lancaster and Paley-Baildon, eds, *Coucher Book of the Cistercian Abbey of Kirkstall*, pp. 22–23.

[79] W. Hillingworth and J. Caley, *Rotuli Hundredorum*, 2 vols (London: Record Commission, 1812–1818), 1, 112.

The Conversi 'Type': Topoi of Recalcitrance and Piety

We find the idea of the disorderly *conversus* in various moral tales and exempla collections in both England and France, predominantly in stories where lay brothers have acted contrary to their specified vocation. At the English monastery of Stratford Langthorne, a lay brother is said to have been denied entry to Paradise when he died because he was not wearing the Cistercian hood at the hour of his death. He had been deprived of that 'distinguishing mark of his status' when he had once apostatized.[80] Another tale from a Clairvaux manuscript tells of a vision of Purgatory, where a lay brother is tied to a pole and immersed in fire because he was always 'acting out of hand' and had been sent from one monastery to another before he eventually apostatized and died in bed with his wife.[81] It would be erroneous to assert that the initially important presence of lay brothers in the grange economy exempted *conversi* in some way from the humility of their devotional situation within the Cistercian order generally. Lay brothers were, of course, still expected to uphold the monastic virtues they had accepted after the period of their novitiate was over. Obedience is frequently stressed as being the virtue *conversi* are most likely to ignore, and particularly in the thirteenth-century exempla tales, the consequences of disobedience are always serious and in some cases terminal. A common tale is the story of the lay brother at a Clairvaux grange who presumed to clean his shoes without the permission of the grange master. While he was doing this, he heard two voices, then felt such a violent blow to his back that he died of his injuries within days. This invisible punishment was for disobedience.[82] Another story

---

[80] London, Lambeth Palace, MS 51. The stories involving Cistercians in this manuscript have been partially transcribed by Christopher J. Holdsworth, 'Eleven Visions Connected with the Cistercian Monastery of Stratford Langthorne', *Cîteaux*, 13 (1962), 185–204. This particular story appears at p. 190. The lay brother eventually gets to heaven by appearing to the abbot and prior in a vision, and persuading them to restore the hood to his corpse.

[81] Paris, Bibliothèque Nationale, MS Lat. 15192. For descriptions of this manuscript, see McGuire, 'Purgatory, the Communion of Saints and Medieval Change'; Brian Patrick McGuire, 'The Cistercians and the Rise of the Exemplum in Early Thirteenth-Century France: A Reevaluation of Paris BN MS Lat. 15192', *Classica et Medievalia*, 34 (1983), 211–67.

[82] Paris, Bibliothèque Nationale, MS Lat. 15912, fol. 30^v: 'Conversus quidam claraevallis colligas suas sine licentia in una grangiarum lavare praesumpserat. Quod dum faceret voces quosdam audivit dicentum ad alterum 'percute, percute, merge in aqua'. Quorum unus respondit 'Nequaquam quo dicto'. Statim sensit se in dorso vehementer percussus. Unde graviter infirmatis infra dies paucos obiit. Ecce quam periculosum professis obedientiam quicquam inobedienter agere.' Note, however the difference between this tale and the story of St Bernard's great humility manifested in the saint cleaning his own shoes. Caesarius of Heisterbach, *Dialogus Miraculorum*, book 4, ch. 7, *De Sancto Bernardo Abbate, quem calcios suos inungentem superbiae spiritus subsannavit*. The issue of lay brothers and shoes reappears in the *Exordium Magnum*, this time in the context of a revolt planned by lay brothers to steal the boots of the choir monks and tear them to shreds. The plan failed when the main conspirator, who was 'non

from the same collection tells of two demons who came to the dormitory one night and on finding a disobedient *conversus* beat him to the gates of the monastery, where he was only saved by the divine intervention of the Blessed Virgin.[83]

Conversi who try to read are seen as particularly perverse. Jocelin of Furness tells of the lay brother Walter at Melrose who, influenced by Satan, learnt how to read 'until he had almost lost the light of reason and preferred the Jewish sect to a belief in Catholic truth'. Ultimately, after being shown visions of heaven, hell, and purgatory, Walter saw the error of his ways, and managed to 'adhere to the rules' for the remainder of his life. It is interesting to note that after Walter had confessed to his grave crimes, his new-found literacy became attributed to wondrous influences and lost its association with Satan.[84] This was not so for the lay brother of Klosterkamp, who learnt to read, and took pleasure in the private ownership of books. He was promised by a demon who came to him in a dream in the form of an angel that if he continued to study, he would end up as bishop of Halberstadt. Instead, the lay brother was hanged as a thief after he stole a horse 'so that he might come to his See with becoming dignity'.[85] A lay brother, says Caesarius of Heisterbach, 'should not speak to learned men'.[86]

By far the greatest number of references to lay brothers in these moral and edificatory tales appear in the thirteenth-century *Dialogus Miraculorum*. Examples of lay brothers being punished for doing the wrong thing include the lay brother who was asleep in choir while the other monks saw the devil in the form of a serpent on his back. Another lay brother was seen to have a cat on his head that made him fall asleep by putting its paws over his eyes. Again, the cat was the devil.[87] A lay brother who ate meat was swept up to the top of the belfry by

---

conversus sed perversus', suddenly died. The story is retold by Donnelly, *The Decline of the Medieval Cistercian Laybrotherhood*, pp. 34–35.

[83] Paris, Bibliothèque Nationale, MS Lat. 15912, fol. 30: 'Quidam frater maiori suo dormitum perrexit. Cui nocte duo demones adstarant. Et dixit unus ad alterum 'Quis est hic?' Alter respondit 'Conversus quidam'. Primus ait, 'Non sed inobediens quidam tunc ille intulit. Ergo arripiamus eum'. Et arripientes eum ceperunt flagellare trahentes per dormitorium. Accurunt fratres vident illum solotenus protrahi. Sed eos qui trahebant non poterant vituere. Quem cum usque ad portam monasterii iam horribiliter pertraxissent beata Virgo affuit et liberavit eum.'

[84] 'Brother Walter, before his vision, had been a very simple man, unintelligent and slow of speech. After it, however, he became clever and eloquent, so much so that he could compose impromptu rhythmic verse [...] Using this gift, he afterwards wrote noble verses in the English tongue [and] his hearers, even the abbots and other important personages, used to be overcome with admiration, and he would often bring tears to their eyes [...]' See McFadden, 'An Edition and Translation of the Life of Waldef', p. 328.

[85] *Dialogus Miraculorum*, book 5, ch. 16, *De converso Campensi, qui promissione Episcopatus Halberstadensis deceptus, suspensus est*.

[86] *Dialogus Miraculorum*, book 4, ch. 79: *De sermone monachi illiterati ad Henricum Cardinalem Albanensem*.

[87] *Dialogus Miraculorum*, book 4, ch. 32: *De Serpente, quem vidit frater Conradus in dorso conversi dormitantis in choro*; and ch. 33: *De Converso, cui cattus visus est oculos claudere, cum in choro dormitaret*.

the devil,[88] while one *conversus* who was not sufficiently humble found himself eating coals instead of the host at Mass.[89] Yet another lay brother found himself unable to swallow the host because he had not given up all of his property.[90] Lay brothers are also dangerous and potentially corrupting. A 'miserable' lay brother, who 'was clothed in the habit but not in the spirit of a religious', learnt magic arts and drove a nun in a nearby convent to suicide.[91] A particularly avaricious lay brother was deposed for failing to be penitent and for spending too much time accumulating land,[92] and another conspiratorial lay brother suddenly died before he could carry out his plot.[93] In Jocelin of Furness's *Life of Waldef*, a lay brother named Sinuin, 'a *conversus* by profession and habit, but perverse in act and intent' was deservedly butchered by a 'man of terrible size, features and bearing', but was miraculously restored by an angel on the prayer of Waldef.[94]

The offences committed by lay brothers in these tales are usually fairly minor transgressions of the monastic rule, such as eating meat, being greedy for private property, or falling asleep in choir. But in the narrative context of these edificatory tales, the lay brother is constructed as a transgressor, and is used to illustrate the folly and humiliation that will follow if the rules are broken. In this sense, the lay brother's behaviour is utilized as the measure of monastic discipline within the Cistercian institution. And the lay brother is the potential vehicle by which the institution may be disrupted on both practical and spiritual levels. This is particularly evident in the stories where the devil is let into the monastery because a lay brother has done the wrong thing. And in the more serious episodes, the lay brother poses an active physical threat to others within and without the abbey.

Not only is the lay brother in many of these stories presumptuous and insolent, but he is perverse, an actor against requisite conduct. Perversity is a threat to the Cistercian order's reputation, which is a clear concern of the General Chapter and of the storytellers,[95] but it is also a threat to the carefully erected systems of hierarchy and power that characterize the monastic milieu. In any system based on sets of written rules, language assumes enormous

---

[88] *Dialogus Miraculorum*, book 4, ch. 85: *De converso in cellario comedante carnes*.

[89] *Dialogus Miraculorum*, book 9, ch. 63: *De Wirico converso qui carbones sumere visus est loco sacramentis*.

[90] *Dialogus Miraculorum*, book 9, ch. 64: *De converso proprietario qui corpus Christi sumere non potuit*.

[91] *Dialogus Miraculorum*, book 4, ch. 42: *De sanctimoniali, quae a quodam maligno converso dementata, in puteum se praecipitavit*.

[92] *Dialogus Miraculorum*, book 4, ch. 62: *De Ulrico Praeposito Steinveldensi, et converso avaro, quem amovit*.

[93] *Dialogus Miraculorum*, book 11, ch. 57: *De subita morte cuiusdam conversi conspiratoris*.

[94] McFadden, 'Life of Waldef', p. 305.

[95] Herbert of Clairvaux, *Liber Miraculorum*, PL 185, col. 1141, 'non conversus sed perversus' was the description of the leader of a conspiracy at Schönau.

significance not only in creating the environment in which those rules are to be acted upon and played out, but also in becoming the instrument in which that environment perpetuates itself. Establishing an adversary in language reinforces the zone of proper conduct. This is why, in the case of the Cistercian lay brother, the exempla collections must be understood as being closely linked to legislation: both are creative and prescriptive, both envisage and correct simultaneously. The lay brother is a necessary part of the Cistercian community, and is not, therefore, entirely 'demonized' as the occasionally sympathetic exempla indicate. It is the possibility of transgression and its consequences that *conversi* illustrate, rather than transgression's inevitability.

But, in some of these stories, *conversi* are represented as devout and worthy recipients of visions. One lay brother was delivered from the attacks of the devil by constantly reciting the *Ave Maria*, according to the *Dialogus Miraculorum*,[96] while in Herbert of Clairvaux's *Liber Miraculorum*, it is said that St Bernard was pleased and surprised to hear that one lay brother had no fear of dying because he was confident of God's mercy, 'since he gave himself to serve'.[97] There is occasionally some awareness here that *conversi* may have been justifiably disenchanted with their conditions. A story from Clairvaux tells of a lay brother at a Clairvaux grange who was weeping because he was not allowed to join in the liturgy for the feast of Mary.[98] While he was crying, Mary appeared to him in a vision and told him that he ought to be a part of the devotions with the choir monks. The lay brother then heard a choir of angels singing the office, and when all this was reported back to St Bernard, he said that he wished that such an office had been sung in church as had appeared to that *felix conversus* in the field. This unusual story, with its ostensible overtones of divine justice, reiterates the message that the lay brother ought to be *felix* in his occupation, and that it is the lay brother who does not presume to alter his station who will be rewarded. There is no hint of rebellion or dissidence here, only sadness and the blessing of Mary as compensation for disenfranchisement. One of the most poignant examples of a lay brother's despair comes once more in the *Dialogus Miraculorum*. A lay brother in a state of grave melancholy was asked why he was so sad. He replied that he was sure of going to hell because he could not say

---

[96] *Dialogus Miraculorum*, book 7, ch. 25: *De converso, qui per Ave Maria ab infestatione diaboli liberatus est*.

[97] *PL*. 185, col. 1141. See also M. Casey, 'Herbert of Clairvaux's Book of Wonderful Happenings', *Cistercian Studies*, 25:1 (1990), 37–64 (p. 57).

[98] Paris, Bibliothèque Nationale, MS Lat. 15192. The story has been partially transcribed and translated by Brian Patrick McGuire, 'The Cistercians and the Rise of the Exemplum' (p. 266, note 101). The story is repeated in the *Exordium Magnum*, with some changes, including the lay brother's attempt to be part of the office by singing the *Ave Maria* over and over to himself. McGuire (p. 243) has proposed that in the *Exordium Magnum*, 'the moral [...] is changed [...] to the lay brother's obedience and acceptance of his rightful place', although it is difficult to see how the stories are essentially distinguishable on those grounds. There is a third version of the story which I have not yet seen. It appears in H. D. Oppel, 'Eine kleine Sammlung Cisterciensischer Mirakel aus dem 13. Jahrhundert', *Würzburger Diozesan Geschichtsblatter*, 34 (1972), 5–28.

his prayers like he used to. The lay brother was admitted to the infirmary, but wandered out one morning, and committed suicide by drowning himself in the fishpond.[99]

Sinnulph, Lay Brother of Fountains Abbey

From the Yorkshire Cistercian houses, the most complete representation of a lay brother is found in Hugh of Kirkstall's *Narratio*. The picture of Sinnulph, a lay brother of Fountains abbey, appears in a section describing the conversion of abbot Ralph to the Cistercian order. This section is the first to explicitly include Hugh of Kirkstall's own reminiscences, rather than to rely solely on the memories of the ninety-nine year old Serlo of Fountains, whom Hugh had interviewed for the early history of the abbey.[100] And as Hugh himself is keen to display, the memories he has of abbot Ralph are extremely fond ones:

> I owe myself altogether to him and always shall, for it is from his holy hand that I received the pattern of this holy institution, under his authority that I lived as a friend, and by him was given me, through God's grace, the opportunity of salvation and the pattern of healthful discipline. No subject would be to me more welcome, no story more pleasant, than one which keeps alive the memory of our holy father and describes his character. Willingly, therefore I approach this task [...].[101]

The pattern of abbot Ralph's own life, however, was in the hands of Sinnulph, a lay brother of Fountains abbey. Sinnulph was 'conspicuous for singular grace and purity. He was simple and illiterate, but God had taught him. For book he had his conscience, for teacher the Holy Spirit'. Sinnulph's influence over abbot Ralph began before the abbot had himself become part of the Cistercian monastery, and Ralph would tell stories about Sinnulph's great piety. When Ralph was weary of his secular life, Sinnulph prayed for him and—in the manner of St Bernard—Ralph felt 'the chains of my heart loosed and the yoke melted from the surface of the oil'. During the following evening, Ralph experienced the presence of God, and on the advice of Sinnulph, the 'Man of God', made the decision to enter the Cistercian order.[102]

One remarkable aspect of the influence of Sinnulph is that this *conversus* was fully able to establish a relationship with a secular person, although he was himself part of the ostensibly enclosed monastery. Ralph was a knight at this time, according to Hugh of Kirkstall, a status which, oddly, seems to have made it easy for Ralph to be transformed into a monk: 'He was once a soldier of this

---

[99] *Dialogus Miraculorum*, book 4, ch. 41: *De converso, qui ex desperatione in piscina se suffocavit*.

[100] For some discussion of Sinnulph, see Derek Baker, 'Heresy and Learning in Early Cistercianism', in *Schism, Heresy and Religious Protest*, ed. by D. Baker, Studies in Church History 9 (London: Cambridge University Press, 1972), pp. 93–107.

[101] *Mem. F.*, p. 117.

[102] *Mem. F*, pp. 118–19.

world and did not loose the girdle of his soldier life, but changed it for a better, joining himself to the camp of the Hebrews, afterwards to be prince among the people of God'.[103]

While Ralph was still a soldier, he would visit Sinnulph for spiritual advice and guidance, impressed by '[t]he piety and moderation of the man, the gravity of his silence, his modest carriage, the discipline of his habits; how ready he was to exhort, how helpful his consolation [...]'.[104] The intimacy of the relationship between knight and lay brother can be seen, in this case, to suggest the quasi-secular nature of the lay brotherhood itself. Not only was Sinnulph able to meet with Ralph, but according to Hugh's narrative, he was also able to converse with him at great length. This is an unusual relationship between two characters both of whom were on the margins, as it were, of monastic life. A comparable relationship, this time between a lay brother and a choir monk, can be found in a collection of early-thirteenth-century visions from the Cistercian monastery of Stratford Langthorne.[105] Roger, also a simple lay brother, told the compiler of this collection about his friendship with the monk Alexander. They would often discuss 'questiones de vita presenti et de morte futura', and ultimately made a pact that whoever of the two should die first would appear to the survivor after death and tell him of the afterlife.[106] Alexander was the one to die first, and appeared to his friend Roger as he had promised. Like the relationship between Ralph and Sinnulph, the tie that bound Roger and Alexander was an intimate friendship; but, as stated above, the relationship between Ralph and Sinnulph ought to be read in the light of a more general relationship between secularized characters.

Also unusual in Hugh's description of Sinnulph is the lay brother's status as a spiritual adviser and as a seer. Although Hugh is careful to stress Sinnulph's illiteracy, as do other representations of pious lay brothers in Cistercian literature, Sinnulph is also depicted as possessing a shrewd and prophetic sort of wisdom. Meeting with Ralph one morning, Sinnulph 'fell suddenly silent and, bursting into abundant tears, gazed on me [Ralph] with a serene countenance. "God has now revealed to me", he said, "what He has himself ordained concerning you. You shall assume the habit of religion in the monastery of Fountains, and, when your course is run, you shall there end your days"'.[107] This revelation remained with Ralph throughout his life; when he fell sick at Clairvaux one year and others despaired of him living, Ralph is said to have declared, 'I need not die at Clairvaux; it is at Fountains I expect the day of my departure. For brother Sinnulph foretold to me that I should live at Fountains and there should end my life'.[108]

---

[103] *Mem. F.*, p. 117. For other Cistercian uses of military imagery, see Newman, *Boundaries of Charity*, pp. 29–37.

[104] *Mem. F.*, p. 118.

[105] Holdsworth, 'Eleven Visions', pp. 197–201.

[106] Holdsworth, 'Eleven Visions', p. 198.

[107] *Mem. F.*, pp. 119–20.

[108] *Mem. F.*, p. 123.

This prophetic knowledge attributed to Sinnulph is reminiscent of exempla tales of devout *conversi* receiving visions or miraculous powers. However, it is important to recognize in all of these tales, and whenever *conversi* are singled out and described as wise or special, that it is always within the precise and demarcated context of their vocation. Sinnulph, like Roger in the Stratford Langthorne collection, and other pious *conversi* in the *Dialogus Miraculorum*, the *Exordium Magnum*, the *Liber Miraculorum*, and Jocelin of Furness's *Life of Waldef*, is simple, illiterate, and although given the influential role of spiritual adviser by Hugh, does not stray from the discipline of his profession. In other words, the 'knowledge of holy things and the spirit of revelation' that Sinnulph possesses in Hugh of Kirkstall's narrative, is developed in order to emphasize the spiritual development of the future abbot of Fountains, rather than to elevate this 'homo simplex et illiteratus'. I argue that even where lay brothers are depicted as positive influences or as upholding monastic virtues, these representations are based on implicit or explicit narrative strategies that effectively 'other' the lay brother. In Hugh of Kirkstall's *Narratio*, for example, the simple *conversus* is used to describe the spiritual development of the abbot. In the *Dialogus Miraculorum*, the stories of lay brothers seeing the Virgin are related to their humility, and in the *Life of Waldef*, the patient lay brother who did not presume to enter the chapel filled with glittering light in which the saint was praying, was rewarded with the vision of Waldef radiant and bathed in glorious light emerging after his prayers were complete.[109] In this way, I believe that the close relationship between Ralph and Sinnulph should not be misconstrued as indicating a more flexible role for lay brothers in Yorkshire or as being essentially different from French or German tales in the representation of a *conversus*. Even as an example of probity, the Cistercian lay brother occupied a very carefully demarcated site in monastic culture.

Within the Cistercian world, lay brothers occupied a particularly ambiguous space. Literary topoi of disorder and piety in the exempla texts of the thirteenth century illustrate the polarization of the *conversus*, who simultaneously served as a measure of unacceptable disorder and necessary humility. Institutionally, too, lay brothers seemed to have been used to test the boundaries of the monastery and its environs. Frequent contact with the secular world, especially commercially, brought with it the taint of worldliness while within the *terra sancta* of the monastic precinct itself, lay brothers were excluded from the most significant of the spiritual areas. The tension between material circumstance and imagined space is perhaps nowhere more evident than in the case of thirteenth-century Cistercian *conversi*, whose use as economic agents and manual labourers allowed them only limited experience of earthly transcendence. Ultimately, lay brothers were of no more use to the Cistercian monasteries of northern England, either economically or like Sinnulph of Fountains abbey, as the occasional conduits of revelation. Irrelevant to the changing Cistercian enterprise by the close of the thirteenth century, the spatial story of Cistercian *conversi* was rapidly drawing to an end, as they were simply and conclusively eliminated from the monastic world.

---

[109] McFadden, 'Life of Waldef', p. 20.

CHAPTER 7

Apostasy and the Contravention of Monastic Space

In 1286, Archbishop John le Romeyn of York included in his register an order to search for a monk of Rievaulx abbey, Godfrey Darel. According to the outraged archbishop, Darel 'had apostasised, and it is said, is wandering around in secular habit [...] deceiving the faithful by sorcery and nefarious incantations [...] and rejecting the church to the injury of his salvation and to the scandal of all orthodox Christians'.[1] 'Oh sadness', intoned the archbishop, before ordering that Darel be found and his case investigated.

The case of Godfrey Darel was one of many during the thirteenth century concerning apostates and fugitives from Cistercian houses. Darel, like other runaway monks, was considered to have contravened one of the most fundamental tenets of monasticism—that a monk should not go outside the enclosure of his monastery without permission, as was decreed in the *Regula S. Benedicti*.[2] Darel was a scandal to Christians not only because of the magic he was said to have been practising—although this was evidently of some concern to the archbishop—but also because he had violated the boundary between the monastic and secular worlds. Darel had abandoned his profession, broken his monastic vow, and acted in a way that completely undermined the monastic expression of Christian faith.

Various meanings were given to the act of monastic apostasy and the responses that it engendered. Representations of apostasy and apostates in monastic literature and in Cistercian legislation reveal that the abandonment of a Cistercian house signified the violation of monastic boundaries in both physical and spiritual ways. During the thirteenth century monastic space was represented as physically confining but at the same time spiritually liberating.

---

[1] *The Register of John Le Romeyn Lord Archbishop of York*, 2 vols (London: Publications of the Surtees Society 123, 1834), 1, 437.

[2] Manning, 'La Règle de S. Benoît', p. 263, ch. 67.

Apostasy was the absolute and unequivocal contravention of these understandings of space. Not only was the apostate rejecting the immediate monastic enclosure, but he was also seen to reject the freedom from the world and the ultimate liberation of the spirit that the monastery represented metaphorically. As a consequence, apostates were frequently described as weak, driven by the desires of the flesh, and vulnerable to the influence of the devil.

Legal recourse to the non-monastic world, as Darel's case shows, was not uncommon, while the scandal that runaways were seen to cause to the integrity of the Cistercian order was a constant preoccupation of the General Chapter and of individual Cistercian abbots. Fugitives from Cistercian houses faced grave consequences both within their monastery and from the secular arm if they were found, while they were also the subjects of vitriolic representation in Cistercian texts. The issue of apostasy itself must also be conceptualized broadly, especially in relation to the many medieval definitions of what constituted an apostate and the act of apostasy itself.

The Fugitives

Violating the Boundaries

The first obstacle to be encountered by a would-be fugitive from a Cistercian house was the walled monastic precinct. Negotiating the physical confines of the monastery was a recurrent theme in the exempla and other Cistercian literature of the twelfth and thirteenth centuries and in these stories, the built environment of the abbey is often represented as being unexpectedly difficult to transverse. Caesarius of Heisterbach, for instance, narrated an episode in which a nun attempted to scale the wall of her convent, but hit her head so badly while climbing, that she fell to earth within the convent walls and decided to stay.[3] Likewise, a Cistercian abbot called Serlo, on a visit to England, was said to have heard during confession the tale of an English nun who unsuccessfully attempted twice to unlock her door at night. On the third attempt, she discovered that the hand of the Virgin held fast the lock, preventing her escape.[4] The attempts of one novice to leave Rievaulx were continually thwarted by the strange transformation of the monastic environment. According to Walter Daniel, this novice was 'a man of no mental stability, always staggering about, now here, now there, from one thing to another, shaken like a reed by the wind of his changeful will'. The novice decided to leave Rievaulx, 'not heeding the counsel of salvation [...] unknowingly and foolishly ignorant '. All day long, it is said, 'he wandered aimlessly about the woods until, shortly before sunset, he

---

[3] *Dialogus Miraculorum*, book 4, ch. 56: *De sanctimoniali, quae dum nocte vellet ire ad saeculum, et caput ostio illideret, a tentatione liberata est.*

[4] London, British Library, MS Add. 15723, fol. 86.

came to the road by which he had left, and suddenly found himself again within the monastic wall'.[5]

This wondrous return encouraged the novice to remain at Rievaulx, but only temporarily. Some time later he attempted to leave again, saying to abbot Aelred: 'Everything here and in my nature are opposed to each other. I cannot endure the daily tasks, the sight of it all revolts me, I am tormented and crushed down by the length of the vigils, the food cleaves to my mouth, more bitter than wormwood, the rough clothing cuts through my skin and flesh down to my very bones'. Again, the novice decided to leave. This time, when the fugitive got to the gate, he felt the 'empty air at the open doors as though it were a wall of iron'. Again and again he tried to break through and could get no further. At last, in intense rage, it was said, he took hold of the hinges of the gate with both hands, and stretching out his leg, tried to put one foot forward. Still he could not reach the boundary, and at last, he resigned himself to living within the monastery permanently.[6]

The walls and doors of a monastery served as practical and symbolic boundaries, as I have argued in the first chapter of this book. Stories which focus on these physical features as barriers to apostasy not only reinforce the idealized division between the monastic and secular worlds, but also illustrate the changeful and potentially transforming nature of certain physical sites. In these tales, man-made structures are appropriated and used by divine and mysterious forces to illustrate the need for both spiritual and corporeal stability. The stone walls that marked the end of the precinct, according to these stories, also miraculously functioned as spiritual guardians and instruments of deterrence to those who attempted to violate the boundaries of the monastery or convent. The runaway's attempt to tackle and overcome these powerful symbols of enclosure always ends in failure in the exempla tales.

Monastic fugitives thus stood at the boundaries of monastic places and at the edge of the secular realm. Fugitives subverted Cistercian disciplinary practices of emplacement or enclosure within the monastery by attempting to relocate their bodies in non-Cistercian spaces. More generally, by merely exerting control over their own bodies, apostates had contravened their monastic status as disciplinary subjects. An example of this can be found in Caesarius of Heisterbach's tale of a novice who ran back to the world while his mentor was praying for him in the church.[7] In this story, the novice runs away before his impressionable soul can be persuaded to stay by the prayers of his master, while the narrative subtext defines the choice made by the novice in terms of space. The novice rejects the devotional site of the Cistercian church for the space of the outside world.

Such renegotiation and rejection of fixed sites may be interpreted as a tactic of empowerment by the movement of people who are otherwise placed in

---

[5] *Life of Ailred*, pp. 24–25.

[6] *Life of Ailred*, pp. 30–31.

[7] *Dialogus Miraculorum*, book 1, ch. 4: *De novicio in Hemmenrode converso, de quo sanctus David dixit: Non est omnibus datum.*

carefully demarcated physical places.[8] In other words, the movement of a subject from one place to another creates an entirely different space over which that subject exercises his or her own autonomy. In the context of apostasy and the violation of monastic boundaries, there are certainly theoretical aspects of such empowerment to be found. On an empirical level, however, the use of space as a device of empowerment becomes more problematic. Although the *effect* of the act of apostasy may have been to merge the monastic and secular worlds into a space defined by the wandering body of the apostate, the reasons for the act of apostasy were not always so easily described or expressed in terms of empowerment and control over space. The next part of this chapter will look more closely at the ways in which reasons for apostasy were represented by medieval Cistercians.

Reasons for Leaving

Myriad reasons have been proposed by historians of monastic apostasy as to why men and women who had made a permanent religious commitment felt unable to continue living in the monastic environment. These reasons range from the burden of the vow of poverty, a celibate life, personality differences with the abbot or superior, sheer boredom, and even fear, together with premature taking of the monastic vow as a novice.[9] The sources themselves are often frustratingly silent on the point of motivation, while the voices of apostates are conspicuously absent from most of the material. Establishing why monks may have left their Cistercian monastery is therefore frequently fraught with speculation.

Monks who witnessed the apostasy of others from their houses provided their own reasons for the discontent of the fugitives. Serlo of Fountains abbey, for instance, tells of the apostasy of two monks in the early days of the abbey's foundation. 'The tempter [...] approached two of the brothers, Gervase and Radulf', reports Serlo, 'who had not armed themselves sufficiently with the shield of faith and prayer'. The devil 'suggested poisonous advice to them, suggested treacherous thoughts, such as the austerity of the order, the horror of solitude, the tenderness of the body [and] the ways of their former life'. The two monks were urged to return to their former ways (as monks of the Benedictine abbey of St Mary's at York), because their new vocation as Cistercians, according to the devil, was too radical a step for these Benedictines to take. 'It is a difficult thing to live in the desert, to give up all that is familiar, to urge the body beyond what is natural, and to remain unmoved by the tortures of the flesh', whispered the devil. Gervase and Radulf succumbed to the venomous suggestions of the 'tempter', and returned to their former lives, much to the

---

[8] De Certeau, *The Practice of Everyday Life*, pp. 91–130.

[9] Logan, *Runaway Religious*, pp. 73–83. Logan also suggests that there were reasons peculiar to women religious that ought to be considered separately from the reasons pertaining to male religious (pp. 83–96); Christopher Harper-Bill, 'Monastic Apostasy in Late Medieval England', *Journal of Ecclesiastical History*, 32 (1981), 1–18.

scandal of their friends, according to Serlo, and to the derision of their enemies. Gervase, however, later realized the error he had made, and found strength enough to return 'like a soldier after flight' to Fountains abbey. Radulf, on the other hand, never returned, but 'made a pact with his flesh and his belly was swallowed into the ground'. 'Alas', mourns Serlo elsewhere, 'no place is safe from temptation while we live here on earth'.[10]

Weakness of the will and of the flesh are the means by which the devil can find a way into the monastery in Serlo's story. Gervase and Radulf are seen to be at least partially responsible for their own decline, as they were not sufficiently faithful nor wary enough to stave off the advances of the devil. Thus Serlo attributes the ultimate act of apostasy to the insufficient and shaky religiosity of the monks themselves. In Serlo's narrative, the bodily desires of these monks are the focus of the devil's suggestions because the body is essentially disorderly, and may easily be influenced, tempted, and coerced into transgression. The difficulties of austerity and solitude and the weakness of the flesh are encountered by all monks in Serlo's description of the early history of Fountains abbey. Radulf and Gervase, however, lacked the spiritual armour to defend themselves against the body's demands. Similar assumptions may be found in the story of the unstable novice in Walter Daniel's *Vita Ailredi*, described above. Like Gervase and Radulf at Fountains, the novice at Rievaulx was concerned with the desires of the body, such as better food, while 'my will', he is alleged to have said, 'is always desiring other things, it longs for the delights of the world, and sighs for its loves and affections and pleasures'.[11]

Matthew of Rievaulx, on the other hand, understood ambition to be a major reason why monks became discontented with their way of life. Like Serlo of Fountains, Matthew of Rievaulx believed that no place was safe from temptation, and evil surrounded all.[12] Ambition for things outside the cloister, according to Matthew, is like a virus or poison spread by the devil, while actually leaving the monastery for this reason means that the monk has been seduced by Satan. 'While wandering outside, Dinah was subjected by wantonness; Shechem deflowered her'; and likewise, analogized Matthew, 'Satan seduces the wandering monk'.[13] Succumbing to the desire to follow the 'exterior way' is the worst error a monk may make, in Matthew's view. The poem concludes with a final warning that ambition causes dissatisfaction with all things, and that is when the cloister will seem like a prison.[14] This is also the point at which apostasy may occur.

---

[10] For the preceding, see *Mem. F.*, pp. 30–31. For very similar vocabulary, see St Bernard's letter to a canon regular who has been tempted to return to the world in *PL* 182, cols 79–87. Bernard views the canon as a knight of Christ who would be deserting combat if he were to forsake his religious commitments.

[11] *Life of Ailred*, p. 30.

[12] Wilmart, 'Mélanges', pp. 64–65, *De Fece Ambitionis*.

[13] Wilmart, 'Mélanges', p. 65.

[14] Wilmart, 'Mélanges', p. 65: '[i]nde/ Lectio displicet hiis, et claustrum carcer habetur'. For the metaphor of monastery and cloister as prison in a more positive context, see Leclercq, 'Le cloître est-il une prison?'; G. Penco, 'Monasterium-Carcer', *Studia*

Matthew of Rievaulx expands these themes in another discourse on ambition, which is a lengthier version of the first poem.[15] In the second version, the monastic vocation and the enclosure of the monastery are more centrally located in the narrative. This creates an essentially spatial frame of reference against which the sin of ambition may be tested. Addressing those who have chosen the monastic life—*claustrales*—Matthew reiterates the seductive and secret work of the devil in encouraging ambition and a desire for external things: 'Flee, brothers!', the poet warns, 'the snake hides in the grass, here the serpent tricked Adam and seduced Eve'. Matthew sets up a clear distinction between interior and exterior worlds. The slow routine of the monastery—'now reading, now an Hour, now as the order demands, labour'—is the way to heaven.[16] The metaphor of cloister as paradise is repeated in this poem in order to underline Matthew's argument that the interior life of the monastery is all that a 'disciple of Benedict' should desire. No monk should exhibit desire for anything else.

The sin of ambition is the destroyer of the division between the monastic and non-monastic worlds, according to Matthew. As with the first poem on ambition, Matthew uses the metaphor of disease and infection to describe the insidious and essentially ruinous nature of ambition—it is a virus, a monster, leprosy. Ambition devours the mind; it is like an internal worm gnawing away at monastic stability. These images of contagion and pollution underpin Matthew's argument that thinking on things outside the monastery is not only wrong, but also dangerous. Ambition represents the weakening of the monk's defence against instability, temptation, and the devil. The image of the monastery wall being punctured by a viper is invoked here.[17] Again, it is the space of the monastery which represents all that is needed and right, while the external world is constructed as threatening, vile, and corrupting. The monk's duty is to address the ambition that lurks within, and send it without.

Thus the precondition for leaving the monastery, for both Serlo and Matthew, is frailty and vulnerability of spirit. Both Serlo's historical narrative and Matthew's poetic diatribe against ambition understand monastic discontent to be rooted in the omnipresence of Satan, who identifies the weak in spirit and then insinuates ideas of either discontent or ambition into their minds. The didactic effect of Matthew's and Serlo's texts is to make the reader or receiver of their

---

Monastica, 8 (1966), 133–43. The language of confinement did not, according to Jean Leclercq, become a common metaphor for the monastic life until the twelfth century, although the use of *solitudo, eremus, carcer, claustrum*, and *clausura* had been used to describe the Christian life for some time. Leclercq gives the example of Tertullian's letter *Ad Martyras*, where the imprisonment of the martyrs is likened to the voluntary seclusion and enclosure of the earliest eremites. Tertullian also comforted the martyrs with the observation that: 'quotiens eam spiritu de ambulaveris, totiens in carcere non eris'.

[15] Paris, Bibliothèque Nationale, MS lat. 15157, fols 53–55, *De ambitione qui est radix omnium malorum*. I am very grateful to Dr James Girsch for making his transcription of this poem available to me.

[16] Paris, Bibliothèque Nationale, MS lat. 15157, fol. 54.

[17] Paris, Bibliothèque Nationale, MS lat. 15157, fol. 53^v: 'Ambitus excruciat, pungit quasi vipera mordens'.

words wary, alert, and cautious of the lurking evil that stalks every Christian. Apostasy is seen to be the manifestation of Satan's success.

Other Cistercian sources are less analytical of runaway monks, although perhaps more direct in reporting the pragmatic motive for flight from the monastery. Legal records are one example. Assize records from the late thirteenth century provide two cases of flight from the Yorkshire Cistercian houses which are quite clear as to the immediate reason for the decision of a monk to apostasize. The first involved a monk of Jervaulx abbey, William de Modwither, who fled from the monastery in 1279 after murdering abbot Philip. The case was brought to the assize as it was initially thought that Thomas, abbot Philip's successor, had been the one to commit the murder. The jury at the assize found that William de Modwither was the culprit and, as he had left the monastery and could not be found, William was outlawed.[18] This case is also reported in the statutes of the General Chapter for that year.[19] Although neither the assize records nor the General Chapter attempt an explanation for the murder of the abbot in this case, it is clear that the committing of this crime was the motivating factor in William de Modwither's apostasy.

Rather less clear is the case of William of Akerton, a monk of Rievaulx. It is reported, again in the assize rolls, that in 1279 William wanted to leave the monastery. He is said to have gone to the prior, Nicholas of York, and claimed to be a leper. As he was leprous, William is supposed to have said, it would be impossible for him to live amongst the monks at Rievaulx, and he should therefore depart to live elsewhere. When another monk, Jordan de Normanton, tried to examine him to find out if he really did have leprosy, William stabbed him and ran off into the woods. The abbot immediately sent two of the brothers to find him and return him to the abbey, which they did successfully. In the process, however, William de Acton was beaten so badly that he died at Rievaulx some days later.[20] This case was brought to the attention of the assize, not because of the apostasy involved, but because of the homicide that ensued. The reason for William's initial attempt to leave Rievaulx remains unknown, although it was evidently desperate enough to make the possible social exclusion faced by a leper seem preferable.[21]

Representations of fugitive monks and the reasons for their apostasy are therefore various. Within narrative and literary sources, the body of the apostate was constructed as being vulnerable to temptation and discontent. The mens rea of the apostate is more difficult to evaluate, given the absence of apostates' voices themselves from the sources. This is why Michel de Certeau's theoretical

---

[18] Reported in *The Victoria History of the County of Yorkshire*, ed. by W. Page, 3 vols (London: A. Constable, 1907–1913), 3, p. 141.

[19] *Statuta* 1279, 22, t. 3.

[20] This is also reported in Page, *VCH Yorks*, vol. 3, p. 151.

[21] See R. I. Moore, *The Formation of A Persecuting Society: Power and Deviance in Western Europe 950–1250* (Oxford: Blackwell, 1987), pp. 59–60 for the effects of the thirteenth-century laws of segregation in England. Moore also notes the existence of a leper house in Ripon—not too far from Rievaulx.

claim—that repositioning of the body outside or beyond fixed and demarcated sites should be seen in terms of empowerment—must be qualified in the case of monastic apostasy. Other Cistercians, however, found ways to explain the act of apostasy, ranging from weakness to the influence of the devil. The instances of fugitive monks I have described above raise some more important questions as to the meaning of the various words used to describe the act of leaving a religious house without permission, and the meanings given to that act. Runaways are referred to as fugitives, as egressors and as apostates. If this is the case, then what did constitute apostasy, both in canon law and in monastic understanding?

Apostasy and its Meanings

Defining the Term

Within the period of the thirteenth century canon law underwent significant transformations, first and foremost with the 1234 collection of the *Decretals* of Gregory IX. Although the *Decretum Gratiani* of c. 1140 had been an attempt to collect and systematize the very disparate and often contradictory collections of canon law prior to this period, Gratian's collection remained legally unofficial, despite its great influence.[22] By 1298, with the *Liber Sextus* of Boniface VIII added to the 1234 collection of Gregory IX, medieval canon law was codified and centralized in a way that facilitated common understanding and common practice of legal, canonical issues. Definitions of apostasy, however, remained broad. Medieval commentators on the decretal collections did distinguish between apostasy that involved the complete abandonment of the Christian faith, and apostasy that referred to the abandonment of one's religious house. In the 1160s, for instance, Stephen of Tournai (commenting on Gratian) stated: 'today it is called apostasy when someone who has taken the religious habit afterwards returns to the world'. This definition is the one utilized by F. Donald Logan, who argues that medieval monastic apostasy referred to a situation where 'a professed religious abandons the religious life and returns to the world'.[23]

Within the Cistercian milieu, canon law was not necessarily perceived as being the most persuasive authority, and in 1188, the General Chapter legislated that the work of Gratian should not be held in the communal book-cupboard of Cistercian houses.[24] This prohibition did not mean that the *Decretum* was

---

[22] For Gratian, see the *Corpus Iuris Canonici* (Leipzig, 1879–81); for the canonist in England, see the classic article by S. Kuttner and E. Rathbone, 'Anglo-Norman Canonists in the Twelfth Century', *Traditio*, 7 (1949–51), 279–358.

[23] Logan, *Runaway Religious*, p. 10.

[24] *Statuta* 1188, 7, t. 1. See also P. C. Bock, 'Les cisterciens et l'étude du droit', *ASOC*, 7 (1951), 3–31.

ignored altogether, or even entirely forbidden; rather, the General Chapter was attempting to confine the reading and study of this 'non-monastic' law to a small and select group of Cistercians, who would then presumably decide how best to interpret and implement it. The semi-prohibition of the reading of canon (and civil) law was repeated in 1240. In relation to the binding force of canon law definitions of apostasy, it would appear that Cistercians were able to create their own meanings and definitions, although the broad paradigm on which such definitions were based was not, of course, at odds with the laws of the universal church. In some cases, such as at Rievaulx abbey in the later-twelfth century, papal decrees were even requested by Cistercian houses in order to deal with potential fugitives and apostates.[25] Decretals specifically concerned with the question of apostasy also indicate that the primary jurisdiction to which a fugitive monk had to answer was that of his order,[26] although abbots and superiors were subject to canonical sanction if the way in which they dealt with a situation strayed too far from a papal decree. It is thus within Cistercian sources themselves that Cistercian understandings of apostasy are most usefully found.

Broad definitions such as those given by medieval canonists and modern historians do not always cover every situation where apostasy might be inferred. The situation of monastic novices is one such example. Was a novice who was still within his year of probation to be charged with apostasy if he left the order? Caesarius of Heisterbach's *Dialogus Miraculorum* addresses this issue in consecutive exempla, where a novice asks: 'Surely it is a grievous sin for a novice to return to the world?' The novice-master first tells of the novice Leo, who was tempted to return to the world after having made his initial profession at Hemmenrode. Leo became ill, then mad, calling out the names of women with whom he had sinned. Finally, despite the attempt of his friends to remedy his madness by cutting up puppies and placing their warm flesh on Leo's head, the apostate novice died in agony.[27] The following story concerned a monk on probation, who 'like a dog to his vomit', returned to the world. After a time, this novice died, and at his death, a storm broke and a murder of crows was seen to hover above the roof of his house. 'See then how they die, those who depart from God', says the novice-master.[28]

The novice questions the retribution of a horrible death exacted on the novice

---

[25] A papal mandate dated by Powicke, *Life of Ailred* (p. 40, note 3), to between 1171–81 and addressed to the parsons of churches in the province of York, is recorded in the Rievaulx cartulary. See *Cart. Abb. de Rievalle*, p. 194, no. 261.

[26] Gregory IX's decretal *Ne religiosi* is one example, where Gregory states that the decision to excommunicate an apostate should lie with the apostate's monastic superior. For an outline of the decisions regarding apostasy in the universal church, see Harper-Bill, 'Monastic Apostasy', passim, and A. J. Riesner, *Apostates and Fugitives from Religious Houses* (Washington: Catholic University of America, 1942).

[27] *Dialogus Miraculorum*, book 1, ch. 14: *De miserabili morte de Leonii novicii apostatantis*.

[28] *Dialogus Miraculorum*, book 1, ch. 15: *De horrenda morte Benneconis novicii, et quod non liceat noviciis redire ad saeculum post votum*.

in the second story, asking for the novice-master to explain the meaning of the *Regula S. Benedicti*'s instruction that 'this is the law under which you desire to enlist; if you can keep it, go forward; if not, go in peace'. 'If it is such a mortal sin for the novice to return to the world', asks the novice, 'what then is the meaning of St Benedict's instruction?' The master replies that the two situations described in the stories are both wrong. However, he says, 'of the two ills the holy father prefers that the novice should depart while still a novice rather than desert as a monk after making his profession'. The novice remains unbound by the place of the monastery, but bound by his vow. This means, according to Caesarius' novice-master, that a novice may be transferred to another, less severe religious house if necessary, although he cannot be given dispensation to return to the secular world. A monk, however, has promised constancy with regard to both place and vow, according to the novice-master, and therefore cannot leave.[29]

Defining the Act

The *Speculum Novitii* attributed by Mikkers to Stephen of Sawley also advises the novice on apostasy. In this text, the author is less concerned with stressing the binding nature of the monastic obligation, than with the effect of apostasy on the novice himself. Apostasy, in this text, is defined as disobedience. The novice is urged not to give in to the desire to depart from the monastery, however temporarily:

> When a desire to ride horseback plagues you remember what happened to Dinah, who had only gone outside to take a walk. If you feel that by your going out some future good will happen to your monastery, remember Esau's departure, who while busily hunting, lost his father's blessing. If you have a strong desire to visit your parents, remember Tamar, David's daughter, who, even though she had her father's permission to leave her chamber to visit her sick brother, was corrupted by him.[30]

The secular world is thus fraught with danger. The instructions then given to the novice to avoid the temptation of experiencing this threatening environment are quite different in tone and form to those given by the German Caesarius of Heisterbach in the exempla tales. The author says intimately:

> Dear son, I warn you that of all the snares of the ancient enemy, you must avoid the following as you would avoid the noonday devil: from the time you entered religion never entertain, cherish or give credence to the suggestion that it is of scant use to remain, that you could be more useful to yourself and others in another way of life, another habit.[31]

It is for the novice's peace of mind, according to the *Speculum*, that he must

---

[29] *Dialogus Miraculorum*, book 1, ch. 15.

[30] Mikkers, 'Speculum', p. 66.

[31] Mikkers, 'Speculum', p. 66.

remain in the Cistercian monastery. Unlike Caesarius of Heisterbach, the author of the *Speculum Novitii* does not dwell on the potential ferocity of divine punishment that will ensue if a novice should apostasize. Rather, this text emphasizes the psychological effects of the temptation to leave the monastery. The novice's heart will become 'like a broken jar which will not hold wisdom', says Stephen, while 'your soul will be torn apart trying to choose between many professions and pursuits'. The division and distraction of the soul is the result of thoughts of apostasy, a state which will leave the novice unable to retain the advice of the master 'or any unction of grace or the seasoning of spiritual flavour'. The consequence of this, the novice is warned, 'will be the danger that not only weariness but a frightful despondency over the good work you have begun will overwhelm you'.[32]

The novice is instructed to ponder on the grace of God that brought him to the Cistercian monastery in the first place. The novice, according to the *Speculum*, has been reinstated to paradise as a result of making the commitment to the monastic life; unlike Adam, who 'almost died of weeping and wailing when he realised that he, who had lived in a place of sunshine and sweet delights, had suddenly been cast out into so dark and terrible a prison, a land of drought and a symbol of death'. The novice's duty is to give thanks that he has been invited 'to the angels' table, to the music and harps of paradise, to the chanting of heaven'. Rather than terrify the novice with grim stories of the retribution that may be exacted on a monk or novice who has already left the abbey, the author stresses the need for the novice to understand and meditate on the reasons for him being in the monastery at all. As I have argued previously, the practice of meditation served a number of disciplinary and pragmatic purposes: in the context of potential apostasy, meditating on the grace that 'opened to you the long-closed entrance to paradise' serves in the *Speculum Novitii* to encourage stability.[33] Apostasy, in the *Speculum*, is thus understood to be instability, dividedness of purpose and rejection of the liberating and eternal space of the monastic world.

Defining the Apostate

Defining what constituted an apostate, even when departure from a monastery had occurred, was not always easy for Cistercians. As already mentioned above, the meanings given to a monk's flight from a monastery were understood and expressed within the context of ostensible motives for that departure. Related to these representations was the definition and identification of apostates themselves. An example may be found in Hugh of Kirkstall's *Narratio*, where Serlo of Fountains recalls the circumstances in which the monastery at Fountains was established. Unlike other English Cistercian houses, Fountains abbey had been founded, not as a 'daughter house' of another Cistercian abbey, but by a group

---

[32] Mikkers, 'Speculum', pp. 66–67.

[33] Mikkers, 'Speculum', p 68.

of discontented monks from the Benedictine abbey of St Mary's at York.³⁴ Serlo of Fountains remembered that some of the Benedictine monks heard of the austerity and purity of the Cistercian order and desired to imitate it themselves—'with all eagerness', says Serlo, 'they longed for the desert, for manual labour [...] it was their pleasure to clothe themselves in sackcloth and hair-shirts, to correct by a more severe judgement the indulgences of a too lax mode of life'. This desire, however, was not well received by other members of St Mary's community, and Serlo reports of 'a rumour spread among the brethren that the prior and associates were moved by the spirit of levity, were teaching secession from their own people and were setting about new and unlawful plans which would lead to the contempt of their order, the disgrace of their house, the confusion and scandal of the whole convent'. Even the abbot of St Mary's believed that this plan to follow the Cistercian observance was tantamount to desertion, and that the monks had no right to break the oath which they had so solemnly taken.³⁵

Nonetheless, with the support of Archbishop Thurstan of York,³⁶ the group of thirteen monks who desired to secede from the Benedictine community did just that, much to the consternation of the abbot of St Mary's, who accused them of apostasy. 'Surely, he said, it was absurd for a monk to transgress the form of his profession, to change the laws of his predecessors, to declare a law unto himself, to cross the bounds which his fathers had fixed from the beginning'.³⁷ Even when the discontented monks proposed the reform of their own house, the abbot, who 'admitted that he was not very quick in comprehension', declared 'that he and the rest of the monks desired to observe no other than their established institutions'.³⁸ The abbot did ultimately promise not to stand in the way of the monks' decision, despite the animosity of the rest of the community. The questions then addressed by Hugh of Kirkstall, the author of the text, are whether the monks who seceded from the abbey of St Mary's could rightly be called apostates, and whether the monastery of Fountains was therefore founded by runaway monks.

The voice adopted by Hugh of Kirkstall to discuss these questions is not that of the Cistercian monk Serlo, but rather that of the Archbishop of York, Thurstan, who is reported to have written a letter to William of Corbeil describing the events.³⁹ Not surprisingly, Hugh of Kirkstall has the archbishop decide that the reformers from St Mary's were not apostates. He concentrates on the

---

³⁴ For these unusual conditions, see Baker, 'The Foundation of Fountains Abbey'; Baker, 'The Genesis of Cistercian Chronicles in England: The Foundation of Fountains Abbey'.

³⁵ *Mem. F.*, pp. 6–7.

³⁶ Who believed, according to Serlo, that men should not remain in circumstances where the grace of God has not given them strength to lead a good life among licentious customs. See *Mem. F.*, p. 18.

³⁷ *Mem. F.*, p. 10.

³⁸ *Mem. F.*, pp. 21 and 24.

³⁹ *Mem. F.*, pp. 11–29.

nature of their monastic vow, deciding that the content or ideals behind the vow were more important than the ties that bind a monk to a particular place. It is the fact that the thirteen Benedictines wanted to observe a stricter and more austere way of life that is the issue, in the author's view, not that the monks were leaving the physical space of the monastery in which they had made their initial profession, nor even that they were leaving one monastic order for another: 'these men should not be regarded as turning away from their vow, but looking forward, since they are leaving a place where the opportunity of sin is too great and desire to serve God in more security'.[40] This is in stark contrast to the case of Gervase and Radulf which I have recounted above, where these two monks are unequivocally described as apostates by Serlo on the basis that they chose to return to a more lax way of life, after having made a commitment to the Cistercian order.

The crucial point, according to Hugh of Kirkstall, is that monks decide to follow, not so much a particular institution or order, but the *Regula S. Benedicti*; in reality, decides Hugh (echoing the words of Odo of Cluny), 'that monk [is] an apostate who in habit, diet, and other ways does not shrink from transgressing the rule of the blessed Benedict'.[41] Men who wish to follow the rule ought not be blamed and censured, dictates Hugh, but should rather be protected. The alleged authoritative tones of the Archbishop of York lend these arguments a decisive weight, as does the inclusion of historical precedent. The letter states that 'we ought to remember that a similar secession of monks from Molesmes founded and established that perfect rule of Cistercian life which is now the marvel of almost the whole church'.[42]

The narrative technique of introducing a powerful and respected figure into the text as a way of expressing the view of the Cistercian author is replicated almost immediately in relation to the same issue of apostasy. St Bernard himself is said to have written letters of support to Richard, the first abbot of the new monastery at Fountains, to Archbishop Thurstan and two letters to Geoffrey, abbot of St Mary's at York, the texts of which are incorporated into Hugh of Kirkstall's *Narratio*.[43] In the first of these letters, Bernard briefly congratulates Richard for the courageous act of going beyond mediocrity to embrace a better life. Bernard speaks of the security of the monks' decision, and affirms their righteousness by saying: 'a most rare bird on the earth is a man who advances even a little from the stage he has once reached in religion'. The second letter is also a congratulatory one, this time to Thurstan of York, whose support for the Cistercians Bernard describes as 'truly episcopal' and 'a notable example of paternal love'.[44]

---

[40] *Mem. F.*, p. 27.

[41] *Mem. F.*, p. 22.

[42] *Mem. F.*, p. 29.

[43] *Mem. F.*, pp. 36–37 for the letter to abbot Richard and his companions; *Mem. F.*, pp. 38–39 for the letter to archbishop Thurstan; *Mem. F.*, pp. 40–45 for the letters to abbot Geoffrey of York.

[44] *Mem. F.*, p. 38.

To the abandoned abbot of St Mary's, Bernard writes of the question of apostasy. First, the return of the monks Gervase and Radulf to the monastery of St Mary's is described by Bernard as a grievous disaster for them. Bernard refuses, however, to condemn these monks as apostates, saying only that: 'I ought not to condemn them. The Lord knows them that are his'. Bernard writes that he can only tell of how he would feel if he were in the situation of Gervase and Radulf:

> I, Bernard, if I had freely, of my own will and deed, turned from good things to better or from dangerous things to greater safety, and afterwards had presumed by some unlawful decision to return to the things I had left, I should be fearful of becoming, not only an apostate, but even unfit for the kingdom of God.[45]

Here, the apostasy of Gervase and Radulf has been turned around to describe the act of fleeing from Fountains abbey, not the act of fleeing with the other monks from St Mary's.[46] Even abbot Geoffrey of St Mary's was concerned about reinstating these apostates into his community, although, paradoxically, they were returning to the place in which their vows had been made.

Thus, in monastic discourse, apostasy did not always present itself in absolutely clear-cut cases, while defining the apostate was by no means the common-sense task that the definitions given at the beginning of this section might suggest. Establishing apostasy depended on the stage at which a monastic vow had been taken, and it depended on how that vow was construed. Apostasy could be described in terms of instability, weakness, divided loyalty and laxity, as well as the violation of monastic enclosure, while Cistercian sources distinguish between apostasy committed in order to live a better life in the eyes of God, and apostasy committed for other reasons.[47] Once apostasy had been established, however, the institutional superstructure of the order and its allies sprang into action.

---

[45] *Mem. F.*, p. 43.

[46] For a similar situation, also described by Bernard, see his *Epistola Prima ad Robertum nepotem*, PL 182, col. 71, a letter to his nephew, who had deserted the Cistercians for the Cluniacs. The General Chapter, too, legislated against changing from the Cistercian order to any other. See for instance, *Statuta* 1223, 12, t. 2.

[47] Modern canon law distinguishes more carefully between an apostate and a fugitive: see *Canon Law: A Text and Commentary*, ed. by T. L. Bouscaren and A. C. Ellis, 3rd rev. edn (Milwaukee: Bruce Publishing, 1957), p. 305 regarding the distinction between one who leaves a religious house without permission but with the intention of returning (a fugitive), and one who leaves with no intention of coming back (an apostate). For an overview of the more haphazard medieval uses of these terms, see Jean Leclercq, 'Documents sur les fugitifs', in *Analecta Monastica*, Studia Anselmiana 54, 7th series (Rome: Pontificium Institutum S. Anselmi; Orbis Catholicus Herder, 1965), pp. 87–145.

Responses to Apostates and Fugitives

The General Chapter

In the Cistercian world, the General Chapter was the forum in which decisions were made as to the treatment and punishment of apostates and fugitives. Throughout the twelfth and thirteenth centuries, the General Chapter issued a number of statutes and codifications which attempted to deal with the question of monastic apostasy. The statutes are primarily concerned with the punishment of apostates, and the means by which they could be reincorporated into the Cistercian house from which they had fled. Within these statutes, there are also examples of the cooperation that existed between secular and monastic jurisdictions. The issue of apostasy often brought the non-monastic world into contact with Cistercian houses, and the legislation of the General Chapter provides some insight into the legal relationships that were forged between the secular and monastic spheres.

First and foremost, however, the statutes deal with the punishment of apostates who had been found and brought back to their monastery, or who had returned voluntarily, or who were attempting to hide in another house. Abbots were bound to punish these monks to the degree determined by the General Chapter. In 1175, for instance, the General Chapter ruled that an abbot who knowingly sheltered an apostate was to fast on bread and water every Friday and was prohibited from celebrating mass. The fugitive himself was subject to a number of different punishments, depending on the nature of the apostasy itself and the other crimes that were seen to accompany it. Apostates were considered to be thieves if they had left the monastery with more than two habits and cowls, while an apostate who had married and produced children during his period of apostasy was to be punished as if he had been found in carnal sin.[48]

As the thirteenth century wore on, punishments faced by an apostate were directed more at the humiliation of his body, while many of the corrections implemented by the General Chapter were to be carried out in public view. In 1221, for example, the General Chapter legislated that an apostate monk should accept the discipline in chapter every Friday for one year, and on these days he should fast on bread and water. The same year, the General Chapter decided that monks or lay brothers who returned to the monastery as a thief should be last in all things, fast on bread and water every Friday for a year, and eat rough bread for forty days. In the case of *conversi*, they were to eat their meals from the floor of the refectory for forty days and be beaten in chapter for a year. In 1231, it was ruled that unless an apostate had been away for less than seven days, he would be last in all things; and in 1263, the General Chapter decreed that an apostate should have no new clothes for three years and was to be refused any administrative office. By 1271, all apostates were to eat their meals from the refectory floor for a year and wear the habit which the abbot decreed for them.

---

[48] *Statuta* 1195, 11, t. 1; *Statuta* 1195, 7, t. 1 and repeated with slight amendments in *Statuta* 1266, 4, t. 3, when the problem had evidently grown; *Statuta* 1195, 21, t. 1.

Monks were to be shaved in the same way as lay brothers for a year as well.[49] The following year, the General Chapter revised the 1271 ruling, stating that monks who had come back to the monastery should be flogged in chapter every Friday for a year, but that *conversi* should be received in their usual habit and should be shaved according to their usual custom. The end of the thirteenth century saw a general amnesty granted for fugitives and apostates: in 1296, all apostates were to give themselves up by Christmas. Any fugitive subsequently discovered was to be found and detained at the expense of his own abbey.[50]

The thirteenth-century codifications added little to existing legislation concerning apostasy. The collections of 1237 and 1257, for instance, served to consolidate the existing decisions of the General Chapter on apostates and fugitives. Abbots were reminded of their duty to search for runaways and to bring them back to the monastery,[51] and other penalties for apostasy were reiterated. The codifications also mention the use of the secular arm for the finding and return of apostates and fugitives from Cistercian houses. Thus the various forms of legislation implemented by the Cistercian General Chapter show that apostasy remained a concern of the order throughout the thirteenth century, while the punishments brought to bear on fugitives tended to focus on the disciplining of the apostate's body via fasting, beating and public humiliation. Forcible enclosure in the form of incarceration within the monastery, too, was an additional punishment introduced during this period.

The Secular Arm

The General Chapter was not the only judicial forum encountered by the fugitive Cistercian monk. From the mid-thirteenth century in England, apostates were signified for arrest by the secular authorities through the issuing of the writ *De Apostata Capiendo*.[52] These writs allowed the secular arm to assume responsibility for the capture of apostate Cistercians (and monks and nuns of other orders), although the monastery itself was the place to which the fugitive was returned and where he was punished. If, however, an apostate had committed a crime while a fugitive from his monastic house, then he answered to the secular realm for that crime. The previously mentioned case of William de Modwither, who fled from Jervaulx in 1279 after murdering abbot Philip, is one example, where the assizes were the apostate's place of trial. It would seem, however, that the issue of apostasy was still addressed by the culprit's monastery once an apostate had been returned.

---

[49] *Statuta* 1271, 3, t. 3.

[50] For the preceding, see *Statuta* 1221, 11, t. 1; 1221, 10, t. 2; 1231, 7, t. 2; 1263, 3, t. 3; 1271, 3, t. 3; 1296, 2, t. 3.

[51] B. Lucet, *Les codifications cisterciennes de 1237 et de 1257*, p. 279, *De fugitivis quibus negatur reversio*. This was also the case at canon law: see also *Decretales Gregorii IX*, 31.24.

[52] These writs are the main focus of Logan, *Runaway Religious*. The extant Cistercian examples are to be found in the London Public Record Office, series C81/1788.

The writ *De Apostata Capiendo* served to bring about the return of the apostate to his religious house by means of the intervention and aid of the secular arm. The surviving examples of this procedure are in the form of letters from the abbots of various religious houses to the King and Chancellor, petitioning them for aid in apprehending specific runaway monks.[53] The petitions were answered by the issue of the writ itself. Within the letters of request, sparse details are provided as to the reasons for the runaway's flight from the monastery, or even his whereabouts. These petitions are predictably formulaic, naming the apostate, stating that he has left the abbey without permission, that he has endangered his salvation and the reputation of the religious order he has abandoned, and that the monastery requests secular aid in his return.

There is only one example of a thirteenth-century request for the issue of *De Apostata Capiendo* that came from a Yorkshire Cistercian house, although there are many examples from other English Cistercian houses, and, after 1300, from the Yorkshire monasteries as well. The lone example comes from Fountains abbey and is dated 1289. It concerns a lay brother, Henry Sampson, who was said to wander around to the detriment of his soul and the great scandal of the Cistercian order.[54] The abbot of Fountains requested that the secular arm be used to capture Sampson for his delivery back to the monastery. This petition does not differ from the other, later examples from Cistercian houses contained within the same series, although other petitions often reveal that the apostate has dressed in secular clothes.[55]

The involvement of secular procedures in solving the problem of apostasy may be understood in a number of ways. First, the use of secular writs such as *De Apostata Capiendo* by religious houses demonstrates a closer relationship between monastic affairs and those of the 'outside world' than may be expected. Although monastic and secular jurisdictions were ostensibly separate, the cooperation revealed in the monastic petitions to Chancellor and King indicates that the boundaries between the two could be renegotiated if necessary. The General Chapter also provides evidence that the secular arm could be brought to bear on runaways. In 1256, for example, the General Chapter reported that a lay brother of Maris Stellae, brother Warner, was to be imprisoned in perpetuity for his indecency and dishonesty. Brother Warner had evidently also apostasized, as the General Chapter decreed that the secular arm could be brought in if necessary in order to effect his punishment.[56]

---

[53] Comparable petitions, although for excommunication, may be found in the Public Record Office's C85 series, which form applications from secular authorities for secular aid to secure excommunicated people. The King and Chancery were therefore heavily involved in the recording and tracking of religious transgression.

[54] PRO, C81/1788.

[55] See for instance PRO, C81/1788 from Fountains abbey, where Robert de Sturtonis was signified for arrest in 1392, having left his abbey fourteen years previously, and dressed in secular garb.

[56] *Statuta* 1256, 7, t. 2.

It should be remembered, nevertheless, that the monastic world only invoked the secular arm when necessary in these cases. Petitions for the issue of a writ for the arrest of an apostate and even the involvement of the secular arm in the recovery of monks such as brother Warner came from the Cistercian houses themselves—not, it seems, at the instigation of the non-monastic world. Although, as mentioned previously, abbots and religious superiors were bound to recover apostates and fugitives who had left their houses, and to receive them back three times,[57] the means by which the recovery occurred was not solely through administrative formulae such as *De Apostata Capiendo*. The case of the pseudo-leper at Rievaulx abbey in 1279 shows that other monks could be sent out to recover an apostate, however infrequently, while the General Chapter's amnesty for apostates at the close of the thirteenth century shows that in other cases, Cistercian houses were happy to wait for the fugitive to return to the monastery voluntarily.[58] Cistercian monasteries did not, therefore, always avail themselves of the secular arm.

More generally, these encounters between secular and monastic worlds reveal some of the seriousness with which apostasy was regarded by Cistercians. Within the seemingly non-descriptive petitions for secular aid, reasons stating the urgency for the return of apostates stress two points in particular. First, fugitive monks were said to be endangering their souls and their prospects for salvation. This is a point that is brought out by some of the other sources describing meanings of and motives for apostasy by Matthew of Rievaulx and the author of the *Speculum Novitii*, as I have mentioned above. This point was reiterated by Gregory IX in *Ne Religiosi Vagandi*.[59] The second point made in almost every petition is that the apostate was causing scandal to the Cistercian order. Henry Sampson, the runaway *conversus* from Fountains, for instance, had caused scandal to the order, while the wandering magician Godfrey Darel, cited at the beginning of this chapter, had gone even further in causing scandal to all Christian people. Caesarius of Heisterbach's *Dialogus Miraculorum* includes the tale of a novice who attempted to leave the monastery, but who was stopped by his abbot, who told the novice, 'Bring me an axe'. When the novice wisely asked why, the abbot replied: 'That your feet may be cut off. For believe me, I would rather keep you without your feet, than to let you go away and bring shame upon our house'.[60] Shame on a Cistercian house and scandal to the order in general were perceived to be as serious as the fact that an apostate was in grave danger of forgoing his own chances of

---

[57] *Statuta* 1221, 8, t. 2.

[58] *Statuta* 1296, 2, t. 3. See also note 55 above, where the abbot of Fountains waited fourteen years before requesting aid for the return of Robert de Sturtonis. Another example of monks returning voluntarily to their monastery may be found in Walter Daniel's *Vita Ailredi*, where Walter says that during the seventeen years of Aelred's abbacy, only four monks left Rievaulx without permission, but that three of them returned. See *Life of Ailred*, p. 40.

[59] Cited in Logan, *Runaway Religious*, p. 143.

[60] *Dialogus Miraculorum*, book 4, ch. 50: *Item de tentatione Reneri successoris eiusdem*.

salvation. The involvement of the secular arm in monastic affairs may be seen as an attempt to avert such potential public scandal.

Other Solutions

Although the proliferation of punishments and the relationship forged between the secular realm and English Cistercian houses over the question of apostasy may imply that the only time an apostate was dealt with was after committing an act of apostasy, other solutions were available to unhappy and discontented Cistercians. As the poems of Matthew of Rievaulx and the *Speculum Novitii* reveal, abandonment of the monastery was seen to be a constant threat to the Cistercian order, while even the General Chapter itself realized the need for preventing the problem of flight from the monastery before the judicial machinery of punishment was invoked.

One such preventive measure was to relax some of the more severe elements of monastic life in order to make the potential runaway's existence within the confines of the abbey a little more comfortable. The novice who unsuccessfully tried to negotiate the suddenly magic precinct walls at Rievaulx was promised better food and softer clothes to tempt him to remain at the abbey.[61] Other more senior monks were able to request that their duties be changed. Maurice of Rievaulx, for instance, was able to resign from the position of abbot at this monastery, as he was 'irked by the burdens of pastoral care' and 'preferred to resume his seat in the cloister'.[62] Maurice's return to the cloister being related to the possibility of his leaving the monastery altogether is unlikely; nevertheless, the example does indicate that, in some circumstances, it was possible for a monk to request a change in his situation.

One example where an unhappy abbot was denied the option of resigning from his abbacy appears in Hugh of Kirkstall's *Narratio*. Richard, the second abbot of Fountains, was said to have accepted the burden of the abbacy unwillingly, as 'he had dedicated himself to spiritual joys and undertook worldly duties with reluctance'.[63] Despite this reluctance, Richard is said by Hugh to have been an excellent abbot, while the monastery at Fountains expanded under his leadership. However, a speech impediment made the performing of public duties extremely difficult for Richard, so much so that he began to tire of his office, deciding after four years that he did not want to continue as abbot. Abbot Richard approached Bernard of Clairvaux three times, requesting that he be allowed to return to the contemplative life in the cloister; yet Bernard refused his plea on the grounds that he should keep the position to which he had been elected, that he was not his own master, and that his duty was to attend to the pastoral care of his community.

Richard persisted, however, and finally managed to procure a letter from St

---

[61] *Life of Ailred*, p. 31.
[62] *Life of Ailred*, p. 33.
[63] For the following, see *Mem. F.*, pp. 74–76.

Bernard allowing his release from his position if the monastic community of Fountains abbey would consent. The letter was shown by Richard to his monks, and he explained to them that he wished for their consent in relieving him from his office. Serlo of Fountains remembered that 'all the brethren were distressed at this announcement, our faces changed and all our joy was clouded by grief'. The community then vigorously refused to allow Richard to resign—'With one voice all shouted together, "We will never consent!"', recollected Serlo. The reason for the community's reluctance to allow their abbot to resign was elucidated by one of the monks, Hugh de Matham, who rose to his feet in the midst of the general clamour, and addressed the unhappy abbot:

> This monastery which you undertook to rule by the gift of God had been known to flourish in such peace from the time of its foundation, that no tempest has upset its internal tranquillity, no malice of the old enemy has bewitched it. It does not know the troubles of schism, it does not know the weeds of dissension; it is holy ground and does not produce the seeds of scandal [...] We wonder at this unholy innovation which would disjoint the head from the body, denigrate our fame and extinguish whatever light of holy religion there is in us. Who would not be scandalized hearing that the abbot of Fountains has deserted his monks, the father his sons, the shepherd the sheep entrusted to him?[64]

Hugh's speech uses the language of apostasy to emphasize the impropriety of abbot Richard's request. Richard would be a deserter if he were to absolve himself from his pastoral responsibilities—Hugh remonstrates with the abbot to be still, to be silent and not to desert 'those whom you cannot desert without scandal'.[65] The implication here is that abbot Richard would be committing almost all the crimes associated with apostasy, without actually leaving the monastic site, should he step down as abbot. The suggestion that Richard was asking for something almost scandalous was enough to prevent the abbot from insisting on his right to resign, while the invocation of the threat of the 'ancient enemy' underlined the inherently improper nature of his request. Without being an apostate, abbot Richard was being threatened with that label. Likewise, the language used by Hugh in this instance underscores the sacred site of the monastery as a holy garden in which constant and vigilant cultivation is needed to prevent these 'weeds' of scandal from choking the carefully tended plants of stability.

Finally, it was also possible for monks to be transferred to another Cistercian house if it was perceived that their current situation was difficult or potentially dangerous to the rest of the monastery. I have already mentioned cases of transfer in the previous chapter, in the context of troublesome *conversi*. On a more general level, transfer from one monastery to another was used as a precautionary measure. Transfer involved a number of formal negotiations between Cistercian houses, including letters of supplication from the monk's present house, reports on his recent behaviour indicating that he could not remain in his monastery without grave scandal, and transfer documents relin-

---

[64] *Mem. F.*, p. 76.

[65] *Mem. F.*, pp. 76–77.

quishing responsibility for the monk.⁶⁶ It was also possible for monks to be transferred to another order entirely. In comparison to other monastic orders, Cistercians seem to have been less enthusiastic about transferring their monks out of the Cistercian world, and even more reluctant to receive transferees from non-Cistercian houses, yet transfer between Cistercian houses was clearly carried out. A monk of Bonaecumbae was exiled to a monastery far from his own house in 1232, for example, while in 1278, a number of lay brothers from a French house were transferred to remote houses permanently.⁶⁷

Institutional responses to the issue of apostasy were therefore predominantly formulated by the legislation of the General Chapter, although the secular arm was invoked if Cistercian houses thought it necessary in the recovery of apostates. Some attempts were also made to prevent the possibility of apostasy altogether, as the examples of change in status and transfer show. These responses to apostasy were all nevertheless directed at the return of the apostate to the Cistercian order, and in the case of imprisonment as punishment, also reinforced ideas of monastic enclosure. A runaway monk was publicized, not only to the secular arm in order to effect his return, but also within the monastery by means of certain punishments inscribed on his body. Even when a difficult or unhappy monk who had not committed apostasy was transferred or relieved of burdensome duties, institutional responses and procedures governing change of status were made public within Cistercian houses. The institutional superstructure thereby served to restore the monk as a disciplinary subject within the site he had attempted to escape.

The ways in which fugitives from Cistercian houses were represented and described by both their monastic brethren and by the jurisdictions to which they were made to answer reveals that the apostate, who existed on the fringes of monastic and secular domains, was subject to the censure of both. The apostate monk and the act of apostasy signified a range of meanings for medieval Cistercians themselves. Apostasy was understood in terms of the contravention of physical space, manifested in the apostate's departure from the site of the monastery. Apostasy was also understood in more abstract ways. Depictions of apostates in Cistercian literature focus on weakness and disorder within the body and soul of a fugitive monk, while at the same time warning that these traits potentially exist in all Cistercians, whether monks or abbots.

Apostates themselves are not given a voice in most of the sources. Reconstructing reasons for apostasy and trying to uncover an apostate's own responses to recapture, punishment, and incarceration remain historically speculative. Returning to the case of Godfrey Darel we may uncover some of the meanings that his apostasy had for the monastery at Rievaulx, the Archbishop of York, and even for the Christian Church. Darel himself, however, remembered and recorded as an apostate, leaves no trace as to what apostasy meant for him. The silence that the runaway monk left in his wake was quickly filled by the clamour

---

⁶⁶ Leclercq, 'Documents sur les fugitifs', pp. 122–23 for a transcription of some of these thirteenth-century letters from a French Cistercian house (Vatican Library, MS Vat. Lat. 7528, possibly from Languedoc).

⁶⁷ *Statuta* 1232, 24, t. 2; 1278, 26, t. 3.

of outrage during the Middle Ages. Almost as soon as he left the monastery, a convicted apostate was deprived of his voice—spiritually demonized, vigorously reviled, and ultimately silenced.

Boundaries between the Cistercian monastery and the secular world were often challenged during the medieval period. Apostates, for instance, defied both physical and imagined or abstract boundaries by choosing to leave the monastery. I want to turn to another instance where the division between physical and abstract space in Cistercian houses was renegotiated—in the places and spaces occupied by the dead. During the thirteenth century, the complex relationship between the living and the dead in Cistercian houses reflected the merging of physical, material places with the imagined or abstract spaces of the other worlds. The final chapter of this book explores the ways in which the boundaries between living and dead were created and maintained.

CHAPTER 8

Sites of Death and Spaces of Memory

The Cistercian monks of Yorkshire encountered the dead in all corners of their world. They knelt for confession on the graves of former abbots in the chapter house; they buried their brethren in cemeteries east of the church; they even stumbled on the tombstones that paved their cloister walks. The sites occupied by the dead in the abbey were places often also utilized by the living; the perennial presence of the supposedly absent dead meant that the living were in constant and very tangible communication with the other world. The reciprocal nature of the relationship between the living and the dead in the medieval period has long been stressed by historians, particularly in terms of intercessionary responsibilities and memorial and commemorative practices.[1]

---

[1] Patrick Geary, *Living with the Dead in the Middle Ages* (Ithaca and London: Cornell University Press, 1994). Geary claims that modern society's avoidance and abhorrence of death stems from monks and canonesses becoming responsible for the dead, either for their burial or for their memory. This set the scene for a severing of the affairs of the dead from the affairs of the living. (p. 2). This modern attitude, says Geary, is absolutely different to medieval attitudes towards death and the dead, where 'the dead did not cease to be members of the human community' (p. 2). Geary's more specific work on *memoria* reinforces this view—see Geary, *Phantoms of Remembrance*. This is also the view of Esther Cohen, *The Crossroads of Justice: Law and Culture in Late Medieval France* (Leiden/New York: Brill, 1993), who describes the punishment of the dead in the later Middle Ages—especially suicides—as growing out of a fundamental belief that the bodies of the dead still exercised power over the realm of the living (pp. 134–45). For recent scholarship on the early medieval origins of many liturgical and commemorative practices concerning the dead, see Megan McLaughlin, *Consorting with Saints: Prayer for the Dead in Early Medieval France* (Ithaca and London: Cornell University Press, 1994); Frederick Paxton, *Christianizing Death: The Creation of a Ritual Process in Early Medieval Europe* (Ithaca and London: Cornell University Press, 1990); and for Anglo-Norman England, see Christopher Daniell, *Death and Burial in Medieval England 1066–1550* (London and New York: Routledge, 1997).

Not only were the living responsible for maintaining the memory of the dead, but the dead were also obliged to provide services for the living. This spiritual economy of exchange and negotiation governed decisions as to the immediate use of physical space in a monastery. Burial sites, for example, were chosen for lay patrons and benefactors on the strength of their donations to the abbey while living, while the performance of prayers for the dead dictated the form of much liturgical activity within the church.

Throughout the course of the thirteenth century, both the use of space by the dead and the memorial practices that recognized them underwent significant changes in the Yorkshire Cistercian houses. Lay people, including women, were buried more frequently in sacred and previously inaccessible areas of the abbey—such as in the chapter house and in the east end of the church. At the same time, the number of liturgical offices for the dead was severely reduced, with individual anniversaries for the dead being replaced in 1273 by twelve general memorial services each year.[2] Cistercian houses formerly empty of shrines or relics began to build tributes in the form of memorial tombs to former abbots, while the church floor itself was littered with lay graves that crept from the galilee porch on the west facade to the very steps of the altar in the holiest site of all. The dead were insistently encroaching on more and more areas of the monastery, although traditional memorial practices seemed to be declining. Interpreting these changes solely in terms of the increasing power and influence of the secular over the monastic world is an attractive but ultimately unsatisfactory temptation. From their foundation in the previous century, all the Cistercian houses in Yorkshire had consistently enjoyed close and beneficial relationships with patrons and benefactors—yet it is not until later that we find the graves of benefactors in the church or chapter house. The proliferation of non-monastic burial plots in Cistercian abbeys shows that association with the Cistercian world was still important for lay people in the north of England during the thirteenth century, yet the liturgical changes within the order itself seem to suggest that the ways of commemorating the dead were in a process of flux.

In this chapter, I argue that the nature of the change in the burial and remembrance of the dead should be associated with changes in the way the spaces occupied by the dead were viewed by both Cistercians and lay people. During the thirteenth century, the dead became more closely associated with specific topographies within the monastery. This was true of both lay patrons and benefactors and the monks themselves, especially abbots. Remembrance of the dead through the usual liturgical offices was scaled down, I believe, because commemoration itself had become more site-specific. Memory itself was subject to new discourses of space that were more firmly anchored in particular sites. These new discourses may be traced in Cistercian representations of dying and death, together with these changes in burial practices for both lay and monastic people.

---

[2] *Statuta* 1273, 2, t. 3. Daniell, *Death and Burial in Medieval England*, p. 179 says that this reduction of memorial services for the dead 'was a remarkable admission that the role of memory had become too much for the living'. I will explore Daniell's claim later in this chapter.

Representations of Dying and the Dead

Abbots

Aelred of Rievaulx's breathing grew slower and quieter over the four days preceding the Ides of January in 1166. He had been ill for over a year; even Mass had become so exhausting for the abbot that afterwards, he could neither speak nor move, but lay as though unconscious in his cell for an hour after the service.[3] When his suffering had become unbearable, Walter Daniel reports that Aelred had wished aloud that 'God may speedily deliver me from this prison and lead me to a place of refreshment'. As the abbot's pain grew, so did the pain of the monks who surrounded him as he lay gasping in his cell—'the following night brought the father much pain, and us most pain of all, because his was only of the body, while ours was the pain of a sorrowful mind, extremely sad because of him'. When Aelred finally died, with almost the whole community in attendance and with Walter Daniel himself cradling Aelred's head in his hands, it was with both joy and terrible grief that the monks of Rievaulx farewelled their abbot—with 'smiles and tears running together'. The monks had time to prepare their abbot for a proper death; some days before he died, Aelred had been anointed by Roger of Byland, and before that, Aelred had delivered his final address to his community.

Walter Daniel's description of Aelred's death places grief and joy at the centre of its narrative theatre. While the monks of Rievaulx crowded around the abbot's bed, Aelred is said to have counted their love as 'the greatest of all blessings, that he should have been chosen by God and men to be so well beloved'. The monks themselves rejoiced that their abbot would soon have 'happy relief', and they marvelled at the whiteness and purity of his corpse as he was washed after death.[4] Aelred's body 'shone [...] was fragrant as incense, pure and immaculate in the radiance of his flesh' and the community marvelled long on 'that sweetness, that beauty, that glory'.[5] Jean Leclercq has challenged the assumption that death in the Cistercian monastery was a fearful experience, on the grounds that narratives like Walter Daniel's demonstrate the absolute serenity of dying in the monastic world. Leclercq argues that Cistercian twelfth-century depictions of death reveal that dying was 'awaited without fear, accompanied by serenity and even by beauty'.[6] St Bernard, for example, had described

---

[3] *Life of Ailred of Rievaulx*, p. 54–61 for the following.

[4] *Life of Ailred*, p. 62. For a similar tale of bodily purity in death, see Stephen of Sawley's *Speculum Novitii* where Stephen tells of the nun who 'held the sign of the holy cross in such reverence that as she was being carried to her grave, the face of the crucified Christ turned around toward her', while her entire body putrified in death but for her thumb, 'because with it, she was wont to make the sign of the cross', in Mikkers, 'Speculum', p. 61.

[5] *Life of Ailred*, p. 62.

[6] Jean Leclercq, 'The Joy of Dying According to St Bernard', *Cistercian Studies*, 25 (1990), 163–74.

the 'moment of jubilation' that occurred as brother Gerard sang while he died, and talked of singing again while Malachy, archbishop of Armagh died on the particularly auspicious All Souls' Day.[7] For Bernard, the day of death—*dies illa*—was not a judgement day to be feared. Rather, according to Leclercq, it was a day of light. Walter Daniel reiterates the theme of light—'My God! He did not die in darkness, as those who have been long dead, not so Lord, but in your light, for in his light, we see your light'.[8]

Nevertheless, death also brought great sadness to Cistercian communities. St Bernard believed that every time one of his monks died, he himself began to die—'ego morior in singulis'.[9] Walter Daniel missed Aelred so much that he wished to lie with his friend and abbot 'withered and dead, covered with stone and constricted by a mass of earth', darkened by dust, sightless and still.[10] The reason for his grief, according to Walter Daniel, was not that Aelred had died, but that Walter himself would no longer see him or hear him; as Jean Leclercq has pointed out, death from the world meant loss for those left behind, rather than resentment directed at those who departed. Excessive grief, however, was seen by Walter Daniel to be self-centred and although he is overwhelmed by sorrow and tears, he is also aware that 'while the sorrowing mind will be caught up in excessive agitation for a long time, there are certain bounds'.[11] Walter Daniel is aware that his grief will not last forever, even though he feels that it might.[12]

Although Walter Daniel's description of Aelred's death is strongly characterized by Walter's own emotional experience of grief and joy, the narrative emphasizes the ceremonial playing out of the abbot's death. Aelred dies in delineated stages, according to Walter. The stage is set with the backdrop of the abbot's illness, while scene by scene, the drama of Aelred's death is allowed to unfold. There was nothing sudden or violent about Aelred's experience of dying; he had time to summon the community to his deathbed and advise them on choice of a successor. He surrounded himself with significant and favourite objects—a glossed Psalter, Augustine's *Confessions*, the text of

---

[7] *Sermo in Cantica Canticorum* 26, cited by Leclercq, 'Joy of Dying', p. 168; *Vita Malachi, SBO,* vol. 3, pp. 375–78.

[8] *Life of Ailred*, p. 63.

[9] Leclercq, 'Joy of Dying', p. 169; *In Obitu Domni Humberti, SBO*, vol. 5, pp. 440–47.

[10] *The Life of Aelred of Rievaulx by Walter Daniel,* tr. F. M. Powicke and intro. M. Dutton, CFS 57 (Kalamazoo: Cistercian Publications, 1994), p. 145. Powicke's 1950 edition did not include Walter Daniel's *Lament for Aelred*, which follows the *Vita Aelredi* in Cambridge, Jesus College, MS QB 7, fols 74^{r-v}. The Latin text has not been transcribed.

[11] *Lament*, p. 146. For other examples of ecclesiastical disapproval of excessive mourning, see Phillipe Ariès, *The Hour of Our Death*, trans. by H. Weaver (New York: Knopf, 1981), pp. 142–45.

[12] Walter Daniel also says that he has never been so stricken as when he heard Aelred beg for a speedy death, *Life of Ailred*, p. 60.

John's Gospel, the relics of certain saints, and a small crucifix which had previously belonged to the archbishop of York. He conversed with angels when he was unable to speak to the monks who crowded around him, and later on, was seen to smile and weep at the reading of the Passion. Throughout, the abbot's breath first gasped, then slowed, with one final burst of clarity the night before he finally died.[13] Walter Daniel's depiction of the abbot's death is thus carefully orchestrated to privilege the beauty of death; certainly, Walter also shows the grief that this entails, but the unfolding of Aelred's death and the precision with which each scene is resolved gives the impression that dying may be perfect, just as the soul is made perfect after death.

Other Cistercian narratives describing the death of abbots reinforce the view that dying may occur at appropriate times and in proper circumstances. Serlo of Fountains remembers that Richard, the first abbot of Fountains, died because 'the Lord had made a better provision for him and rewarding the labours of his faithful servant, took him from pilgrimage to his home, from toil to the longed for rest of God. On the way home from Rome, he was seized with a fever, and finishing his life in the way of obedience, ran his course and was perfected in peace'.[14] Abbot Richard the second waited until 'all the due ceremonies had been performed around him by our holy father Bernard' before he 'put off the outer man' and died at Clairvaux,[15] and abbot Richard the third saw that 'all things were at peace without and within according to his prayer, he saw his sons flourishing in grace and walking in love. So in a good old age, surrounded by a company of holy men, full of days and rich in virtues, he departed to the Lord'.[16] Abbot William, too, experienced a restful death—'having finished his course he fell asleep in peace in a holy old age, leaving to posterity a grateful memory in grace and blessing'.[17] Likewise abbot Ralph, who, as prophecy had foretold, died at Fountains with 'all due ceremonies [...] departed rich in merits to the Lord'.[18] Only abbot Robert seems to have died unexpectedly—'In the ninth year, as he was returning from the general chapter, he fell ill on the journey. At Woburn he grew worse and met his appointed end. His body was brought to Fountains with due honour and laid in the sepulchre of the fathers'.[19] In Serlo's account, the words 'dying' or 'death' are infrequently used; it is more usual for

---

[13] *Life of Ailred*, pp. 57-61.

[14] *Mem. F.*, p. 72.

[15] *Mem. F.*, p. 78. That succession could be a difficult issue may be discerned in Bernard of Clairvaux's letter to the community of Fountains after the death of abbot Richard the second. Bernard sent Henry of Clairvaux to take over the vacant abbacy, together with a letter asking the monks of Fountains to receive Henry without schism or dissent and to accept his regulations and reform for the community as if they had been directed by Bernard himself.

[16] *Mem. F.*, p. 113.

[17] *Mem. F.*, p. 116.

[18] For Ralph's death, see *Mem. F.*, p. 125.

[19] *Mem. F.*, p. 115.

Serlo to talk of 'falling asleep' or *dormitione*, than it is for *mors* to be explicated directly.[20]

Monks

Sources describing the deaths of ordinary monks and lay brothers are not as abundant or detailed as those which describe and commemorate the death of abbots. Despite their paucity, however, these sources often provide glimpses of Cistercian experiences of dying that are not framed within narratives emphasizing beauty and release. The Yorkshire Assize Rolls of 1218–19, for example, reveal that monks from the Yorkshire houses sometimes died unexpectedly, violently and accidentally. William le Brun of Kirkstall, for instance, was killed by his own cart, which appears to have fallen on him. Another case tells that 'a brother of Meaux was found drowned in the Ribble and he fell from his horse. 'No one is suspected', found the jury, and the judgement was handed down as 'misadventure'. The Assize investigated cases where other monks or lay brothers had fallen from their horses and died. Brother Roger of Rievaulx was found drowned in the river Aire after falling from his horse. Again, the judgement was misadventure, while brother William Wulsy, also of Rievaulx, was found drowned, fallen from his horse into the river Rye in similar circumstances.[21]

Other texts show a variety of experiences of dying. Like the descriptions of abbots' deaths, these tend to relay particular acts of meanings depending on the narrative context. In the *Speculum Novitii*, for example, the death of the monk Gerard is described in relation to monastic obedience. In this story, death is presented as a state which can be negotiated. As Gerard lay dying, he experienced a vision of Christ, who revealed to Gerard that no Cistercian would perish if he loved the Order—'he is cleansed at the moment of death, or shortly thereafter'.[22] This statement serves a twofold purpose. In the context of the *Speculum*, the idea that the Cistercian is predestined for eternal life reinforces the need for unequivocal devotion and obedience to the Cistercian order. The promise of life-everlasting is thus contingent on the disciplined monk. Second, the author's reporting of the vision suggests to the novice that death is an experience over which the novice himself may exert some influence. If the novice is mindful of death, then he may avoid its trials. In other parts of the *Speculum*, the idea of death is introduced into the novice's quotidian experience. 'When going to bed, think of Christ's burial and your own',[23] the novice is

---

[20] For other idealized representations of dying, see Michael Casey, 'Herbert of Clairvaux's Book of Wonderful Happenings', *Cistercian Studies*, 25 (1990), 37–64 (p. 59) who says that Herbert of Clairvaux transforms death into a 'scene of extraordinary beauty'.

[21] For the preceding cases, see Clay, *Three Yorkshire Assize Rolls*, pp. 207, 249, 260, and 350.

[22] Mikkers, 'Speculum', p. 63.

[23] Mikkers, 'Speculum', p. 64.

advised, and 'compare your bed to your grave, just as if you were entering it for burial'.[24] The *Speculum* teaches the novice that death is not only omnipresent, but that the novice ought to reflect on it daily.

Death and dying in the Cistercian monastery were therefore often represented in idealized and didactic terms, as shown in the stories recounted above. The 'perfect death' of abbots such as Aelred and monks such as Gerard and Malachy of Clairvaux, are reflections of their great devotion and purity while alive. Other Cistercians, however, did not enjoy such a smooth and happy transition from this world to the next. Visions and prophecies of death from various Cistercian houses reveal that death was also expressed in terms of justice and punishment. Those who had been troublesome or sinful during life were likely to suffer deservedly difficult deaths. An example may be found among the various tales of miracles and prophecies in the *Vita Ailredi*. One story tells of the mysterious death of a vituperative Cistercian abbot, who had insulted and reviled Aelred with 'arrows of cursing' and the 'spittle of lies'.[25] Aelred himself became resentful of the abbot's malice, and appealed to God that the wrongful accusations heaped on him should come to an end. Whereupon the abusive abbot 'barely crossed the threshold of his own house' when he became ill and died 'miserably' and in agony seven days later.[26] The possibility of an extended period of tribulation in purgatory plays a role in other visions, such as those relating to disobedient *conversi*, while other stories emphasize the reparations that must be made in death for sins or transgressions committed in the Cistercian monastery in life. An instance of the latter may be found in Paris, Bibliothèque Nationale, MS Lat. 15192, where a dead abbot is said to have suffered in death every time his living brethren disobeyed the *Rule*, as he had failed to uphold sufficient discipline in his monastery while he was alive.[27] Integral to these stories is the notion that the boundaries between life and death are not always distinct. This may be more readily illustrated by the insistent return of the dead into the world of the living.

The Return of the Dead

The dead did not confine themselves to the space of the afterlife. They returned to the monastery in order to instruct, correct, strike fear into those who had transgressed, or to ask for intercessionary favours. The dead also reappeared by invitation, as a result of pacts and bargains made with fellow monks during life. The relationship between the living and the dead was therefore consolidated by continual contact.[28] The stories and visions which recount the reappearance of

---

[24] Mikkers, 'Speculum', p. 62.

[25] *Life of Ailred*, p. 44.

[26] *Life of Ailred*, p. 45.

[27] At fol. 81ʳ.

[28] See Jean-Claude Schmitt, *Les revenants: les vivants et les morts dans la société médiévale* (Paris: Editions Gallimard, 1994), esp. chapter 3, 'L'Invasion des revenants',

the dead in the monasteries are concerned with the miraculous aspects of these visits. At the same time, however, these tales stress the presence of the dead in the more day-to-day experience of Cistercian monks. The dead were continually remembered within the monastery, not only by means of liturgical commemoration, which I shall discuss later, but also by virtue of their continued physical engagement with the living. As Patrick Geary has stated, 'the dead did not cease to be members of the human community'.[29] Many meanings can be found in the reporting of contact between the dead and living, as the following exempla and miracles illustrate.

In several Cistercian stories from the thirteenth century and some from the twelfth, the bridge between the living and the dead is expressed in terms of friendship. This was the case with Benedict, a monk from Fountains abbey, who prophesied his own death and promised to his friend the infirmarer that once this prophecy came to pass, Benedict would return to Fountains and tell the infirmarer all about death.[30] A similar story tells of another monk from Fountains abbey, who heard of Benedict's reappearance and, doubting that a corpse could reappear, had a vision of the dead Benedict showing him how the dead could get up out of their place of burial.[31] In both these stories, the living monk shows curiosity as to what the afterlife will bring for him; in the first, the infirmarer asks if he could feel the fires of purgatory to know what that heat would be like, while in the second tale, the living monk asks the dead Benedict to enquire of Mary if he was a good monk. The dead friend provides the means by which these questions may be answered.[32]

Another example of the communication between living and dead friends comes from a late-twelfth century manuscript from the southern English abbey of Stratford Langthorne, in which the story is told of a bargain made between two members of the monastery, Roger and Alexander.[33] The two monks agreed that whoever died first should come back to the other and tell the survivor about the afterlife. The agreement was even confirmed by a cyrograph. The monk Alexander was the first to die and, as the agreement had stated, he came back twice. The first appearance was thirty days after his death, and the second was a

---

pp. 77–98 for the literary typologies containing twelfth- and thirteenth-century visions of the dead and pp. 151–58 for 'l'apport cistercien' and in particular Caesarius of Heisterbach's *Dialogus Miraculorum*.

[29] Geary, *Living with the Dead in the Middle Ages*, p. 2.

[30] Paris, Bibliothèque Nationale, MS Lat. 15912, fol. 126ʳ. Also cited in McGuire, 'The Cistercians and the Rise of the Exemplum in Early Thirteenth-Century France'.

[31] Paris, Bibliothèque Nationale, MS Lat. 15912, fol. 126ʳ. In reply to the living monk, Benedict says, echoing the liturgy: 'Ecce hoc est corpus meum, et hec est cuculla et tunic in qua sepultus sum'.

[32] For a more detailed explication of monastic friendship, see McGuire, *Friendship and Community*.

[33] London, Lambeth Palace, MS 51, fols 124–26. See chapter 6 for some discussion of the friendship between lay brothers and choir monks. The story is transcribed and edited by Holdsworth, 'Eleven Visions', pp. 197–201.

year later. The first time, Alexander stated that he had returned as their agreement had dictated. Roger, like the brothers of Fountains abbey mentioned previously, was most interested in precisely where his friend had gone and where exactly he would go when his soul exited his mortal body. Alexander was not initially forthcoming with detailed information, but did reveal that souls went to a particular place immediately. More than that, Alexander would not impart.[34]

Alexander's second visit occurred a year after his death. In this, the bodily, material Alexander was absent, and what appeared to Roger was a great luminosity and a rush of sweet fragrance.[35] The voice of Alexander was audible, however, and to it, Roger addressed similar questions to those he had asked his dead friend a year before—'Are you in paradise?', 'Are you in eternal glory?', 'In what state of glory are you?'. When Alexander was able to provide his friend with more specific replies, Roger went on, excitedly: 'Are there many saints there?', 'Is Mary there?', 'Is Saint Benedict there?', 'Are our families there?', 'Is brother William Rufus who died before you there?', and so on. Alexander described the various hierarchies of *gloria*, together with the people who populated each level, revealing that the Virgin, for example, was in a higher paradise within the sight of God. He told Roger that there was no way to reach these glorious places hurriedly, and that Roger should remember that the trials of the world would be compensated for in life everlasting. 'Let your heart be comforted', said Alexander, 'that you will come to be with us in our mansion in the future'. The experience ended when the other brethren were about to begin Matins, and the light disappeared. Roger told the compiler that he had seen the light and spoken to his friend—'amicus cum amico'.[36]

The story of Roger and Alexander shows that advice from the dead was worthy of attention, while the ties that bound monks in friendship were not broken by physical separation. The tale also reveals some of the Cistercian monk's primary concerns about dying. Most prominent of these concerns is the need for the spatial definition of the afterlife. Roger's questions about the exact whereabouts of his friend, together with his interest in the presence of familiar figures among the dead provide the means by which a reassuring view of death can be elucidated. The comforting presence of a friend can also provide important intercessionary support. In the *Exordium Magnum*, Conrad of Eberbach tells of a vision in which a host of dead Cistercians returned to their abbey in procession in order to collect one of their brethren who had been ill, thus ensuring protection and inclusion in the Cistercian afterlife.[37] Similarly, Caesarius of Heisterbach's *Dialogus Miraculorum* narrates the story of a pious lay brother, who was visited by a dead friend unable to bear the pains of purgatory. The dead man asked that the lay brother organize Cistercian masses to be said for

---

[34] Holdsworth, 'Eleven Visions', pp. 198–99.

[35] Holdsworth, 'Eleven Visions', p. 200–01 for the following.

[36] Holdsworth, 'Eleven Visions', p. 201.

[37] McGuire, 'An Introduction to the *Exordium Magnum*', pp. 292–93.

him, which the *conversus* duly did, thus preventing any more suffering for his friend.[38]

The Formality of Dying

Dying was also subject to a number of very formal and ritualized practices within the Cistercian world. These practices ensured the dying monk a safe and certain transition from this life to the next, and are generally to be found in the *Ecclesiastica Officia*. In this text, there is a series of chapters devoted to the rituals and procedures that should be followed from the time a monk approaches death, to his burial, and then to the commemorative offices that maintained the link between the dead individual and his living community. I shall discuss the issues of burial and memory in more detail below. At present, I focus on the formality surrounding the process of dying and the hour of death together with the Cistercian emphasis on community that the procedures involved.

When it became apparent that one of the brethren was about to die, he was placed on a cloth on the ground. The cloth was itself laid upon ashes in the shape of a cross.[39] The arranging of the body was probably organized by the infirmarer, who was then obliged either to ring the bell in the church or to knock on the *tabula* in the cloister a number of times in order to let the rest of the community know of the impending death. On hearing the sound of the bell or tabula, the monastic community hastened to the dying man, saying the *Credo* two or three times. The sacristan and the cantor brought the stola, the crozier, the collectaneum, the holy water and the cross to the abbot who, with the rest of the brethren, stood around the dying monk. Depending on the time of the day, the sound of the tabula or bell brought different responses. If the brethren were at *collatio*, for example, the reader was to interrupt his reading with the words 'Tu autem domine', while the abbot was to say 'Adiutorium nostrum'. More importantly, if the community was at Mass or singing one of the canonical Hours in the church when the summons to the dying was heard, all but the abbot or prior were to remain in the church unless ordered to go to the dying man. It was therefore possible that a dying monk might not always be surrounded by the entire community on his death bed, although the *Ecclesiastica Officia* stresses communal presence.

The death of a monk did not always occur at a liturgically convenient time. Thus, the *Ecclesiastica Officia* provides a number of different instructions for the community, all depending on how long the brother took to die and at what time of day death occurred. In the church for example, many of the brothers remained to complete the canonical hour as mentioned. However, the offices could be altered, according to the progress of the dying monk. If the monk was still living at the end of the recitation of the litany, then the brothers were to say

---

[38] Cited in McGuire, 'Purgatory, the Communion of Saints and Medieval Change', p. 68. The reference is to Caesarius of Heisterbach, *Dialogus Miraculorum*, book 12, ch. 33.

[39] Griesser, 'Ecclesiastica Officia', p. 257 for the following.

the seven Penitential Psalms. If he still had not died at the end of these, the community was to depart from the church, leaving the cross and holy water in readiness for the body to be brought to the church. If the *tabula* was heard during the office of the dead, it was permitted that the office could be abandoned temporarily and recommenced when convenient. Likewise, if a chapter meeting was interrupted by the imminent or actual death of a monk, the community should return to complete it later, and the dead monk should be absolved on their return. And in the refectory, a meal should be completed after the death if it had been interrupted, while the reader should resume his reading at the point at which it had stopped.[40] The practice of devotion within the community thus continued despite the interruption of death.

For the dying monk, unction was given as soon as the infirmarer had judged it necessary. Again, this rite was essentially administered by the whole community, who processed from the church singing 'Domine Ne In Furore Tuo'. The priests followed the abbot, who was led by the cross, the light, and the holy water.[41] The sacristan was in charge of carrying the unction oil, although the abbot himself gave the unction prayers, while the sick monk could also receive communion at this time. After the sick monk had been anointed and had taken communion (witnessed by the community), the brethren left to continue their usual duties and to wait for the monk's imminent death.[42]

The moment of death heralded a whole new set of procedures governing the treatment of the corpse. If a monk died during the recitation of a canonical hour, those of the community who were with the body should stand around the dead man in the order they kept in choir.[43] The office of the dead was then said, followed by the Psalter. The body was then washed, usually by the prior, and then taken to the church in a procession. The procession was headed by more senior members of the monastery, including the abbot, who brought the holy water, the thurible, the light, and the cross. The rest of the community followed. The corpse itself was carried to the church by four monks, while the procession finished with the lay brothers. Once in the church itself, the procession placed the body in the choir. The *Ecclesiastica Officia* is careful to state that this should only be done if the corpse were not fetid, in which case it should not be brought to the church at all. If the corpse—like that of Aelred described above—was pristine, then it was arranged where the abbot directed, while the abbot himself made the commendation of the dead man's soul. The body was

---

[40] Griesser, 'Ecclesiastica Officia', p. 258.

[41] Griesser, 'Ecclesiastica Officia', pp. 256–57.

[42] Frederick Paxton, 'Anointing the Sick and the Dying in Christian Antiquity and the Early Medieval West', in *Health, Disease and Healing in Medieval Culture*, ed. by S. Campbell, B. Hall, D. Klausner (New York: St Martin's Press, 1992), pp. 93–102. Also H. B. Porter, 'The Origin of the Medieval Rite for Anointing the Sick and the Dying', *Journal of Theological Studies*, n.s. 7 (1956), 211–25.

[43] Griesser, 'Ecclesiastica Officia', p. 258.

the focus of these rituals, and the *Ecclesiastica Officia* also directs that, until interred, the body should never be left alone.[44]

These rituals served to ensure the transition from life to life-everlasting, while the immediate focus was on the transformation of the earthly body. The earthly body, as Caroline Walker Bynum has shown,[45] was not a useless shell to be discarded at the moment of death. Rather, the material body was needed in order to ensure a complete and whole resurrection. Thus the dying body was prayed over, oiled, and laid on cloth and ashes, while the dead body was washed, laid out, and also prayed over until a suitable time for burial. Until the body had been interred, the Cistercian community was still able to influence its progression from life to death, even once physical death had occurred. During this transitional period, the body was under the complete control of the monastery. As custodians of the corpse, Cistercian monks asserted their own authority over the dead and exerted some spiritual influence over the course that the body and soul would take. One way that this was done was by means of liturgical ritual as demonstrated in the *Ecclesiastica Officia*. Authority over bodies and the spaces occupied by the dead was also exerted through burial practices.

Burial

Patrick Geary has warned that scholars of medieval burial practices should not interpret burial sites and grave locations as proof of medieval 'monolithic' belief systems, where everything to do with death is an established ritual, and where potentially shifting cultural and historical meanings are ignored. Death, says Geary, has been particularly subject to totalizing theories of belief, theories which barely recognize the actions of people and which privilege the primacy of doctrines of the church. Geary believes that 'medieval religion was not believed but danced',[46] a statement which may seem extraordinary in the context of the monastic world, but which gives, I believe, proper weight to the significance of human action in creating and giving meaning to experiences such as death. In the Cistercian monastery, burial practices reveal many such layers of meaning. The demarcation of burial zones in Cistercian houses was at once highly regulated and continually changing during the thirteenth century, while what constituted an appropriate burial site for a lay person also changed. The liturgical rites that accompanied a funeral had been set out by early legislation, but were subject to shifting interpretations and practices. How and where Cistercians buried the dead, therefore, can be read as reflecting the changing concerns of the living.

---

[44] Griesser, 'Ecclesiastica Officia', p. 258.
[45] Bynum, *Resurrection of the Body*, pp. 175–76.
[46] Geary, *Living with the Dead in the Middle Ages*, p. 44.

Regulation of Burial and Burial Rites

A monastic funeral began in the church, where the corpse lay. Those who carried the cross, the light, the incense, and the holy water stood at the head of the dead man, while the abbot stood behind them, ready to process from the church to the place of burial.[47] The body was incensed at the end of each collect, and when the recitation of the collectaneum had been completed, the cantor would begin the antiphon to the psalm, and the procession began. Not everyone was involved in the procession to the grave—the *Ecclesiastica Officia* ruled that someone should remain to look after the church and the cloister. Four brothers wearing the *scapula* carried the body to its place of burial, while the abbot led the procession and placed himself at the head of the grave once the tomb had been reached. It is worth noting here that there is no mention in the *Ecclesiastica Officia* of burials potentially taking place within the church; all the liturgical and processional instructions direct the corpse and the community out of the church, and presumably into the cemetery at the east of the church building.

The body was placed on the ground by the four monks who had carried it from the church, while the rest of the community arranged themselves around the grave. The grave itself was prepared while the community was standing around it. Once ready, the grave was asperged, as was the body. Incense was also waved over the grave-digger and the body. The body was then lowered into the ground with the abbot beginning *Temeritatis quidem*, whereupon the community arranged themselves into processional form to begin the journey back to the church. The seven Penitential Psalms were recited by the monks, who proceeded to the choir in the inverse order—lay brothers first, and then the rest of the monastery in ascending order of seniority. Once inside the church, the abbot divested himself of his sacerdotal vestments and prostrated himself at the altar steps. The funeral ended with the *Requiem eternam* and collect being recited after the psalms.[48]

The *Ecclesiastica Officia* is also careful to make sure that others within and without the community were aware of the death and burial. Lay brothers were part of the funeral processions and part of the funeral services as well, although they retained their usual position in their choir or the retrochoir after the burial. Monks who were in the infirmary were encouraged to participate in the burial liturgy if they were able, while death notices were distributed to pilgrims by the gatekeeper.[49] For thirty days after the burial, extra collects were said at the Offices and at masses for the dead man, while on the thirtieth day, the soul of

---

[47] Griesser, 'Ecclesiastica Officia', pp. 260–01 for the following.

[48] Griesser, 'Ecclesiastica Officia', p. 261.

[49] Griesser, 'Ecclesiastica Officia', p. 261. Mortuary rolls for abbots and others were also taken to other houses to be signed. See S. Hasquenoph, 'La mort du moine au moyen âge (Xe–XIIe siècles), *Collectanea Cisterciensia*, 53 (1991), 215–32; also Moulin, *La vie quotidienne des religieux*, p. 67. No mortuary roll from the Yorkshire Cistercian houses survives, although London, British Library, MS Royal 15 A x, dated *c.* 1216, is a mortuary roll for the abbot of Thorney, signed by the abbots of Fountains, Jervaulx, Rievaulx, Byland, and Kirkstall *inter alia*.

the monk was absolved in the chapter house and the collect *Omnipotens sempiterne deus cui numquam* was said. This signified the release of the soul from thirty days of purgation. Each priest was obliged to say three private masses for the dead man, each cleric was to recite the Psalter, those monks who did not know the Psalter were obliged to recite the *Miserere Mei Deus* one hundred and fifty times, while those who did not know the *Miserere* were to recite the *Pater Noster*.[50]

Other regulations determining burial procedures may be found in the *Statuta*. Here we discover that other factors aside from liturgical ones were at work in the process of burial, especially in relation to the location of graves. There are a number of cases of abbots who were reprimanded and punished for what the General Chapter evidently saw as violation of the usual practice of burial. In 1217, for example, the body of a monk of Montpeyroux was ordered to be reinterred within the cemetery confines and the abbot who had neglected to do this was severely punished.[51] Other bodies were disinterred if the General Chapter thought it necessary. The bones of the abbot of Locedio, for instance, were relocated to the chapter house in 1218,[52] and, even earlier in 1200, there is mention of the *extumulato* of a *conversus* at Aiguebelle.[53] Other monastic bodies were excluded from burial in the consecrated ground of the cemetery. This was the case with a lay brother who had not confessed to having money in his possession and who had not made confession before receiving communion. When he died suddenly, the General Chapter ordered that he be buried outside the cemetery.[54] In 1193, there was debate about whether a monk of Bonneval who had been an apostate could be buried in the abbey. Once the monk reverted to the Cistercian order on his death bed, it was ruled that he could be buried in the cemetery.[55] And in Yorkshire, abbot Robert Ayling of Meaux resigned in 1280, leaving the abbey in such huge debt that he was denied burial with the other abbots in the chapter house.[56]

Sometimes particular corpses were in demand by more than one monastery. At Meaux abbey around 1210–20, the chronicler reports that the nuns of nearby Swine convent stole the body of Amandus Pincerna from the Meaux cemetery, buried him in their cemetery, and then laid claim to the land with which

---

[50] Griesser, 'Ecclesiastica Officia', p. 261.

[51] *Statuta* 1217, 49, t. 1. For the report to the General Chapter the following year and the punishment of the abbot, see *Statuta* 1218, 34, t. 1.

[52] *Statuta* 1218, 36, t. 2.

[53] *Statuta* 1200, 36, t. 1.

[54] *Statuta* 1228, 20, t. 2. For the implications of burial outside consecrated ground, see Daniell, *Death and Burial in Medieval England*, Ariès, *The Hour of Our Death*, pp. 29–92; W. L. Wakefield, 'Burial of Heretics in the Middles Ages', *Heresis*, 5 (1985), 29–32.

[55] *Statuta* 1193, 27, t. 1.

[56] Bond, ed., *Chronicon Monasterii de Melsa*, 2, p. 157: 'Sed sepulcrum eius inter sepulchra praedecessorum seu successorum suorum minime reperitur, nec recolitur locus ubi fuerat tumulatus'.

Amandus had endowed the house.[57] John of Ford's early thirteenth-century *Life of Wulfric of Haselbury* tells of a similar tussle in the south of England. When Wulfric died at Ford, his corpse was immediately subject to dispute between the monks of Ford and the monks of Montacute, both of whom believed that Wulfric should be interred in their houses. The Montacute monks burst into the Ford church, grabbing the body to take it away, but had not counted on the monks of Ford locking them into the church. The enterprising monks of Montacute nevertheless managed to shove Wulfric's corpse through a window to more Montacute monks waiting outside. The Ford monks started a brawl in which the body of Wulfric was pushed to and fro through the window, and much blood was shed. Ultimately, the Ford monks were successful in shaming the invaders out of the church. Wulfric was buried peacefully in the Ford church, albeit secretly in the inferior, western part of the church, in order that his body would not be stolen from its resting place.[58]

Other unforeseen and irregular circumstances could also play a role in how and where burial took place. During the early years of the thirteenth century in England, the interdict affected usual burial customs. At Meaux abbey, for instance, the chronicler reported that the interdict made it impossible for the monks to bury their dead in the monastic cemetery, in the cloister, or in any of the other places used for grave sites, while the Divine Office was said behind closed doors. The seven monks, sixteen lay brothers, the novice, the abbot of Hovedoe, the priest, and other people who died during this time were therefore interred either in an apple orchard on the outskirts of the precinct, or outside the monastery altogether. The remains of people buried outside the monastery grounds, according to the chronicler, lay in a heavily wooded place near the bridge to Northgrange, a place where the ghostly light of a cross could be seen by men crossing the bridge at night.[59] Sites for burial were thus occasionally selected on the basis of necessity, rather than monastic policy.

---

[57] Bond, ed., *Chronicon Monasterii de Melsa*, 1, p. 356 et seq. for the protracted litigation that ensued.

[58] An editon of the life of Wulfric is found in M. Bell, *Wulfric of Haselbury, by John, Abbot of Ford* (Somerset: Somerset Record Society 47, 1933); a more recent translation of the fight for Wulfric's body is in *The Cistercian World: Monastic Writings of the Twelfth Century*, ed. by P. Matarasso (Harmondsworth: Penguin Books, 1993), pp. 271–73.

[59] *Chronicon Monasterii de Melsa*, 1, p. 343: 'Nam reliqui familiares nostri et famuli ac alii compatriotae nostri ex terminos monasterii, ex opposito loci ubi capella in bosco situatur, ultra fossatum supra quod est pons qui ducit ad Northgrangiam versus boream, in loco tunc vasto modo nemoroso, parvo et quadrangulato, prope pontem a sinistris transeuntibus ad Northgrangiam, terrae traditi fuerunt. Ubi in angulo prope dictum pontem crucem quandam ligneam vidi stantem, a diu ibidem erectam, sed tamen cornibus carentem, in signum ipsius sepulturae. Ubi homines veterani noctanter transeuntes, aliquoties lumen quoddam insolitum, miraculose ut credebant, se conspicere in dubitanter asserebant'. For another Yorkshire Cistercian response to the Interdict, see Matthew of Rievaulx's poem to the pope in Paris, Bibliothèque Nationale, MS Lat. 15157, fol. 46^v, transcribed in E. Faral, *Les arts poétiques du XIIe et XIIIe siècles* (Paris: Champion, 1924), pp. 24–26.

The greatest number of references to burial in the proceedings of the Cistercian General Chapter relates to lay burials within the monasteries. These statutes reveal that from the start of the thirteenth century, the issue of lay burial in Cistercian houses was seen as increasingly important. At the same time, there seems also to have been a tremendous variation in the practices and policies of individual Cistercian monasteries. Abbots who allowed secular people to be buried within the monastery were subject to censure early in the century, although it seems that the location of the burial sites within the monastery was the subject of concern rather than the presence of seculars per se. In 1205, for instance, the abbot of Vallis Sanctae Mariae was punished for burying a nobleman in the church of his abbey, while in 1219, the abbot of Bebenhausen spent six days in 'levi culpa' for (among other offences) allowing the burial of a nobleman in the chapter house.[60] Thus abbots could be punished if lay people were buried in inappropriate places,[61] or if inappropriate lay people were buried in holy places. It appears that women were not initially to be buried in Cistercian houses at all. The abbot of Vieuville was punished for burying a woman in 1201,[62] although by 1215, the implication seems to have been that certain women could be buried in the monastery but only in particular places and not in the church.[63] In 1217 the General Chapter ruled broadly that secular persons could be received for burial in Cistercian monasteries.[64]

The Yorkshire Houses

In what way do the Yorkshire monasteries comply with the rulings on lay burial set forth by the General Chapter? First, it is clear that the northern English houses were by no means exempt from requests for lay burial, especially from patrons and benefactors. A cross-section of the upper echelons of northern society requested to be buried at abbeys such as Fountains. Joan Wardrop has found that ninety people requested to be buried at Fountains abbey between 1132–

---

[60] *Statuta* 1205, 15, t. 1; *Statuta* 1219, 19, t. 1.

[61] Another example of this is the case of the Duke of Lotharingia, whose corpse was the subject of great dispute from 1214, when the General Chapter decided that it should be moved to another abbey from Clairlieu (*Statuta* 1214, 21, t. 1). The following year, the same request was made, the monks and abbot of Clairlieu having failed to comply with the General Chapter's first order (*Statuta* 1215, 26, t. 1). In 1216, a sentence of interdict was passed on Clairlieu for still failing to relocate the duke's body (*Statuta* 1216, 7, t. 1), while the following year the abbot was deposed by the General Chapter for continually refusing to move the duke (*Statuta* 1217, 75, t. 1). A formal investigation was underway by 1218 (*Statuta* 1218, 21, t. 1), while the saga ended in 1220, by which stage the duke's remains were no longer referred to as a body, but as bones (*Statuta* 1220, 63, t. 1).

[62] *Statuta* 1201, 15, t. 1.

[63] *Statuta* 1215, 31, t. 1.

[64] *Statuta* 1217, 3, t. 1: 'Mortui saeculares qui in coemeteriis nostris sepulturam sibi eligunt, si de licentia sacerdotum suorum hoc faciant, recipiantur'.

1300.⁶⁵ Of these, there are none that the charters indicate as having requested a particular burial site. Only nine women made individual requests to be buried, and two of these were included in their husband's request.⁶⁶ All except one of these women had some association with the house—Christiana de Arncliffe, for example, had a brother who had been received as a *conversus* there.⁶⁷ Eleanor and Lambinus de Studley were so keen to be buried at Fountains that they threatened 'malediction [and] the wrath of God' if any attempt was made to infringe their will to which they had attached their seals.⁶⁸ Many burial requests took the form of grants of land or property made in conjunction with the grant of the benefactor's body. Geoffrey Haget, for example, gave his body to be buried at Fountains and bequeathed it with his property and the village of Thorpe.⁶⁹ And at Kirkstall between 1200–08, the less wealthy Stephen of Hamerton granted the monks of Kirkstall twenty loads of hay, together with his chattels at his death, as well as his body.⁷⁰

During the late twelfth century, the sites of these graves tended to be in the cloister walks and, occasionally in the monastic cemetery itself. At Fountains abbey, for example, the grave of William de Percy, who died in 1174, lay in the eastern cloister walk outside the sacristy.⁷¹ By 1203 at Fountains, the monastic church had become the focus of patronal burial, although secular grave plots were confined to the outskirts of the church, and in particular, to the galilee porch. The tombs of William de Stuteville and Matilda, countess of Warwick occupied part of the galilee porch at Fountains abbey at the western entrance to the church. At Rievaulx abbey in the middle of the thirteenth century, the galilee porch was still being used as a burial site—Isabella de Roos was interred there in 1264 while the traces of a further seven graves may be discerned in the same place (Fig. 8.1). The western end of the church was a more secular space than the eastern end of the building. The presence of lay brethren in the west part of the nave consolidated the quasi-secular nature of the site, while the galilee porch's distance from the hub of liturgical activity at the altar meant that lay use of this area did not encroach on more sacred topography.

Proximity to the church was evidently important for those lay people who chose to be buried near the western façade, and it may be useful to interpret the thirteenth-century burials here in the context of lay understandings of that particular space. The geography of burial in Cistercian houses was, of course, significantly determined by the availability of a particular site; the absence of

---

⁶⁵ Wardrop, *Fountains Abbey and Its Benefactors*, p. 261. London, British Library, MS Cotton Tib. C xii, fols 279ᵛ–80 says that Stephen de Catton is buried in the *cimiterium*, but hardly any other charters actually say where they were interred.

⁶⁶ Wardrop, *Fountains Abbey*, p. 264.

⁶⁷ London, British Library, MS Cotton Tib. C xii, fols 132ʳ–132ᵛ.

⁶⁸ Wardrop, *Fountains Abbey*, p. 270.

⁶⁹ Wardrop, *Fountains Abbey*, p. 225.

⁷⁰ Lancaster and Paley-Baildon, eds, *Coucher Book of the Cistercian Abbey of Kirkstall*, pp. 200–01.

⁷¹ Coppack, *Fountains Abbey*, p. 66.

Fig. 8.1: Rievaulx abbey, tombs in the Galilee porch (Photos, author)

twelfth-century lay graves in the eastern end of the church, for example, strongly implies that lay people were simply not allowed to occupy this space in death, just as they had not been allowed to occupy it in life. Yet, in relation to the galilee porch, it is also possible that this may have been a preferred site for interment. Christopher Daniell has indicated that liminal zones, such as the boundaries or edges of cemeteries, were sometimes chosen specifically as sites for burial, as these spaces 'may equate with the soul crossing the boundary from the earth to the afterlife'.[72] The galilee porch may be understood as one of these 'liminal' areas, not entirely within the church space, yet not without it either. Requests for burial, such as that made by Robert Colcerellus of Coldona to the monastery of Meaux, may have been understood to refer to a liminal zone such as the site of the galilee porch. The Meaux chronicler simply reports that Robert give his body to be buried 'near us', together with a gift of land.[73]

Other burial locations are not as easy to identify, especially from written sources alone. Hugh of Kirkstall's *Narratio* for instance, mentions that Robert de Sartis and his wife Raganilda were both buried in 'the sepulchre of the just' at Fountains and, according to the aged Serlo who recalls them, 'their memory is held blessed among us'.[74] Where precisely this 'sepulchre of the just' might be, is not clear. Also buried at Fountains was Serlo of Pembroke, a young man of the king's household who died at Fountains after granting the village of Cayton to the monastery, and was interred 'among the saints'.[75] This is a similarly vague description of what must have been a specific site. Canon Serlo of York and canon Tosti of York also both died at Fountains abbey and were presumably interred there, although in this case, Hugh of Kirkstall does not include any mention of the location of these graves.[76]

Later on in the thirteenth century, other sites within the Cistercian monastery were used by lay people as places of burial. In 1298, Roger de Mowbray was buried in the church at Fountains abbey—possibly in the north arcade;[77] and by 1315, the grave of Henry, Lord of Alnwick and member of the distinguished Percy family, was situated immediately in front of the high altar in the presbytery itself.[78] At other Cistercian houses in the north of England, the same trend may be seen. The presence of thirteenth-century effigies of lay people in Cistercian churches shows that the practice of burying secular people in Cistercian oratories was carried out in Yorkshire earlier than may be expected from the prohibitions of the General Chapter. At Jervaulx abbey, for instance, the effigy

---

[72] Daniell, *Death and Burial in Medieval England*, p. 100.

[73] Bond, ed., *Chronicon Monasterii de Melsa*, 1, p. 366.

[74] *Mem. F.*, pp. 54–55.

[75] *Mem. F.*, p. 56.

[76] *Mem. F.*, pp. 53–54.

[77] Coppack, *Fountains Abbey*, p. 66.

[78] Coppack, *Fountains Abbey*, p. 66.

of Ralph FitzHenry, a lay patron, was placed under the crossing of the church.[79] At Sawley, the bodies of Lord William de Percy, who died in 1244, and his grandson Henry were placed in adjoining graves in the choir of the church, although their tombs do not survive.[80]

A number of points may be made about the move toward burying lay people within Cistercian churches. First of all, it is important to recognize that the encroachment of lay people on monastic space took place at a different rate in different houses. At Rievaulx, for instance, as mentioned previously, the galilee porch was still being used as a lay burial site in the middle of the thirteenth century, while at the same time at Sawley, for instance, the Percy family was buried in the oratory choir. Nicola Coldstream has argued that the demand for burial in the monasteries of England was an 'insidious pressure', and that monasteries such as Meaux were 'cajoled into accepting lay people for burial in exchange for grants of land'. Thus, poorer houses may have been the first to accept lay burials in particularly sacred sites such as the church, as financial relief and land came with bodies.[81] Second, the shift toward accepting women for burial in Cistercian houses may usefully illustrate the status and influence of women as benefactors within Yorkshire society. That women such as Isabella de Roos could be interred at Rievaulx and that women such as Alice de Gant could receive the same masses and psalms as a monk after her death,[82] suggests that lay burials were often discretionary in practice, and often reflected Cistercian interest in familial and patronal ties rather more than concerns relating to gender.[83]

More generally, the merging of lay and secular space in the site of burial may be related to changes in the ways in which commemoration was understood by both lay people and monks. The scaling down of traditional liturgical offices for the dead occurred at the same time as increases in private masses and at the same time as the presence of lay graves in Cistercian churches became marked.

---

[79] W. St John Hope and H. Brakspear, 'Jervaulx Abbey', *Yorkshire Archaeological Journal*, 21 (1911), 303–44 (p. 313); Coldstream, 'Cistercian Architecture from Beaulieu to the Dissolution', p. 157.

[80] J. Walbran, 'On the Recent excavations at Sawley Abbey in Yorkshire', *Mem. F.*, p. 171.

[81] Coldstream, 'From Beaulieu to the Dissolution', p. 157. See also David Postles, 'Monastic Burials of Non-Patronal Lay Benefactors', *Journal of Ecclesiastical History*, 47:4 (1996), 620–37 for burial grants *cum corpore*. Also Brian Golding, 'Burials and Benefactions: An Aspect of Monastic Patronage in Thirteenth-Century England', in *Symposium on England in the Thirteenth Century: Harlaxton Conference Proceedings*, ed. by M. Ormrod (Nottingham: University of Nottingham, 1985), pp. 64–75.

[82] Wardrop, *Fountains Abbey*, p. 251. For other manifestations of deathbed piety, see Gougaud, *Devotional and Ascetic Practices in the Middle Ages*, pp. 131–45.

[83] Another example of this may be found in Lancaster and Paley-Baildon, eds, *Coucher Book of Kirkstall Abbey*, where Herbert de Arches confirmed the grant of 12 acres of land in Shadwell made by his mother 'pro sepultura sua'. The editors note that the Arches family was 'an offshoot of the more important family of the same name which [...] held a considerable estate' (p. 133).

The final part of this chapter explores the ways in which traditional liturgical practices of commemoration altered during the thirteenth century, and how these changes related to new understandings of the sites and spaces of memory.

Commemoration

Remembrance of the dead began as soon as a body had been interred. The *Ecclesiastica Officia* directed that—as described previously in this chapter—masses and other prayers were said for a dead monk by all the members of the community. The *Ecclesiastica Officia* also set out the manner in which the Office of the Dead, a more general service for the memory of monastic and non-monastic people, should be undertaken. This office was to be said almost every day in the church. As at other monastic services, the Office of the Dead took the form of antiphon and response, with one of the community leading the service. The service began with the antiphon and response, then the psalm, followed by the *Pater Noster*. After a number of responses, one side of the choir sang 'Exultabunt domino ossa'. The community ended with the *Laudate Dominum de celis*, followed by the *Requiem Eternam*, which was repeated after the *Confiteor*, and again after the psalm *De Profundis*, as with the other psalms. At the end of the service, after the collects, the priest said 'Dominus vobiscum' and 'Requiescat in pace', before the brothers responded with 'Amen'.[84]

The principal offices for the dead were described separately. 'In the commemoration of all the faithful and our parents, brothers and sisters and on the anniversaries of abbots', dictates the *Ecclesiastica Officia*, 'the psalm should be sung while standing and in the same way at Vespers and at Vigils'. These Offices of the Dead were to be sung with all the members of the community present unless a suitable reason prevented someone from attending. In other ways, these 'principal offices' took a very similar form to the usual Office for the Dead. The collects to be said for the dead were specific: *Deus cuius miseratione* for those resting in the cemetery, then *Omnipotens sempiterne deus cui nunquam* for the brothers of the congregation, families and other dead. The *Quesumus domine pro tua pietate* was recited for 'our mothers and sisters and other women', likewise for the other dead. For the faithful in general, the collect *Fidelium deus omnium* was said. On the anniversary of one of the brother's deaths, or that of another dead person, *Presta Domine quesumus* should always be said, followed by *Omnipotens sempiterne deus* if the community remembered a man and *Quesumus domine* if a woman.[85]

Thus, some forms of remembrance were always collective in Cistercian monasteries.[86] However, other ways of remembering the faithful departed were

---

[84] Griesser, 'Ecclesiastica Officia', pp. 214–15 for the preceding.

[85] Griesser, 'Ecclesiastica Officia', pp. 216–17.

[86] For another example, see Griesser, 'Ecclesiastica Officia', pp. 261–62, where directions are given for the formal and collective remembering of all Cistercian dead at the General Chapter meeting at Cîteaux each year.

more individually focussed. This is especially true of the prayers and masses that were to be said for specific people after their deaths. I have already described some of the prayers that were to be said for monks. The *Ecclesiastica Officia* had stipulated in the mid-twelfth century that there should be three private masses said for each departed within the monastery.[87] Lay people, however, were more complex subjects for remembrance, while lay requests for burial within Cistercian houses from the later twelfth century onward show that lay people felt able to dictate to the monastery exactly how many masses ought to be said for them, and in some cases, what other forms of commemoration ought to take place as well.[88]

One example may be found in the Meaux chronicle, which tells of the foundation of a chapel in 1238 near the monastery of Meaux by Sir Peter Mauley, whose wife Isabella had requested burial within the monastic precinct. With the gift of a substantial endowment of land, the monks of Meaux were required to maintain the chapel and two secular priests and two clerks, who were to say a mass for the dead, the hours, and a mass of the Virgin Mary daily, as well as a mass for the anniversary of Isabella's death.[89] This example is important, as the request was not only for burial, but also for a very particular commemorative practice that was to be the monks' responsibility even though it took place outside the church space. Another example of a very specific request for spiritual benefit is the case of Robert de Hesding, who gave land and other rights to Fountains abbey, so that when he decided to enter the Cistercian order he would be welcomed at Fountains, but if he died a layman, then the souls of his parents and ancestors could still avail themselves of the prayers and benefits of the Cistercians 'forever'.[90] The expectation that commemorative prayers could carry on 'forever' gives some indication of the potential liturgical pressure faced by Cistercian monasteries in this context.

Other examples of requests for remembrance are found in the many charters and grants of land that the Yorkshire houses accumulated during the twelfth and thirteenth centuries. These tend to follow a standard form—many include the conventional 'pro anima mea et uxoris mei, pro animabus patris et matris mee et omnium antecessorum et heredum meorum', while others ask for more specific consideration. Robert, son of Helewise de Burdon, for instance, confirmed a grant of land made by his mother, saying that the grant was made for the prayers and other benefits of Kirkstall in which the monks would receive Robert and his brother, their heirs and the souls of their parents and predecessors.[91] The idea of

---

[87] Griesser, 'Ecclesiastica Officia', p. 261.

[88] Royal requests for spiritual support came fairly frequently as some of the *Statuta* show, for example, *Statuta* 1271, 75, t. 3, where the King of France asked for prayers from Cistercians to aid the recovery of the Holy Land.

[89] *Chronicon Monasterii de Melsa*, 2, pp. 59–61.

[90] London, British Library, MS Cotton Tib. C. xii, fols 248^{r-v}. Cited also in Wardrop, *Fountains Abbey*, p. 252.

[91] Lancaster and Paley Baildon, eds, *Coucher Book of Kirkstall Abbey*, p. 89. Another example may be found in Bond, ed., *Chronicon Monasterii de Melsa*, 1, p. 361, where John, son of Peter of Meaux, gave his servant Odo together with Odo's descendants to

receiving souls was not uncommon; Baldwin Fitzralph and his wife gave their son Adam to Kirkstall and simultaneously entrusted the monks with the souls 'omnium predecessorum nostrorum participes fecerunt omnium beneficiorum que fiunt in sua domo et in ordine suo'.[92] That the monks were in charge of souls illustrates not only the very fundamental place of intercessionary prayer in the life of the monastery, but also reveals just how many souls expected commemoration as a result of one request.[93] Again, the potential for unmanageable numbers of names being included in the collective Office for the Dead and in individual masses was acute.

Evidence that the Cistercian order in general attracted large amounts of requests for prayers and masses for the dead may be found in the *Statuta*. In 1225, the General Chapter ruled that only one mass each year should be said for a specific soul.[94] This ruling coincided with a period of architectural development in Cistercian churches, where the number of individual altars in these buildings increased. This is the case with Rievaulx abbey, where the lay brothers' western end of the church was appropriated in part for the inclusion of altars, and at Fountains abbey, where the eastern end of the church was entirely remodelled to include the Chapel of Nine Altars from the start of the thirteenth century. This indicates, as mentioned previously, that intercessionary sites were becoming extremely important in the monastic spiritual world. Related to the need for more altars is the growing number of lay requests for this form of intercession and commemoration. Not only were lay people concerned to be interred in the sacred topography of the monastery, but, as the charters and grants of land and gifts demonstrate, lay people also expected to be able to dictate to the monastery what other forms of commemoration ought to be carried out. In this context, the General Chapter's attempt to scale down the number of private masses said for an individual can be read as an attempt to control the heavy liturgical demands made on the Cistercian order.

In 1273, the General Chapter went even further, ordering that the commemoration of all individual anniversaries should cease. These were to be replaced with twelve general services a year instead.[95] Again, we may infer that the sheer number of anniversaries was becoming a burdensome liturgical demand on monastic communities. Yet, deeper meanings may be also extracted from these

---

the monastery in exchange for the saying of a private Mass.

[92] Lancaster and Paley Baildon, eds, *Coucher Book of Kirkstall Abbey*, p. 95.

[93] Wardrop, *Fountains Abbey*, has found that the charters requesting prayers and masses 'pro salute anime' run into the hundreds for the period 1132–1300 (p. 243).

[94] *Statuta* 1225, 6, t. 2: 'Statuitur et firmiter praecipitur observari, ut [...] annuatim una missa privatim tantummodo celebretur'.

[95] *Statuta* 1273, 2, t. 3: 'Quoniam propter multiplicitatem anniversariorum personis pluribus a Capitulo generali concessorum, Ordo multipliciter oneratur, statuit et ordinat Capitulum Generale ut in qualibet abbatia Ordinis singulis mensibus unum anniversarium celebretur die vel hebdomada qua abbas quilibet in domo propria viderit oportere; et praeferatur in eodem anniversario quaecumque persona quam abbas quilibet voluerit, adiunctis personis aliis quibus anniversarium ab Ordine est concessum'.

changes to liturgical and intercessionary practice—especially in relation to the changes in burial practices occurring in Cistercian houses at the same time. The connection between increased lay burials and the Cistercian institutional attempt to control the prayers and masses expected by patrons, benefactors and others, should not only be read as evidence for the significance of the growth of intercessionary responsibilities among monks. These changes should also be understood in terms of shifting understandings and expressions of memory.

Formal ways of remembering the dead fell into two main categories in the Cistercian monasteries of the thirteenth century: liturgical naming of the dead in masses, and remembrance evoked by the topographical presence of graves, tombs, and effigies. The reduction in the number of former commemorative practices at the same time as the pervasive expansion of physical places of commemoration, reveals that memory in the Cistercian world had become extremely site-specific. This was not only the case with Cistercian remembrance of lay people: the thirteenth century was also a period in which shrines for abbots became more popular. Two examples of this are found at Rievaulx abbey, where a shrine to the first abbot, William, was erected between the chapter house and the eastern cloister walk, and where a shrine to Aelred was established within the church.[96] Defining the space to be occupied by the dead, who occasionally still reappeared in the monastery as physical entities and whose post-mortem locations were thus potentially always unfixed, was associated with ways of anchoring memory more firmly to precise topographical sites.

Why the dead and their memories should be subject to spatial definition and topographical confinement may be related to the broader issues raised in this chapter. Death was at once an idealized state, as seen in representations of perfected deaths, and an experience that required the earthly apparatus of formal liturgical guidance and negotiation. The focus in many of these liturgical rites was the body, which was represented both as subject to the authority of the monastery and as potentially disorderly in its violation of the boundaries between life and death. On one level, the creation of a place for the dead on earth may be understood as a Christian attempt to exert some control over the body. On another level, the relationship between burial and commemoration may be seen in more mundane, pragmatic terms. Cistercians were evidently under siege from secular requests for prayer and masses. The combining of these individual liturgical services with the unspoken service of burial in the holy earth of the Cistercian church may simply show that the Cistercian order was attempting to retain an element of unworldliness in their liturgy. In other words, allowing seculars into the monastic church was a compromise.

The places and spaces inhabited by the dead were defined and expressed by the actions of the living. Physical sites of death were spaces associated with memory and commemoration, an association that strengthened as the thirteenth century wore on. Life and death were linked not only in words and prayers, but

---

[96] Fergusson and Harrison, *Rievaulx Abbey: Community, Architecture, Memory*, pp. 166–68.

also in understandings of how space could be given meaning by the bodies that attempted to fix its limits.

Epilogue

The modern visitor to the ruined monasteries of northern England sees a landscape of many layers. Walking through the western entrance of Fountains abbey, for instance, the visitor's gaze is first drawn to the imposing façade of the long, low western range, with the Galilee porch to the church at one end. Surveying the green lawns that sweep up from the gatehouse to the monastery is the church tower, remarkably intact after some 500 years (Fig. 9.1). It is easy for the visitor to then walk through the main body of the ruined monastery, through the cloister and past heaps of masonry lying to the east of the cloister walk. On the other side of the precinct, the visitor may turn around, surveying the great east window of the church. At this point, the visitor's journey around Fountains abbey has only just begun, although the medieval portion of the main site has already ended. The rest of the site of Fountains abbey is a landscaped garden. It is an idyllic eighteenth-century setting—of half-moon ponds, follies, trim parterres, and mysterious pockets of woodland overlooking the valley (Fig. 9.2). The Cistercian abbey is part of the fantasy—like the follies and the garden, the monastery itself is part of the romanticism of post-medieval aesthetics—and former owners even relocated the medieval masonry where it was thought to be unsightly and at odds with the vision of an earthly paradise.

Since the eighteenth century, Fountains abbey has undergone some changes, most notably a series of archaeological investigations of the monastery site. Now in the hands of English Heritage the ruins at Fountains, although still part of the landscaped gardens, have been trimmed of foliage, pathways cleared through the stones, and portions of the surviving buildings clearly labelled for the benefit of the droves of tourists who now pass through the site each year. Recording the site for posterity is also an important part of English Heritage's mission, while pedagogical factors are at work in the site museum and the visitors' centre which allow the tourist to 'experience' medieval life by witness

Fig. 9.1: Western façade of Fountains abbey

ing pictorial displays, mannequins dressed as monks, and brief histories of the building and its post-medieval transformations. Discourses of paradise have given way to twentieth-century agendas of conservation, education, and tourism.

Something of an earthly paradise may still be found at Jervaulx abbey, where ivy and wildflowers are left to creep over the twelfth-century walls, and there is no attempt to reconstruct the monastery as a medieval place in which people lived, worked, and died. The site is designed to be a garden, like Fountains abbey, but a garden for picnickers, with garden benches replacing the cloister seats, and with fallen stonework heaped up into untidy mounds where it was found (Figs. 9.3). Jervaulx, unlike the majority of monastic sites in England, is still in private hands and has not, therefore, been subject to the archaeological attentions of bodies such as the National Trust or English Heritage. Neat and archaeologically correct display are not a priority at this site. The owners of Jervaulx, although welcoming to visitors, have decided to leave the ruins of their monastery picturesque and overgrown as, presumably, they themselves found it. There are no labels attached to the walls of the monastic building to tell the visitor about the original function of the site. Jervaulx is a place where smiling tourists may sit in the chapter house, dreaming up their own meanings for the spaces that surround them (Fig. 9.4).

No such fantasy exists at Kirkstall abbey in Leeds, where the original monastic precinct is carved in two by the A65 motorway. Access to the site of the monastic church and cloister is restricted, while the church itself is closed to the public, with bars across the doorways (Fig. 9.5). The tower of the church itself at Kirkstall is showing signs of irreparable damage caused, according to local rumour, by the lads of Leeds who break into the church and dare each other to scale the masonry on Saturday nights (Fig. 9.6). At Kirkstall, the trimness of the ruins at Fountains and the serenity of the rambling gardens covering the walls at Jervaulx have given way to an agenda of prohibition. Without permission it is impossible to enter the ruined medieval buildings. The site is now in an urban environment, and the stonework is stained with pollution from the cars and buses that pass twenty-four hours a day. The former guest house to the monastery is now the abbey museum, which is devoted to displays of nineteenth-century Yorkshire life. The aesthetic challenges faced by the now urban monastery of Kirkstall have meant that it is not a popular tourist site, like the landscaped Fountains abbey, or the overgrown Jervaulx. Rather, Kirkstall— for the casual visitor—is a site which seems grim, inaccessible, and austere.

To the modern visitor, these ruined monasteries stand in landscapes defined by the owners of those ruins. At Fountains, the site is essentially a formal garden, while there is a strong educational agenda to the presentation of the monastery itself and the visitors' centre. At Jervaulx, the visitor is encouraged to wander about in a less structured way, and allow imagination to work in thinking about the ruins. At Kirkstall, the visitor is actively discouraged from anything but the most peripheral of visits, and a walk through the ruins is a quick and not immediately informative event.

Digging back through the landscape of the past, it becomes clear that such manifestations of imagination and regulation, of order and freedom, of enclosure and liberty, have a long history in the monastic world of northern England. The spaces of the Yorkshire Cistercian abbeys tell many different stories, from the mid-twelfth century to the present day. I have concentrated on one of these stories, focusing on the ways in which medieval Cistercians created and defined their physical and mental landscapes during the thirteenth century. Monastic sites served both to order or regulate physical spaces and to describe the possibilities of transcendence present in those spaces. Space was itself topographically fixed and simultaneously imagined. The relationship between sites of enclosure and concepts of liberation was expressed in the many meanings ascribed to Cistercian monastic space, meanings which developed and changed over time.

Abstract Spaces: Memory, Metaphor, and Freedom

The division between monastic space and secular space was formed by physical boundaries, such as walls. However, intangible boundaries were also constructed to effect divisions between the monastic and non-monastic worlds. The transitional spaces occupied by Cistercian novices during their probationary year and the violation of monastic boundaries by apostates demonstrate Cistercian anxieties over the construction and destruction of such boundaries. One way of fixing boundaries was to articulate them in terms of abstract space as well as in terms of fixed sites or topographies.

Fig. 9.2: The gardens of Studley Royal (Photos, author)

Fig. 9.3: West range stairs at Jervaulx abbey and entrance to west range (Photos, author)

Fig. 9.4: The cloister at Jervaulx abbey; stacks of masonry at Jervaulx abbey (Photos, author)

Epilogue

Memory was one tool used by Cistercians to define the otherworldly spaces signified by the monastery and the monastic vocation. This worked in productive and prohibitive ways. As the thirteenth-century *Speculum Novitii* shows, novices were encouraged to divest themselves of their subjective, experienced memories and to instead adopt collective Christian memories, marking the transition from secular world to Cistercian abbey. In the context of space, the reformation of the novice's memory also created a mental boundary between the monastery and the world. Memory was important in the articulation of community, too, as the commemorative functions of the chapter house show. Liturgical uses of memory in the recitation of the Office of the Dead together with visual reminders of the dead in the form of tombs served to affirm collective notions of responsibility, community, and vocation. The changes to tomb topographies during the thirteenth century, particularly in the church, together with the reduction in services for the lay dead during this time, reveals a shift in Cistercian understandings of memory. The material sites occupied by the dead served more and more as the predominant spaces in which memory was fixed, while liturgical commemoration became less important.

The distinctive nature of Cistercian monastic space was also expressed metaphorically, especially in the identification of cloister and paradise. Abstract space—or the *imagined* space of paradise—was performed in liturgical processions and rituals conducted in the cloister, as well as by literary representations of the cloister site signifying the heavenly paradise. The metaphor of cloister as paradise was extended to describe the monastic landscape in general, as well as the very specific landscapes of the Yorkshire houses. Implicit in Cistercian representations of the cloister site was the idea of symbolic transcendence of earthly or physical space, an idea that may also be traced in the Cistercian churches of the thirteenth century. These buildings were reconstructed during the thirteenth century to facilitate visualization and imagination of the space of heaven. This may be seen in the redesigning of the east end of Yorkshire Cistercian churches and in the use of light within those churches. Cistercian devotional tracts that used the spiritual idea of *meditatio* to encourage the formulation of mental images also stressed the importance of visualization in prayer and singing.

Concepts of abstract space which were voiced in liturgy, imagined in prayer, and represented by the written word, also depended on the physical fabric of buildings and sites. In processional uses of the cloister, the four arcades marked precise moments when metaphor might be activated. In the church, iconographies of light and vision were denoted by windows and acoustics. In the case of wavering Cistercian novices and potential apostates, precinct walls functioned as stark physical boundaries between the monastery and the secular world. Imagined spaces were signified and affirmed by such material sites.

Spatial Regulation, the Body, and Discipline

Relationships between material sites and imagined spaces were also expressed in Cistercian institutional discourses surrounding the monastic body. As a site in

Fig. 9.5: Kirkstall abbey, east end of church and tower (Photo, author)

itself, the body was perceived and represented as potentially disorderly, yet potentially stable. The body was subject to regulation in various ways, while specific areas within the monastery were associated with particular bodies. Chapter house rituals, for instance, emphasised community and commemoration, but at the same time publicized the transgressive, subjective bodies of individual monks in liturgies of correction and punishment. In the infirmary, the disorder of the cosmos was mirrored in the tempestuous bodies of sick monks. Lay brothers, too, were represented in Cistercian sources as disorderly—both spiritually and institutionally—while the areas of the monastery and its precinct which they inhabited were very carefully demarcated.

Epilogue

Fig. 9.6: Kirkstall abbey, barred doorway (Photo, author)

Dealing with transgressive bodies was one way in which medieval Cistercians stabilized their place within landscape and cosmos. This can be discerned in the traditional practice of bloodletting, which functioned within the monastery as both a disciplinary practice and as a way of ordering the turbulent cosmology of the body itself. In broader terms, the relationship between the body and natural world replicated the sometimes tumultuous experience of life within the monastery, as Matthew of Rievaulx expressed. The body was one site wherein union with God could be extremely difficult, as pain, fatigue, and weakness conspired to subvert an easy monastic vocation.

The ambiguous position of Cistercian *conversi* also reveals monastic anxiety about stability and vocation. Although *conversi* had always been subject to segregation within the monastic precinct, it is during the thirteenth century—as they began to be less useful agents of labour—that demarcation of their roles became more and more fixed. Changes in lay brother status may be linked to the polarization of the *conversi* 'type' in both legislative material and in the *exempla* collections of the thirteenth century. Transgressive bodies were also the subject of attention in the chapter house, where institutional strategies for making the body 'docile', in Michel Foucault's terms, are also evident. The rituals of accusation, confession, and punishment in the chapter house reinforced the need of stability in the Cistercian world in the same way as cautionary tales of recalcitrant lay brothers. The site of the chapter house, already associated with collective manifestations of community in its other liturgies, thus signified not only community, but also discipline.

Disciplinary practices associated with the body helped to construct the Cistercian body as an ambiguous site. Regulation of the body in all its guises, whether it be the tonsuring of choir monks, the wearing of beards by the *conversi*, the beating of the body in the chapter house or the draining of bad blood in the infirmary, were all cultural practices which focused on the body's propensity for disorder. In terms of Cistercian understandings of space, disciplining the body reinforced the special place of the monk within his community, within the landscape and within the imagined space of heaven. Regulation and order within the material world allowed for the simultaneous existence of abstract space.

The Blurring of Spatial Divisions

Divisions between material and abstract space were not always neatly effected or envisaged. During the thirteenth century we find frequent concerns in the Cistercian world about the violation of space, or the blurring of boundaries. Apostates, for example, contravened both Cistercian principles of enclosure and Cistercian principles of spiritual freedom by violating the very boundaries which defined those spheres. The dead, too, by revisiting the Cistercian monastery through invitation or invasion, represented the fluidity of the division between this world and the next. Cistercian understandings of material and imagined space reinforced the notion that the domain of the living was also the dominion of the dead. Cistercian affirmations of the proximity of the afterlife can be seen in the Chapel of Nine Altars at Fountains, in the chapter house tombs at Byland and Jervaulx, in the shrine to abbot William at Rievaulx and in the graveyards east of the monastic churches.

The Cistercians of medieval Yorkshire negotiated the tension between freedom and enclosure by ascribing a range of meanings to material and abstract spaces. These meanings were expressed through practices of order, discipline, and regulation, together with ideas of memory, metaphor, and transcendence. Thirteenth-century Cistercians thus perceived space to be both anchored to the earth and liberated from the confines of the earth—like Hamlet, they were 'bounded in a nutshell' but could count themselves 'king[s] of infinite space'.[1] Space was meaningful to medieval Cistercians in both physical and abstract ways and confinement and discipline were tempered by the liberating propensities of transcendence. Likewise, the possibility of transcendence was dependent on principles of enclosure and order. The novices, choir monks, lay brothers, abbots, apostates, and ghosts who inhabited the Cistercian monasteries of northern England were creators and subjects of a spatial world in which boundaries were fixed, transgressed, transmuted, and blurred; in which bodies were disciplined, tended, and remembered; and in which this life and the next coalesced in the quiet Yorkshire earth.

---

[1] Hamlet, II. 2.

Bibliography

Primary Sources

Manuscripts

Cambridge, Corpus Christi College, MS 66
Cambridge, Corpus Christi College, MS 139
Cambridge, Jesus College, MS Q.B. 7
Cambridge, Jesus College, MS Q.B. 17
Cambridge, Sidney Sussex, MS 85
Cambridge, Trinity College, MS 1054
Cambridge, Trinity College, MS 1214
Cambridge, University College, MS Ff. i. 27
London, British Library, Additional MS 15723
London, British Library, Additional MS 46203
London, British Library, Additional MS 62130
London, British Library, Arundel MS 346
London, British Library, Arundel MS 248
London, British Library, Cotton MS Tiberius C XII
London, British Library, Cotton MS Titus D XXIV
London, British Library, Egerton MS 832
London, British Library, Egerton MS 2823
London, British Library, Harley MS 3173
London, British Library, Harley MS 2851
London, British Library, Royal MS 15 A X
London, Lambeth Palace, MS 51
London, Public Record Office, C81/1788
Manchester, John Rylands Library, Latin MS 153

Manchester, John Rylands Library, Latin MS 365
Oxford, Bodleian Library, Ashmole MS 750
Oxford, Bodleian Library, Ashmole MS 1398
Oxford, Bodleian Library, Ashmole MS 1437
Oxford, Bodleian Library, Bodley MS 514
Oxford, Bodleian Library, Bodley Rolls, 22
Oxford, Bodleian Library, Lyell MS 8
Oxford, Bodleian Library, Laudian Miscellany MS 527
Oxford, Bodleian Library, Laudian Miscellany MS 722
Paris, Bibliothèque Nationale, Latin MS 15157
Paris, Bibliothèque Nationale, Latin MS 15912

Printed Editions

Abstracts of the Charters and Other Documents Contained in the Chartulary of the Cistercian Abbey of Fountains, ed. by W. T. Lancaster, 2 vols (Leeds: [J. Whitehead & Son, printers], 1915).
Aelred of Rievaulx, *De Institutione Inclusarum*, in *Patrologiae Cursus Completus, Series Latina*, [hereafter *PL*] ed. by J.-P. Migne, 221 vols (Paris, 1844–64), 32, cols 1451–74.
Aelred of Rievaulx, *De Sanctimoniali Wattun*, *PL* 195, cols 780–96.
Aelred of Rievaulx, *Speculum Caritatis*, *PL* 195, cols 500–54.
Aelred of Rievaulx, *The Mirror of Charity*, trans. by E. Connor and intro. by C. Dumont, Cistercian Fathers Series 17 (Kalamazoo: Cistercian Publications, 1990).
Analecta Hymnica Medii Aevi, ed. by C. Blume und G. M. Dreves (Frankfurt am Main: Minerva, 1961).
Calendar of Inquisitions Miscellaneous (Chancery) preserved in the Public Record Office, (London: Published by Authority of His Majesty's Principal Secretary of State for the Home Dept, 1916–).
Calendar of Patent Rolls, Henry III, 1216–25 (London: HMSO, 1901–).
Calendar of the Charter Rolls preserved in the Public Record Office, Part 1: Henry III. A.D. 1226–1257, ed. by R. D. Trimmer and C. G. Crump (London: HMSO, 1903).
Calendar of the Charter Rolls preserved in the Public Record Office, Part 2: Henry III–Edward I. A.D. 1257–1300, ed. by R. D. Trimmer and C. G. Crump (London: HMSO, 1906).
Calendar of the Fine Rolls preserved in the Public Record Office, prepared under the Superintendence of the Deputy Keeper of the Record (London: HMSO, 1911–62), vol. 1.
Cartularium abbathiæ de Rievalle: Ordinis cisterciensis fundatæ anno MCXXXII, ed. by J. C. Atkinson, Publications of the Surtees Society vol. 83 (Durham: published for the Society by Andrews & Co., 1889).
Charters of the Honour of Mowbray 1107–1191, ed. by D. E. Greenaway, British Academy Record of Social and Economic History, new series 1 (London: Oxford University Press for the British Academy, 1972).

Charters of the Vicars Choral of York Minster: City of York and its Suburbs to 1536, ed. by N. Tringham, Yorkshire Archeological Society Record Series 148 (York: Yorkshire Archeological Society, 1993).

The Chartulary of the Cistercian Abbey of Salley in Craven, ed. by J. McNulty, Yorkshire Archaeological Society Record Series vols 87 and 90 (Wakefield: Yorkshire Archaeological Society, 1933–34).

The Chronicle of St. Mary's Abbey York, ed. by H. H. E. Craster, Publications of the Surtees Society 148 (Durham and London: published for the Society by Andrews and Co. and B. Quaritch, 1934).

Chronicon Monasterii de Melsa, ed. by E. A. Bond, Rerum Britannicarum Medii Aevi Scriptores, 3 vols (London: Longmans, Green, Reader, and Dyer, 1866–68).

Close Rolls of the Reign of Henry III preserved in the Public Record Office (London: HMSO, 1902–38).

Les codifications cisterciennes de 1237 et de 1257, ed. by B. Lucet (Paris, Editions du Centre National de la Recherche Scientifique, 1977).

Corpus Iuris Canonici, 2 vols (Leipzig: ex officina Bernhardi Tauchnitz, 1879–81), I, pars I. *Decretum magistri Gratiani*.

The Coucher Book of the Cistercian Abbey of Kirkstall, in the West Riding of the County of York: Printed from the Original Preserved in the Public Record Office, ed. by W. T. Lancaster and W. Paley Baildon, Publications of the Thoresby Society vol. 8 (Leeds: [J. Whitehead & Son, Printers], 1904).

Curia Regis Rolls of the Reign of Henry III preserved in the Public Record Office, 8 vols (London: HMSO, 1938–72).

Early Yorkshire Charters, ed. by W. Farrer, C. T. Clay, and E. A. Clay, 13 vols (York: Yorkshire Archeological Society Record Series, 1914–65).

Les Ecclesiastica Officia cisterciens du XIIème siècle: texte latin selon les manuscrits édités de Trente 1711, Ljubljana 31 et Dijon 114, version française, annexe liturgique, notes, index et tables, trans. by D. Choisselet and P. Vernet, Documentation cistercienne 22 (Reiningue: La Documentation Cistercienne, 1989).

Exordium Magnum Cisterciense, ed. by B. Griesser, Series Scriptorum S. Ordinis Cisterciensis 2 (Rome: Editiones Cistercienses, 1961).

Griesser, B., 'Die "Ecclesiastica Officia Cisterciensis Ordinis" des Cod. 1711 von Trient', *ASOC,* 12 (1956), 10–288.

Guillelmi Duranti Rationale divinorum officiorum I–IV, ed. by A. Davril and T. M. Thibodeau, Corpus Christianorum. Continuatio mediaevalis 140 (Turnhout: Brepols, 1995).

Herbert of Clairvaux, *Liber Miraculorum, PL* 185, cols 1273–1384.

Hockey, S. F., *The Beaulieu Chartulary*, Southampton Record Series 17 (Southampton: Southampton University Press, 1974).

Honorius Augustodunensis, *De Claustro, PL* 172, col. 590.

Lucet, B., *La codification cistercienne de 1202 et son évolution ultérieure* (Rome: Editiones cistercienses, 1964).

Manning, E., 'La Règle S. Benoît selon les MSS cisterciens (texte critique)', *Studia Monastica,* 8 (1966), 213–66.

Matthaei Parisiensis, monachi Sancti Albani, Chronica majora, ed. by H. R. Luard, 7 vols (London: Longman & Co. [Published by the Authority of the Lords

Commissioners of Her Majesty's Treasury, under the Direction of the Master of the Rolls], 1872–1883).

Memorials of the Abbey of St. Mary of Fountains, ed. by J. R. Walbran, Publications of the Surtees Society vols 42, 67, and 130 (London: published for the Society by Andrews & Co., 1863–1918).

Mikkers, E., 'Un Speculum Novitii inédit d'Etienne de Sallai', *COCR,* 8 (1946), 17–68.

——, 'Un traité inédit d'Etienne de Sallai sur la psalmodie', *Cîteaux,* 23 (1972), 245–88.

Monasticon Anglicanum, ed. by W. Dugdale, 8 vols (London: Printed for Longman, Hurst, Rees, Orme, & Browne, 1817–30).

Monasticon Eboracense: and the Ecclesiastical History of Yorkshire, ed. by J. Burton (York: printed for the author, by N. Nickson,1758).

Les monuments primitifs de la règle Cistercienne, ed. by Ph. Guignard (Dijon: Imprimerie Darantière, 1878).

Noschitzka, C., 'Codex Manuscriptus 31 Bibliothecae Universitatis Labacensis', *ASOC,* 6 (1950), 1–124.

Papal Letters: 1198–1304[–1342], ed. by W. H. Bliss, Calendar of Entries in the Papal Registers, Part 1 and 2 (London: HMSO, 1893–95).

Les plus anciens textes de Cîteaux: sources, textes et notes historiques, Cîteaux. Commentarii cisterciences: Studia et documenta, vol. 2, ed. by J. De La Croix Bouton and J. B. Van Damme (Achel: Abbaye cistercienne, 1974).

Register of John Le Romeyn, Lord Archbishop of York, 1286–1296, ed. by W. Brown, Publications of the Surtees Society 123, 2 vols (Durham: published for the Society by Andrews & Co., 1913).

Rolls of the Justices in Eyre Being the Records of Pleas and Assizes for Yorkshire in 3 Henry III (1218–19), ed. by D. M. Stenton, The Publications of the Selden Society 56 (London: B. Quaritch, 1937).

Rotuli Hundredorum, ed. by W. Hillingworth and J. Caley (London: Record Commission, 1812–18).

The Rule of St. Benedict, trans. by J. McGann (London: Sheed and Ward, 1976).

Sancti Bernardi Opera, ed. by J. Leclercq, H. Rochas, and C. H. Talbot, 8 vols (Rome: Editiones Cistercienses, 1957–1977).

Statuta Capitulorum Generalium Ordinis Cisterciensis ab anno 1116 ad annum 1786, ed. by J. M. Canivez, 8 vols (Louvain: Bureaux de la Revue d'Histoire Ecclésiastique, 1933–41).

Stephen of Sawley Treatises, ed. and trans. by J. O'Sullivan, Cistercian Fathers Series 36 (Kalamazoo: Cistercian Publications, 1984).

Three Yorkshire Assize Rolls for the Reign of King John and Henry III, ed. by C. T. Clay, Yorkshire Archaeological Society Record Series 46 (Leeds: Yorkshire Archaeological Society, 1911).

Walter Daniel The Life of Ailred of Rievaulx, ed. and trans. by F. M. Powicke (London: Thomas Nelson and Sons, 1950).

Wilmart, A., 'Le Triple Exercice d'Etienne de Sallai', *Revue d'Ascétique et de Mystique,* 11 (1930), 355–74.

——, 'Les Méditations d'Etienne de Sallai sur les Joies de la Vierge', *Revue d'Ascétique et de Mystique,* 10 (1929), 268–415.

―――, 'Les Mélanges de Matthieu, Préchantre de Rievaulx au début du XIIIe siecle', *Revue Bénédictine,* 52 (1940), 15–84.

Secondary Sources

Age of Chivalry. Art in Plantagenet England 1200–1400, ed. by J. Alexander and P. Binski (London: Royal Academy of Arts in association with Weidenfeld and Nicholson, 1987).

Arbuckle, G. A., 'Planning the Novitiate Process: Reflections of an Anthropologist', *Review for Religious*, 43 (1984), 532–46.

Archaeological Approaches to Medieval Europe, ed. by K. Biddick, Studies in Medieval Culture, vol. 18 (Kalamazoo: Medieval Institute Publications, 1984).

The Archaeology of Rural Monasteries, ed. by R. Gilchrist and H. Mytum, British Archaeological Reports, British Series 203 (Oxford: BAR, 1989).

Ariès, P., *The Hour of Our Death*, trans. by H. Weaver (London and Harmondsworth: Penguin Books, 1981).

Armstrong Kelly, G., *Politics and Religious Consciousness in America* (London and New Brunswick: Transaction Books, 1984).

Astley, H., 'Roche Abbey', *Yorkshire Archaeological Journal*, 10 (1904), 199–220.

Auberger, J.-B., *L'Unanimité cistercienne primitive: mythe ou réalité?*, Cîteaux, Studia et Documenta 3 (Achel: Administration de Cîteaux, 1986).

Aubert, M., 'Existe-t-il une architecture cistercienne?', *Cahiers de Civilization Médiévale*, 1 (1958), 153–58.

Backaert, E., 'L'Evolution du calendrier cistercien', *COCR*, 12 & 13 (1950 & 1951), 81–94; 108–27.

Bailey, T., *The Processions of Sarum and the Western Church* (Toronto: Pontifical Institute of Mediaeval Studies, 1971).

Baker, D., 'Heresy and Learning in Early Cistercianism', in *Schism, Heresy and Religious Protest*, ed. by D. Baker, Studies in Church History 9 (Cambridge: Cambridge University Press, 1972), pp. 93–107.

Baker, D., 'The Foundation of Fountains Abbey', *Northern History*, 4 (1969), 29–43.

Sanctity and Secularity, the Church and the World, ed. by D. Baker, Studies in Church History 10 (Oxford: Oxford University Press, 1973).

Baker, L. G. D., 'The Genesis of Cistercian Chronicles in England: the Foundation of Fountains Abbey I', *Analecta Cisterciensia*, 25 (1969), 14–41.

―――, 'The Genesis of Cistercian Chronicles in England: the Foundation of Fountains Abbey II', *Analecta Cisterciensia*, 31 (1975), 179–212.

Bandmann, G., *Mittelalterliche Architektur als Bedeutungsträger* (Berlin: Gebr. Mann Studio-Reihe, 1951).

Barraclough, G., *The Medieval Papacy* (London: Thames and Hudson, 1968).

Barthes, R., *Image, Music, Text*, trans. by S. Heath (London: Fontana Press, 1977).

Baschet, J., 'Fécondité et limites d'une approche systématique de l'iconographie médiévale', *Annales ESC*, 46:2 (1991), 375–80.

Bataillon, L.-J., 'Approaches to the Study of Medieval Sermons', *Leeds Studies in English*, 11 (1980), 19–35.

———, 'Similitudines et exempla dans les sermons du XIIIe siècle', in *The Study of the Bible in the Middle Ages: Essays in Memory of Beryl Smalley*, ed. by K. Walsh and D. Wood, Studies in Church History, Subsidia 4 (Oxford: Oxford University Press, 1985), pp. 191–205.

Bauer, G., *Claustrum Animae: Untersuchungen zur Geschichte der Metaphor vom Herzen als Kloster* (Munich: W. Fink, 1973).

Beck, B., 'Les salles capitulaires des abbayes de Normandie, et notamment dans les diocèses d'Avranches, Bayeux et Coutances', *L'Information d'Histoire de l'Art*, 18 (1973), 204–15.

Bedos-Redak, B., 'Towards a Cultural Biography of the Gothic Cathedral; Reflections on History and Art History', in *Artistic Integration in Gothic Buildings*, ed. by V. C. Raguin, K. Bush, and P. Draper (Toronto: University of Toronto Press, 1995), pp. 262–74.

Bell, D. N., 'The Books of Meaux Abbey', *Analecta Cisterciensia*, 40 (1984), 25–83.

———, 'The English Cistercians and the Practice of Medicine', *Cîteaux*, 40 (1989), 139–74.

———, 'The Measurement of Cistercian Space', in *L'Espace cistercien*, ed. by L. Pressouyre (Paris: Comité des Travaux Historiques et Scientifiques, 1994), pp. 253–56.

———, 'The Siting and Size of Cistercian Infirmaries in England and Wales', in *Studies in Cistercian Art and Architecture*, vol. 5, ed. by M. Parsons Lillich (Kalamazoo: Cistercian Publications, 1998), pp. 211–38.

———, *Index of Authors and Works in Cistercian Libraries in Great Britain*, Cistercian Studies Series 130 (Kalamazoo: Cistercian Publications, 1992).

———, *The Libraries of the Cistercians, Gilbertines and Premonstratensians*, Corpus of British Medieval Library Catalogues 3 (London: The British Library in association with the British Academy, 1992).

Benson, R. and G. Constable, *Renaissance and Renewal in the Twelfth Century* (Cambridge, MA: Harvard University Press, 1982).

Berman, C., *The Cistercian Evolution: The Invention of a Religious Order in Twelfth-Century Europe* (Philadelphia: University of Pennsylvania Press, 2000).

Bernard de Clairvaux. histoire, mentalités, spiritualité, œuvres complètes I, Colloque de Lyon-Cîteaux-Dijon, Sources Chrétiennes 380 (Paris: Editions du Cerf, 1992).

Bernardus Magister, ed. by J. R. Sommerfeldt (Spencer, MA: Cistercian Publications, 1992).

Bernstein, A. C., 'Esoteric Theology: William of Auvergne on the Fires of Hell and Purgatory', *Speculum*, 57 (1982), 517–21.

Bethell, D., 'The Foundation of Fountains Abbey and the State of St. Mary's York 1132', *Journal of Ecclesiastical History*, 17 (1966), 11–207.

Biddick, K. 'Decolonizing the English Past: Readings in Medieval Archaeology and History', *Journal of British Studies*, 32 (1993), 1–23.

———, 'Malthus in a Straitjacket? Analyzing Agrarian Change in Medieval England', *Journal of Interdisciplinary History*, 20:4 (1990), 623–35.

———, 'People and Things: Power in Early English Development', *Comparative Studies in Society and History*, 32:1 (1990), 3–23.

Binns, A., *Dedications of Monastic Houses in England and Wales* (Woodbridge: Boydell Press, 1989).

Bishop, T. M., 'Monastic Granges in Yorkshire', *English Historical Review*, 51 (1936), 193–214.

Bloch, M., *Feudal Society*, trans. by L. A. Manyon, 2 vols (London: Routledge and Kegan Paul, 1962).

Bloomfield, M., *Incipits of Latin Works on the Virtues and Vices 100–1500 A.D.: Including a Section of Incipits of Works on the Pater Noster* (Cambridge, MA: Medieval Academy of America, 1979).

Bock, P. C., 'Les cisterciens et l'étude du droit', *ASOC,* 7 (1951), 3–31.

Boitani, P., *English Medieval Narrative in the Thirteenth and Fourteenth Centuries*, trans. by J. Hall (Cambridge: Cambridge University Press, 1982).

Bonde, S., and C. Maines, 'The Archaeology of Monasticism: A Survey of Recent Work in France 1970–1987', *Speculum,* 63 (1988), 794–825.

Bony, J., 'French Influences on the Origins of English Architecture', *Journal of the Warburg and Courtauld Institutes,* 12 (1949), 1–15.

Bossy, J., 'A Social History of Confession in the Age of the Reformation', *Transactions of the Royal Historical Society,* 5th series 25 (1975), 21–38.

Boswell, J., *The Kindness of Strangers: The Abandonment of Children in Western Europe from Late Antiquity to the Renaissance* (Harmondsworth: Penguin Books, 1988).

Bouchard, C., 'Cistercian Ideals versus Reality: 1134 reconsidered', *Cîteaux,* 39 (1988), 217–31.

Bouchard, C., *Holy Entrepreneurs: Cistercians, Knights and Economic Exchange in Twelfth-Century Burgundy* (Ithaca: Cornell University Press, 1991).

Bourdieu, P., *Algeria 1960*, trans. by R. Nice (Cambridge and Paris: Cambridge University Press, 1979), pp. 133–53.

Braert, H., and W. Verbeke, *Death in the Middle Ages*, Medievalia Lovanensia, series 1, Studia 9 (Leuven: Leuven University Press, 1982).

Bredero, A. H., 'Le moyen âge et la purgatoire', *Revue d'Histoire Ecclésiastique,* 78 (1983), 429–52.

Brett, M., 'Canon Law and Litigation: the Century before Gratian', in *Medieval Ecclesiastical Studies in Honour of Dorothy M. Owen*, ed. by M. J. Franklin and C. Harper-Bill, Studies in the History of Religion 7 (Woodbridge and Rochester, NY: Boydell Press, 1995), pp. 21–40.

Britnell, R., 'Boroughs, Markets and Trade in Northern England, 1000–1216', in *Progress and Problems in Medieval England: Essays in Honour of E. Miller*, ed. by R. Britnell and J. Hatcher (Cambridge: Cambridge University Press, 1996), pp. 46–67.

Brontë, E., *Wuthering Heights*, ed. by D. Daiches (Harmondsworth: Penguin Books, 1985).

Brooke, C., 'Reflections on the Monastic Cloister', in *Romanesque and Gothic: Essays for George Zarnecki*, ed. by N. Stratford, 2 vols (Woodbridge: Boydell Press, 1987), 1, pp. 19–25.

Brown, P., *The Body and Society: Men, Women and Sexual Renunciation in Early Christianity* (New York: Columbia University Press, 1988).

Bruzelius, C., 'Cistercian High Gothic: The Abbey Church of Longpont and the Architecture of the Cistercians in the Early Thirteenth Century', *Analecta Cisterciensia,* 35 (1979), 1–204.

Bucher, F., 'Cistercian Architectural Purism', *Comparative Studies in Society and History,* 3 (1960), 98–105.

Burke, P., 'European Ideas of Decline and Revival c.1350–1500', *Parergon,* 23 (1979), 3–8.

Burton, J., 'The Abbeys of Byland and Jervaulx and the Problems of the English Savignacs 1134–1156', in *Monastic Studies II*, ed. by J. Loades (Bangor: Headstart History, 1991), pp. 119–31.

——, 'The Foundation of the British Cistercian Houses', in *Cistercian Art and Architecture in the British Isles*, ed. by C. Norton and D. Park (Cambridge: Cambridge University Press, 1986), pp. 24–39.

——, *Monastic and Religious Orders in Britain 1000–1300* (Cambridge: Cambridge University Press, 1994).

——, *The Yorkshire Nunneries in the Twelfth and Thirteenth Centuries* (York: Borthwick Institute of Historical Research, 1979).

Butler, L., 'Cistercian Abbots' Tombs and Abbey Seals', in *Studies in Cistercian Art and Architecture* 4, ed. by M. Parsons Lillich (Kalamazoo: Cistercian Publications, 1993), pp. 78–88.

Bynum, C. W., 'Why All the Fuss about the Body? A Medievalist's Perspective', *Critical Inquiry,* 22:1 (1995), 3–33.

——, *Docere Verbo et Exemplo; An Aspect of Twelfth-Century Spirituality*, Harvard Theological Studies 31 (Missoula, MT: Scholars Press, 1979).

——, *Fragmentation and Redemption: Essays on Gender and the Human Body in Medieval Religion* (New York: Zone Books, 1992).

——, *Jesus as Mother: Studies in the Spirituality of the High Middle Ages* (Berkeley: University of California Press, 1982).

——, *The Resurrection of the Body in Western Christianity 200–1336* (New York: Columbia University Press, 1995).

Calhoun, C. H., 'Community: Towards a Variable Conceptualization for Comparative Research', *Social History,* 5 (1980), 105–29.

Camille, M., 'Mouths and Meanings: Towards an Anti-Iconography of Medieval Art', in *Iconography at the Crossroads: Papers from the Colloquium Sponsored by the Index of Christian Art*, ed. by B. Cassidy (Princeton: Index of Christian Art, 1993).

——, 'Seeing and Reading', *Art History,* 8 (1985), 26–49.

——, 'Visionary Perception and Images of the Apocalypse in the Late Middle Ages', in *The Apocalypse in the Middle Ages*, ed. by R. K. Emmerson and B. McGinn (Ithaca and London: Cornell University Press, 1992).

Camporesi, P., *The Fear of Hell: Images of Salvation and Damnation in Early Modern Europe,* trans. by L. Byatt (Cambridge: Polity Press, 1991).

Canivez, J. M., 'Le rite cistercien', *Ephemerides Liturgicae,* 63 (1949), 276–311.

Cantor, N., 'The Crisis of Western Monasticism 1050–1130', *American Historical Review,* 66 (1960–1), 47–67.

Carruthers, M., *The Book of Memory: A Study of Memory in Medieval Culture* (Cambridge: Cambridge University Press, 1990).

Casey, M., 'Herbert of Clairvaux's Book of Wonderful Happenings' *Cistercian Studies,* 25:1 (1990), 37–64.

——, 'In Communi Vita Fratrum: St. Bernard's Teaching on Cenobitic Solitude', *Analecta Cisterciensia,* 46 (1990), 243–61.

——, 'The Dialectic of Solitude and Communion in Cistercian Communities', *Cistercian Studies,* 23 (1988), 273–309.

——, *Athirst For God: Spiritual Desire in Bernard of Clairvaux's Sermons on the Song of Songs*, Cistercian Studies Series 77 (Kalamazoo: Cistercian Publications, 1988).

Cassidy, M., '*Non Conversi Sed Perversi*: The Use and Marginalisation of the Cistercian Lay Brother', in *Deviance and Textual Control: New Perspectives in Medieval Studies*, ed. by M. Cassidy, H. Hickey, and M. Street, Conference Series 2 (Melbourne: Department of History, University of Melbourne, 1997), pp. 34–55.

Cassidy-Welch, M., 'Confessing to Remembrance: Stephen of Sawley's *Speculum novitii* and Cistercian Uses of Memory', *Cistercian Studies Quarterly*, 35:1 (2000), 13–27.

——, 'Incarceration and Liberation: Prisons in the Cistercian Monastery', *Viator*, 32 (2001), forthcoming.

Catalogi Codicum Manuscriptorum Bibliothecae Bodleianae (Oxford: Oxford University Press, 1860).

Catalogue of Additions to Manuscripts in the British Museum (London: The British Museum, 1841–1945).

Caviness, M. H., 'Artistic Integration in Gothic Buildings: A Post-Modern Construct', in *Artistic Integration in Gothic Buildings*, ed. by V. C. Raguin, K. Bush, and P. Draper (Toronto: University of Toronto Press, 1995), pp. 249–61.

Certeau, M. de, *The Practice of Everyday Life*, trans. by S. Rendall (Berkeley and Los Angeles: University of California Press, 1984).

Chadd, D. F. L., 'Liturgical Music and the Limits of Uniformity', in *Cistercian Art and Architecture in the British Isles*, ed. by C. Norton and D. Park (Cambridge: Cambridge University Press, 1986), pp. 299–314.

Chemin, A., 'Life Behind Clairvaux's Walls', *Guardian Weekly*, 28 August 1994, p. 15.

Chibnall, M., 'Monks and Pastoral Work: A Problem in Anglo-Norman History', *Journal of Ecclesiastical History*, 18 (1967), 165–72.

Cistercian Art and Architecture in the British Isles, ed. by C. Norton and D. Park (Cambridge: Cambridge University Press, 1986).

Cistercian Ideals and Reality, ed. by J. R. Sommerfeldt (Kalamazoo: Cistercian Publications, 1978).

The Cistercian World: Monastic Writings of the Twelfth Century, ed. by P. Matarasso (Harmondsworth: Penguin Books, 1993).

Clanchy, M. T., *From Memory To Written Record: England 1066–1307*, 2nd edn (Oxford and Cambridge, MA: Basil Blackwell, 1993).

Clay, R. M., *The Medieval Hospitals of England* (London: Frank Cass, 1966).

Cohen, A. P., *The Symbolic Construction of Community* (London: Tavistock Publications, 1985).

Cohen, E., *The Crossroads of Justice: Law and Culture in Late Medieval France* (Leiden and New York: E. J. Brill, 1993).

Cohn, N., *The Pursuit of the Millennium*, rev. edn (Oxford and New York: Oxford University Press, 1970).

Coldstream, N., 'Cistercian Architecture from Beaulieu to the Dissolution', in *Cistercian Art and Architecture in the British Isles*, ed. by C. Norton and D. Park (Cambridge: Cambridge University Press, 1986), pp. 139–59.

Coleman, E., 'Nasty Habits: Satire and the Medieval Monk', *History Today*, 43:6 (1993), 36–42.

Coleman, J., *Ancient and Medieval Memories: Studies in the Reconstruction of the Past* (Cambridge: Cambridge University Press, 1992).

Colker, M. L., *Trinity College Dublin: Descriptive Catalogue of the Medieval and Renaissance Latin Manuscripts*, 2 vols (Aldershot and Brookfield, VT: Scolar Press, 1991).

Collins, S., 'Monasticism, Utopias and Comparative Social Theory', *Religion*, 18 (1988), 101–39.

Comito, T., ed., *The Idea of Garden in the Renaissance* (Hassocks: Harvester Press, 1979).

Conant, K. J., 'Medieval Academy Excavations at Cluny', *Speculum*, 38 (1963), 1–45.

———, *Carolingian and Romanesque Architecture 800–1200*, rev. edn (Harmondsworth: Pelican Books, 1990).

Constable, G., 'Aelred of Rievaulx and the Nun of Watton', in *Medieval Women*, ed. by D. Baker (Oxford: Basil Blackwell, 1978), pp. 205–26.

———, 'The Vision of a Cistercian Novice', *Studia Anselmiana*, 40 (1956), 95–98.

———, 'The Vision of Gunthelm and Other Visions Attributed to Peter the Venerable', *Revue Bénédictine*, 66 (1956), 92–114.

Cook, G. H., *English Monasteries in the Middle Ages* (London: Phoenix House, 1961).

Coppack, G., *Fountains Abbey* (London: B. T. Batsford, 1993).

———, and P. Fergusson, *Rievaulx Abbey* (London: English Heritage, 1994).

Coxe, H. O., *Bodleian Library Quarto Catalogues II: Laudian Manuscripts* (Oxford: The Bodleian Library, 1973).

Croix Bouton, J. de la, *Histoire de l'Ordre de Cîteaux* (Westmalle: Abbaye cistercienne, 1959).

Crossley, P., 'English Gothic Architecture', in *Age of Chivalry. Art in Plantagenet England 1200–1400*, ed. by J. Alexander and P. Binski (London: Royal Academy of Arts in association with Weidenfeld and Nicholson, 1987), pp. 60–73.

———, 'Medieval Architecture and Meaning; the Limits of Iconography', *Burlington Magazine*, 130: 1019 (1988), 116–21.

Cummings, C., *Monastic Practices* (Kalamazoo: Cistercian Publications, 1986).

Daniell, C., *Death and Burial in Medieval England 1066–1550* (London and New York: Routledge and Kegan Paul, 1997).

Darnton, R., *The Great Cat Massacre and Other Episodes in French Cultural History* (New York: Basic Books, 1984).

Davis, G. R. C., *Medieval Cartularies of Great Britain: A Short Catalogue* (London: Longman, Green, 1958).

De La Mare, A. C., *Catalogue of the Collection of Medieval Manuscripts Bequeathed to the Bodleian Library by James P. Lyell* (Oxford: Clarendon Press, 1971).

Dekkers, E., *Clavis Patrum Latinorum*, Sacris Erudiri: Jaarboek voor Godsdienstwetenschappen 3, 2nd edn (Steenbrugge: In Abbatia Sancti Petri, 1961).

Delumeau, J., *Sin and Fear: The Emergence of a Western Guilt Culture 13th–18th Centuries*, trans. by E. Nicholson (New York: St Martin's Press, 1990).

Denton, J., 'From the Foundation of Vale Royal to the Statute of Carlisle: Edward I and Ecclesiastical Patronage', in *Thirteenth Century England IV, Proceedings of the Newcastle-Upon-Tyne Conference*, ed. by P. R. Coss and S. D. Lloyd (Woodbridge: Boydell Press, 1991), pp. 123–37.

Derrida, J., *Margins of Philosophy*, trans. by A. Bass (Chicago: University of Chicago Press, 1982).

——, *Positions*, trans. by A. Bass (Chicago: University of Chicago Press, 1987).

Dies Illa: Death in the Middle Ages, Proceedings of the 1983 Manchester Conference, ed. by J. M. Taylor (Liverpool: F. Cairns, 1984).

Dimier, M.-A., 'Architecture et spiritualité cisterciennes', *Revue du Moyen Age Latin*, 3 (1947), 255–74.

——, 'Infirmeries cisterciennes', in *Mélanges à la mémoire du Père Anselme Dimier: 1/2*, ed. by B. Chauvin (Pupillin, Arbois: B. Chauvin, 1982), pp. 804–25.

——, 'Le mot *locus* employé dans le sense de *monastère*', *Revue Mabillon*, 58 (1972), 133–54.

——, 'Violence, rixes et homicides chez les Cisterciens', *Revue des Sciences Religieuses*, 46 (1972), 38–57.

——, *L'Art cistercien* (Paris: Zodiaque, 1982).

Donkin, R. A., 'Settlement and Depopulation on Cistercian Estates during the Twelfth and Thirteenth Centuries, Especially in Yorkshire' *Bulletin of the Institute of Historical Research*, 33 (1960), 141–65.

——, 'The Cistercian Grange in England in the Twelfth and Thirteenth Centuries, with Special Reference to Yorkshire', *Studia Monastica*, 6 (1964), 95–144.

——, 'The Cistercian Order and the Settlement of Northern England', *Geographical Review*, 54 (1969), 403–16.

——, 'The Disposal of Cistercian Wool in England and Wales during the 12th and 13th Centuries', *Cîteaux in de Nederlanden*, 8 (1957), 109–31; 181–202.

——, 'The Site Changes of Medieval Cistercian Monasteries', *Geography*, 44 (1959), 251–58.

——, 'The Urban Property of the Cistercians', *ASOC*, 15 (1959), 104–31.

——, *A Check List of Printed Works Pertaining to the Cistercian Order in General and to the Houses of the British Isles in Particular*, Documentation Cistercienne 2 (Rochefort: Abbaye ND de S. Remy, 1969).

——, *The Cistercians: Studies in the Geography of Medieval England and Wales* (Toronto: Pontifical Institute of Mediaeval Studies, 1978).

Donnelly, J., 'Changes in the Grange Economy of the English and Welsh Cistercian Abbeys', *Traditio*, 10 (1954), 399–458.

——, *The Decline of the Medieval Cistercian Laybrotherhood* (New York: Fordham University Press, 1949).

Draper, P., 'Architecture and Liturgy', in *Age of Chivalry. Art in Plantagenet England 1200–1400*, ed. by J. Alexander and P. Binski (London: Royal Academy of Arts in association with Weidenfeld and Nicholson, 1987), pp. 83–91.

——, 'The Nine Altars at Durham and Fountains', in *Medieval Art and Architecture at Durham Cathedral*, The British Archaeological Association Conference Transactions for the Year 1977 (Leeds: British Archaeological Association, 1980), pp. 74–86.

Dubois, J., 'The Laybrother's Life in the Twelfth Century: A Form of Lay Monasticism', *Cistercian Studies*, 7 (1972), 161–213.

Duby, G., *Rural Civilisation and Country Life in the Medieval West*, trans. by C. Postan (London: Edward Arnold, 1968).

——, *Saint Bernard: l'art cistercien* (Paris: Flammarion, 1976).

——, *The Three Orders: Feudal Society Imagined,* trans. by A Goldhammer (Chicago: University of Chicago Press, 1980).

Duggan, C., *The Twelfth-Century Decretal Collections and their Importance in English History* (London: Athlone, 1963).

Dumont, C., 'Experience in the Cistercian Discipline', *Cistercian Studies,* 10 (1975), 135–38.

——, 'L'Hymne 'Dulcis Jesu memoria': le 'Jubilus' serait-il d'Aelred de Rievaulx?', *Collectanea Cisterciensia,* 55 (1993), 233–43.

Dumville, D. N., 'The Corpus Christi Nennius', *The Bulletin of the Board of Celtic Studies,* 25:4 (1974), 369–80.

Durham, B., 'The Infirmary and Hall of the Medieval Hospital of St. John the Baptist at Oxford', *Oxoniensia,* 56 (1991), 17–75.

Durkheim, E., *The Elementary Forms of the Religious Life,* trans. by J. W. Swain (New York: Collier Books, 1961).

Dutton, M., 'Intimacy and Erudition: The Humanity of Christ in Cistercian Spirituality', in *Erudition at God's Service,* ed. by J. Sommerfeldt, Cistercian Studies Series 98 (Kalamazoo: Cistercian Publications, 1987), pp. 33–69.

——, 'The Cistercian Source: Aelred, Bonaventure and Ignatius', in *Goad and Nail: Studies in Medieval Cistercian History 10,* ed. by E. R. Elder (Kalamazoo: Cistercian Publications, 1985), pp. 151–78.

Dyer, J., 'Monastic Psalmody of the Middle Ages', *Revue Bénédictine,* 99 (1989), 41–74.

Dynes, W., 'The Medieval Cloister as Portico of Solomon', *Gesta,* 12 (1973), 61–69.

Easting, R., 'Purgatory and the Earthly Paradise in the Tractatus de Purgatorio Sancti Patricii', *Cîteaux,* 37 (1986), 23–48.

Eckenrode, T., 'The English Cistercians and their Sheep during the Middle Ages', *Cîteaux,* 24 (1973), 250–66.

Eco, U., *Art and Beauty in the Middle Ages,* trans. by H. Bredin (London and New Haven: Yale University Press, 1986).

Edwards, G. R., 'Purgatory: Birth or Evolution?', *Journal of Ecclesiastical History,* 36 (1985), 634–46.

English Lyrics Before 1500, ed. by T. Silverstein, York Medieval Texts (York: University of York, 1971).

Eriksen, T. H., *Small Places, Large Issues: An Introduction to Social and Cultural Anthropology* (London and East Haven: Pluto Press, 1995).

L'Espace cistercien, ed. by L. Pressouyre (Paris: Comité des Travaux Historiques et Scientifiques, 1994).

Evans, G. R., *Old Arts and New Theology: The Beginnings of Theology as an Academic Discipline* (Oxford: Clarendon Press, 1980).

——, *The Mind of St. Bernard of Clairvaux* (Oxford: Clarendon Press, 1983).

Faral, E., *Les arts poétiques du XIIe et XIIIe siècles* (Paris: Champion, 1924).

Farmer, H., 'A Letter of St. Waldef of Melrose concerning a Recent Vision', in *Analecta Monastica : Cinquième Série : Textes et Études sur la Vie des Moines au Moyen Age: Cinquième Série, Studia Anselmiana, Philosophica, Theologica; Fasc. 43.,* ed. by Y. Congar (Rome: Herder, 1958), pp. 91–101.

——, 'Stephen of Sawley', *The Month,* 29 (1963), 332–42.

Ferguson, G., *Signs and Symbols in Christian Art* (New York: Oxford University Press, 1966).

Fergusson, P., *Architecture of Solitude: Cistercian Abbeys in Twelfth-Century England* (Princeton: Princeton University Press, 1984).

——, 'Porta Patens Esto: Notes on Early Cistercian Gatehouses in the North of England', in *Medieval Architecture and its Intellectual Context: Studies in Honour of Peter Kidson*, ed. by E. Fernie and P. Crossley (London: Hambledon Press, 1990), pp. 47–59.

——, 'The Cistercian Churches in Yorkshire and the Problem of the Cistercian Crossing Tower', *Journal of the Society of Architectural Historians,* 29 (1970), 211–21.

——, 'The South Transept Elevation of Byland Abbey', *Journal of the British Archaeological Association,* 3rd series 38 (1975), 155–76.

——, 'The Twelfth-Century Refectories at Rievaulx and Byland Abbeys', in *Cistercian Art and Architecture in the British Isles*, ed. by C. Norton and D. Park (Cambridge: Cambridge University Press, 1986), pp. 160–80.

——, and S. Harrison, *Rievaulx Abbey: Community, Architecture, Memory* (London and New Haven: Yale University Press, 1999).

——, and S. Harrison, 'The Rievaulx Abbey Chapter House', *The Antiquaries Journal,* 74 (1994), 211–55.

Fernie, E., 'Archaeology and Architecture: Recent Developments in the Study of English Medieval Architecture', *Architectural History,* 32 (1989), 18–29.

Ferrante, J. M., 'Images of the Cloister—Haven or Prison?', *Medievalia,* 12 (1989), 57–66.

Flemming, P., 'The Medical Aspects of the Medieval Monastery in England', *Proceedings of the Royal Society of Medicine,* (1928), 25–36.

Fletcher, B., *A History of Architecture*, 19th edn (London: Butterworths, 1987).

Forey, A., *The Military Orders* (Toronto: University of Toronto Press, 1992).

Foucault, M., *Dits et écrits 1954–1988*, 4 vols (Paris: Editions Gallimard, 1994), 4, pp. 270–85.

——, *Discipline and Punish: The Birth of the Prison*, trans. by A. Sheridan (Harmondsworth: Penguin Books, 1977).

——, *The History of Sexuality. Volume One: An Introduction*, trans. by R. Hurley (Harmondsworth: Penguin Books, 1990).

Freeland, J. P., 'A Fifteenth-Century Cistercian Processional,' in *Cistercian Ideals and Reality*, ed. by J. Sommerfeldt, Cistercian Studies Series 60 (Kalamazoo: Cistercian Publications, 1978), pp. 344–51.

Freeman, E., 'Aelred of Rievaulx's *De Bello Standardii* and Medieval and Modern Textual Controls', in *Deviance and Textual Control: New Perspectives in Medieval Studies*, ed. by M. Cassidy, H. Hickey, and M. Street, Conference Series 2 (Melbourne: Department of History, University of Melbourne, 1997), pp. 78–102.

French, D. R., 'Ritual, Gender and Power Strategies: Male Pilgrimage to St. Patrick's Purgatory', *Religion,* 24:2 (1994), 103–15.

Freud, S., *Civilization and its Discontents*, trans. by J. Strachey (New York: W. W. Norton, 1962).

Funkenstein, A., *Theology and the Scientific Imagination from the Middle Ages to the Seventeenth Century* (Princeton: Princeton University Press, 1986).

Furniss, D., 'The Monastic Contribution to Medieval Medical Care: Aspects of an Earlier Welfare State', *Journal of the Royal College of General Practitioners*, 15 (1968), 244–50.

Gamber, K., *Liturgie und Kirchenbau: Studien zur Geschichte der Meßfeier und des Gotteshauses in der Frühzeit*, Studia Patristica et Liturgica, fasc. 6 (Regensburg: Pustet [in Komm.], 1976).

Geary, P., 'History, Theory and Historians', *Exemplaria*, 7:1 (1995), 93–98.

———, 'L'Humiliation des saints', *Annales ESC*, 34 (1979), 27–42.

———, *Living With the Dead in the Middle Ages* (Ithaca and London: Cornell University Press, 1994).

———, *Phantoms of Remembrance; Memory and Oblivion at the End of the First Millennium* (Princeton: Princeton University Press, 1994).

Geertz, C., *The Interpretation of Cultures: Selected Essays* (New York: Basic Books, 1973).

Gehl, P. F., 'Competens Silentium: Varieties of Monastic Silence in the Medieval West', *Viator*, 18 (1987), 125–60.

Gellrich, J., *The Idea of the Book in the Middle Ages* (Ithaca and London: Cornell University Press, 1985).

Getz, F., 'Charity, Translation and the Language of Medical Learning in Medieval England', *Bulletin of the History of Medicine*, 64 (1990), 1–17.

———, 'Medical Practitioners in Medieval England', *Social History of Medicine*, 3:2 (1990), 245–83.

Gibbon, E., *The History of the Decline and Fall of the Roman Empire* (London: W. Strahan and T. Cadell, 1776).

Gilchrist, R., 'Community and Self: Perception and Use of Space in Medieval Monasteries', *Scottish Archaeological Review*, 6 (1989), 55–64.

———, *Gender and Material Culture The Archaeology of Religious Women* (London and New York: Routledge, 1994).

Gilson, E., *La théologie mystique de Saint Bernard*, 4th edn (Paris: J. Vrin, 1980).

Gilyard-Beer, R., 'Fountains Abbey: the Early Buildings 1152–60', *Archaeological Journal*, 125 (1968), 313–19.

———, 'The Graves of the Abbots of Fountains', *Yorkshire Archaeological Journal*, 59 (1987), 45–50.

———, *Fountains Abbey Yorkshire* (London: HMSO, 1970).

Girsch, J., 'Matthew of Rievaulx on Select Vices: Poetry and Incidental Prose from Paris BN MS Lat. 15157' (unpublished essay, University of Toronto, 1983).

Given, J. B., *Society and Homicide in Thirteenth-Century England* (Stanford: Stanford University Press, 1977).

Goffman, E., *Asylums: Essays on the Social Situation of Mental Patients and other Inmates* (Chicago: Aldine, 1962).

Golding, B., 'Burials and Benefactions: An Aspect of Monastic Patronage in Thirteenth-Century England', in *England in the Thirteenth Century: Proceedings of the 1984 Harlaxton Symposium*, ed. by W. M. Ormrod (Woodbridge: Boydell Press, 1986), pp. 64–75.

Gougaud, L., 'La pratique de la phlébotomie dans les cloîtres', *Revue Mabillon*, 2:13 (1924), 1–13.

―――, *Devotional and Ascetic Practices in the Middle Ages*, trans. by G. C. Bateman (London: Burns Oates and Washbourne, 1927).

Grant, L., 'Gothic Architecture in Southern England and the French Connection in the Early Thirteenth Century', in *Thirteenth Century England III, Proceedings of the Newcastle-upon-Tyne Conference 1989*, ed. by P. R. Coss and S. D. Lloyd (Woodbridge: Boydell Press, 1991), pp. 113–26.

Graves, C. C., 'The Economic Activities of the English Cistercians in Medieval England (1128–1307)', *ASOC*, 13 (1957), 3–60.

Green, G., *Imagining God: Theology and the Religious Imagination* (San Francisco: Harper and Row, 1989).

Greene, J. P., *Medieval Monasteries* (Leicester: Leicester University Press, 1992).

Greenia, C., 'The Laybrother Vocation in the Eleventh and Twelfth Centuries', *Cistercian Studies*, 16 (1981), 38–45.

Gurevich, A., 'Popular and Scholarly Medieval Cultural Traditions: Notes in the Margin of Jacques Le Goff's Book', *Journal of Medieval History*, 9 (1983), 71–90.

―――, *Categories of Medieval Culture*, trans. by L. G. Campbell (London and Boston: Routledge and Kegan Paul, 1985).

Gurr, T. R., 'Historical Trends in Violent Crime: A Critical Review of the Evidence', *Crime and Justice; An Annual Review of Research*, 3 (1981), 295–353.

Hahnloser, H. R., *Villard de Honnecourt: Kritische Gesamtausgabe des Bauhüttenbücher ms. fr. 19093 der Pariser Bibliothek* (Vienna: Schroll, 1935).

Hallier, A., *The Monastic Theology of Aelred of Rievaulx* (Kalamazoo: Cistercian Publications, 1969).

Hallinger, K., 'Woher kommen die Laienbrüder?', *ASOC*, 12 (1956), 1–105.

Hamburger, J. M., 'The Visual and the Visionary: The Image in Late Medieval Monastic Devotions', *Viator*, 20 (1989), 161–82.

Hamilton-Thompson, A., *Roche Abbey Yorkshire* (London: HMSO, 1957).

Hammond, E. A., 'Physicians in Medieval Religious Houses', *Bulletin of the History of Medicine*, 32 (1958), 105–20.

Harding, A., *England in the Thirteenth Century* (Cambridge: Cambridge University Press, 1993).

Harpan, G., *The Ascetic Imperative in Culture and Criticism* (Chicago: University of Chicago Press, 1987).

Harper-Bill, C., 'Monastic Apostasy in Late Medieval England' *Journal of Ecclesiastical History*, 32 (1981), 1–18.

Harvey, B., *Living and Dying in Medieval England 1100–1540: The Monastic Experience* (Oxford: Clarendon Press, 1993).

Hasquenoph, S., 'La mort du moine au Moyen Age (X^e–XIIe siècles), *Collectanea Cisterciensia*, 53 (1991), 215–32.

Hearn, M. F., 'Villard de Honnecourt's Perception of Gothic Architecture', in *Medieval Art and Architecture and its Intellectual Context: Studies in Honour of Peter Kidson*, ed. by E. Fernie and P. Crossley (London: Hambledon Press, 1990), pp. 127–36.

Hefele, C. H., *Histoire des Conciles*, 11 vols (Paris: Letouzey et Ane, 1907–1952).

Hill, B. D., *English Cistercian Monasteries and their Patrons in the Twelfth Century* (Urbana: University of Illinois Press, 1968).

Hillaby, J., '"The House of Houses": The Cistercians of Dore and the Origins of the Polygonal Chapter House', *Transactions of the Woolhope Naturalists' Field Club,* 46:2 (1989), 209–45.

Hillier, W., and J. Hanson, *The Social Logic of Space* (Cambridge: Cambridge University Press, 1984).

History and Ethnicity, ed. by E. Tonkin, M. McDonald, and M. Chapman (New York: Routledge and Kegan Paul, 1989).

Holdsworth, C. J., 'Eleven Visions Connected with the Cistercian Monastery of Stratford Langthorne', *Cîteaux,* 13 (1962), 185–204.

———, 'Hermits and the Powers of the Frontier', *Reading Medieval Studies,* 16 (1990), 55–76.

———, 'Royal Cistercians: Beaulieu, Her Daughters and Rewley', in *Thirteenth Century England IV, Proceedings of the Newcastle-upon-Tyne Conference 1991*, ed. by P. R. Coss and S. D. Lloyd (Woodbridge: Boydell Press, 1992), pp. 139–150.

Holman, J., 'Stephen of Sawley: Man of Prayer', *Cistercian Studies Quarterly,* 21:2 (1986), 109–22.

Holmes, R., 'Paulinus of Leeds', *Miscellanea*, Publications of the Thoresby Society 4 (Leeds: Publications of the Thoresby Society, 1895), pp. 209–25.

Hontoir, C., 'La dévotion au sacrament chez les premiers Cisterciens', *Studia Eucharistica,* (1946), 132–56.

Hope, W. St John, 'Fountains Abbey', *Yorkshire Archaeological Journal,* 15 (1899), 269–402.

———, and H. Brakspear, 'Jervaulx Abbey', *Yorkshire Archaeological Journal,* 21 (1911), 303–44.

Hordern, P., 'A Discipline of Relevance: The Historiography of the Later Medieval Hospital', *Social History of Medicine,* 1:3 (1988), 359–74.

Horn, W., 'On the Origins of the Medieval Cloister', *Gesta,* 12 (1973), 13–52.

———, and E. Born, *The Plan of St. Gall: A Study of the Architecture and Economy of and Life in a Paradigmatic Carolingian Monastery,* 3 vols (Berkeley and Los Angeles: University of California Press, 1979).

Hoste, A., *Bibliotheca Aelrediana* (Steenbrugge: In Abbatia Sancti Petri, 1962).

Huizinga, J., *The Waning of the Middle Ages: A Study of the Forms of Life, Thought and Art in France and the Netherlands in the Fourteenth and Fifteenth Centuries*, trans. by F. Hopman (Harmondsworth: Penguin Books, 1955).

Hunnisett, R. F., 'The Medieval Coroner's Rolls', *American Journal of Legal History,* 3 (1959), 95–125; 205–21; 324–59.

Hyams, P. R., *Kings, Lords and Peasants in Medieval England: The Common Law of Villeinage in the Twelfth and Thirteenth Centuries* (Oxford: Clarendon Press, 1980).

Iconography at the Crossroads: Papers from the Colloquium Sponsored by the Index of Christian Art, ed. by B. Cassidy (Princeton: Index of Christian Art, 1993).

The Iconography of Heaven, ed. by C. Davidson (Kalamazoo: Medieval Institute Publications, 1994)

Index of Manuscripts in the British Library, 10 vols (Cambridge and Teaneck, NJ: Chadwyck-Healey, 1984–86).

Ivanka, E. Von, 'La structure de l'âme selon S. Bernard', *ASOC,* 9 (1953), 202–08.

Jacquart, D., 'A l'aube de la renaissance médicale des XIe-XIIe siècles: L' "Isagoge Johannitii" et son traducteur', *Bibliothèque de l'Ecole des Chartes*, 144 (1986), 209-40.

———, and C. Thomasset, *Sexuality and Medicine in the Middle Ages*, trans. by M. Adamson (Princeton: Princeton University Press, 1988).

James, M. R., *A Descriptive Catalogue of the Latin MSS in the John Rylands Library at Manchester* (Manchester: Manchester University Press, 1921, repr. Munich: Kraus, 1980).

———, *A Descriptive Catalogue of the Manuscripts in the Library of Corpus Christi College Cambridge*, 2 vols (Cambridge: Cambridge University Press, 1912).

———, *A Descriptive Catalogue of the Manuscripts in the Library of Jesus College, Cambridge* (London: Clay, 1895).

———, *The Western Manuscripts in the Library of Trinity College, Cambridge. A Descriptive Catalogue*, 4 vols (Cambridge: Cambridge University Press, 1900-04).

Jezler, P., et al., *Himmel, Hölle und Fegefeuer. Das Jenseits in Mittelalter* (Zurich: Schweizerisches Landesmuseum, 1994).

Jordan, M. D., 'Medicine as Science in the Early Commentaries on Johannitius', *Traditio*, 43 (1987), 121-45.

Kay, S., and M. Rubin, *Framing Medieval Bodies* (Manchester and New York: Manchester University Press, 1994).

Kealey, E. J., *Medieval Medicus: A Social History of Anglo-Norman Medicine* (Baltimore: Johns Hopkins University Press, 1981).

Ker, N. R., 'The Migration of Manuscripts from Medieval Libraries', in A. G. Watson, ed., *Books, Collectors, and Libraries: Studies in the Medieval Heritage* (London: Hambledon Press, 1977).

———, *Medieval Libraries of Great Britain: A List of Surviving Books* (London: Offices of the Royal Historical Society, 1964).

———, *Medieval Manuscripts in British Libraries*, 4 vols (Oxford: Clarendon Press, 1969-92).

Kerridge, I,. 'Bloodletting: The Story of a Therapeutic Technique', *Medical Journal of Australia*, 163: 11-12 (1995), 631-33.

Kinder, T., *L'Europe cistercienne* (La Pierre-qui-Vire: Editions Zodiaques, 1997).

King, A. A., *Liturgies of the Religious Orders* (London: Longman, 1955).

Knowles, D., *The Historian and Character* (Cambridge: Cambridge University Press, 1963).

———, and R. N. Haddock, *Medieval Religious Houses England and Wales* (London: Longman, 1953).

———, and J. K. S. St Joseph, *Monastic Sites from the Air* (Cambridge: Cambridge University Press, 1952).

———, et al., *The Heads of Religious Houses England and Wales* (Cambridge: Cambridge University Press, 1972).

Krautheimer, R., 'Introduction to the Iconography of Medieval Architecture', *Journal of the Warburg and Courtauld Institutes*, 5 (1942), 1-33.

Kristeller, P. O., 'Bartholomaeus, Musandinus and Maurus of Salerno and Other Early Commentators of the Avicella with a Tentative List of Texts and Manuscripts', *Italia Mediovale e Umanistica*, 19 (1976), 57-87.

——, *Latin Manuscript Books Before 1600: A List of the Printed Catalogues and Unpublished Inventories of Extant Collections*, 4th edn (Munich: Monumenta Germaniae Historica, 1993).
Kuttner, S., and E. Rathbone, 'Anglo-Norman Canonists in the Twelfth Century', *Traditio*, 7 (1949–51), 279–358.
Lackner, B. K., 'Early Cistercian Life as described by the *Ecclesiastica Officia*', in *Cistercian Ideals and Reality* ed, by J. Sommerfeldt, Cistercian Studies Series 60 (Kalamazoo: Cistercian Publications, 1978), pp. 62–79.
Ladner, G., 'Homo Viator: Medieval Ideas on Alienation and Order', *Speculum*, 42 (1967), 233–59.
——, 'Medieval and Modern Understandings of Symbolism: A Comparison', *Speculum*, 54:2 (1979), 223–56.
Laqueur, T., *Making Sex: Body and Gender from the Greeks to Freud* (Cambridge, MA, and London: Harvard University Press, 1990).
Latham, R. E., *Dictionary of Medieval Latin from British Sources* (London: published for the British Academy by Oxford University Press, 1975–[*c.* 1989]).
Lawrence, A., 'Cistercian Decoration: Twelfth-Century Legislation on Illumination and Its Interpretation in England', *Reading Medieval Studies*, 21 (1995), 31–52.
Lawrence, C. H., *Medieval Monasticism: Forms of Religious Life in Western Europe in the Middle Ages* (London and New York: Longman, 1984).
Le Goff, J., *History and Memory*, trans. by S. Rendall and E. Claman (New York: Columbia University Press, 1992).
——, *Intellectuals in the Middle Ages*, trans. by T. L. Fagan (Cambridge, MA: Basil Blackwell, 1993).
——, *L'Imaginaire médiévale: essais* (Paris: Editions Gallimard, 1985).
——, *The Birth of Purgatory*, trans. by A. Goldhammer (Chicago: University of Chicago Press, 1984).
Leask, G. W., *Irish Churches and Monastic Buildings* (Dundalk: Dundalgan Press, 1955).
Leclercq, J., 'Aspects de la vie cistercienne au XIIIe siècle. A propos d'un livre recent', *Studia Monastica*, 20 (1978), 221–26.
——, 'Comment vivaient les frères convers?', *Analecta Cisterciensia*, 21 (1965), 239–58.
——, 'Documents sur les fugitifs', in *Analecta Monastica*, Studia Anselmiana 54, 7th series (Rome: Pontificium Institutum S. Anselmi; Orbis Catholicus Herder, 1965), pp. 87–145.
——, La vie parfaite. Points de vue sur l'essence de l'état religieux (Paris/Turnhout: Brepols, 1948).
——, 'Le cloître est-il une prison?', *Revue d'Ascétique et de Mystique*, 47 (1971), 407–20.
——, 'Le cloître est-il un paradis?', in *Le message des moines à notre temps* (Paris: A. Fayard, 1958), pp. 141–59.
——, 'Lettres de vocation à la vie monastique', *Studia Anselmiana 37*, Analecta Monastica 3rd series (Rome: Pontificium Institutum S. Anselmi: Orbis Catholicus Herder, 1955).
——, 'The Joy of Dying According to St Bernard', *Cistercian Studies*, 25 (1990), 163–74.

——, *Receuil d'études sur Saint Bernard et ses écrits*, 3 vols (Rome: Edizioni di Storia e Letteratura, 1962–69).
——, *The Love of Learning and the Desire for God: A Study of Monastic Culture*, trans. by C Misrahi (New York: Fordham University Press, 1974).
——, 'The Imitation of Christ and the Sacraments in the Teaching of St. Bernard', *Cistercian Studies*, 9 (1974), 36–54.
Lefebvre, H., *The Production of Space*, trans. by D. Nicholson-Smith (Oxford: Basil Blackwell, 1991).
Lehmann-Brockhaus, O., *Lateinische Schriftquellen zur Kunst in England, Wales und Schottland vom Jahre 901 bis zum Jahre 1307* (Munich: Prestel, 1956–60).
Lekai, L., 'Ideals and Reality in Early Cistercian Life and Legislation', in *Cistercian Ideals and Reality*, ed. by J. Sommerfeldt, Cistercian Studies Series 60 (Kalamazoo: Cistercian Publications, 1978).
——, *The Cistercians: Ideals and Reality* (Kent, OH: Kent State University Press, 1977).
Leroux-Dhys, J.-F., *Cistercian Abbeys: History and Architecture* (Köln: Konemann, 1998).
Lescher, B., 'Laybrothers: Questions Then, Questions Now', *Cistercian Studies*, 23 (1988), 63–85.
Levi-Strauss, C., *The Savage Mind* (London: Weidenfeld and Nicolson, 1972).
Leyser, H., *Hermits and the New Monasticism* (New York: St. Martin's Press, 1984).
Little, L. K., *Religious Poverty and the Profit Economy in Medieval Europe* (Ithaca and New York: Cornell University Press, 1978).
Locke, F. M., 'A New Date for the Composition of the Tractatus de Purgatorio Sancti Patricii', *Speculum*, 40 (1965), 641–46.
Logan, F. D., *Runaway Religious in Medieval England c.1240–1540* (Cambridge: Cambridge University Press, 1996).
——, *Excommunication and the Secular Arm in Medieval England: A Study in Legal Procedure from the Thirteenth to the Sixteenth Century* (Toronto: Pontifical Institute of Mediaeval Studies, 1968).
Louth, A., *The Origins of the Christian Mystical Tradition From Plato to Denys* (Oxford: Clarendon Press, 1981).
Madan, F., *Summary Catalogue of Western MSS in the Bodleian Library*, 7 vols (Oxford: Clarendon Press, 1895–1953).
Madden, E., 'Business Monks, Banker Monks, Bankrupt Monks: The English Cistercians in the Thirteenth Century', *The Catholic Historical Review*, 49:3 (1963), 341–64.
Mahn, J.-B., *L'Ordre cistercien et son gouvernement des origines au milieu du XIIIe siècle 1098–1265*, 2nd edn (Paris: Boccard, 1951).
Mâle, E., *The Gothic Image. Religious Art in France of the Thirteenth Century*, trans. by D. Nussey (London: Fontana Library/Collins, 1961).
Mansfield, M. C., *The Humiliation of Sinners: Public Penance in Thirteenth-Century France* (Ithaca and London: Cornell University Press, 1995).
Marks, R., 'Cistercian Window Glass in England and Wales', in *Cistercian Art and Architecture in the British Isles*, ed. by C. Norton and D. Park (Cambridge: Cambridge University Press, 1986), pp. 211–27.
Markschies, C., *Gibt es eine "Theologie der Gotischen Kathedrale"?* (Heidelberg: Abhandlungen der Heidelberger Akademie der Wissenschaft, 1995).

Marosszeki, S., 'Les origines du chant cistercien' *ASOC,* 8 (1952), 41–46.

McCaffrey, H., 'The Meaning of Maundy According to St. Bernard', in *The Chimera of His Age: Studies in Bernard of Clairvaux*, ed. by E. R. Elder and J. R. Sommerfeldt, Cistercian Studies Series 63 (Kalamazoo: Cistercian Publications, 1980), pp. 140–46.

McClung, W. A., *The Architecture of Paradise, Survivals of Eden and Jerusalem* (Berkeley: University of California Press, 1983).

McGinn, B., 'Freedom, Formation and Reformation: The Anthropological Roots of St. Bernard's Spiritual Teaching', *Analecta Cisterciensia,* 46 (1990), 91–122.

——, *Three Treatises on Man: A Cistercian Anthropology* (Kalamazoo: Cistercian Publications, 1977).

McGuire, B. P., 'An Introduction to the Exordium Magnum Cisterciense', *Cistercian Studies,* 27:4 (1992), 277–97.

——, 'Purgatory, the Communion of Saints and Medieval Change', *Viator,* 20 (1989), 61–84.

——, 'Self Denial and Self Assertion in Arnulf of Villers', *Cistercian Studies,* 28:3/4 (1993), 241–59.

——, 'The Cistercians and the Rise of the Exemplum in Early Thirteenth Century France: A Reevaluation of Paris BN MS Lat. 15192', *Classica et Medievalia,* 34 (1983), 211–67.

——, 'The Cistercians and the Transformation of Monastic Friendship', *Analecta Cisterciensia,* 37 (1981), 3–62.

——, *Friendship and Community; The Monastic Experience 350–1250* (Kalamazoo: Cistercian Publications, 1988).

McLaughlin, M., *Consorting with Saints: Prayer for the Dead in Early Medieval France* (Ithaca and London: Cornell University Press, 1994).

McNulty, J., 'Stephen of Eston, Abbot of Sawley, Newminster and Fountains', *Yorkshire Archaeological Journal,* 31 (1934), 49–64.

Meehan, B., 'Durham Twelfth Century MSS in Cistercian Houses', in *Anglo-Norman Durham*, ed. by M. Harvey and M. Prestwich (Woodbridge: Boydell Press, 1994).

Meeks, W., *The Origins of Christian Morality: The First Two Centuries* (New Haven and London: Yale University Press, 1993).

Memory, History, Culture and the Mind, ed. by T. Butler (Oxford and New York: Blackwell, 1989).

Meyvaert, P., 'The Medieval Monastic Claustrum', *Gesta,* 12 (1973), 53–59.

Milbank, J., *Theology and Social Theory: Beyond Secular Reason* (Oxford and Cambridge, MA: Basil Blackwell, 1990).

Milis, L. J., *Angelic Monks and Earthly Men: Monasticism and its Meaning to Medieval Society* (Woodbridge: Boydell Press, 1992).

Miller, E., 'Farming in Northern England during the Twelfth and Thirteenth Centuries', *Northern History,* 11 (1975), 1–16.

——, and J. Hatcher, *Medieval England—Rural Society and Economic Change 1086–1348* (London: Longman, 1978).

Miller, T., 'The Knights of St. John and the Hospitals of the Latin West', *Speculum,* 53 (1978), 709–33.

Minois, G., *History of Old Age*, trans. by S. Hanbury Tenison (Chicago and Cambridge: Polity Press, 1989).

Mohanty, J., 'Foucault as a Philosopher', in *Foucault and the Critique of Institutions*, ed. by J. Caputo and M. Yount (Philadelphia: Pennsylvania State University Press, 1993), pp. 27–40.

Moore, R. I., *The Formation of A Persecuting Society: Power and Deviance in Western Europe 950–1250* (Oxford: Basil Blackwell, 1987).

Moorhouse, S., 'Monastic Estates, their Composition and Development', in *The Archaeology of Rural Monasteries*, ed. by R. Gilchrist and H. Mytum, British Archaeological Reports, British series 203 (Oxford: BAR, 1989), pp. 29–81.

——, and S. Wrathmell, *Kirkstall Abbey vol. 1. The 1950–64 Excavations: A Reassessment* (Wakefield: West Yorkshire Archaeology Service, 1987).

Morant, R. W., *The Monastic Gatehouse and other Types of Portal of Medieval Religious Houses* (Lewes: Book Guild, 1995).

Morson, J., 'Cistercian MSS from the Collection of Sir Sydney Cockerell', *Collectanea Cisterciensia*, 21 (1959), 330–33.

——, 'The English Cistercians and the Bestiary', *Bulletin of the John Rylands Library*, 39 (1956), 146–90.

Moulin, L., *La vie quotidienne des religieux du moyen âge* (Paris: Hachette, 1978).

Mozely, J. H., 'Susanna and the Elders: Three Medieval Poems', *Studi Medievali*, 3 (1930), 27–52.

Mullin, F. A., *A History of the Work of the Cistercians in Yorkshire* (Washington: Catholic University of America, 1932).

Murray, A., 'Confession Before 1215', *Transactions of the Royal Historical Society*, 6:3 (1993), 51–81.

Murray-Jones, P., *Medieval Medical Miniatures* (London: British Library in association with the Wellcome Institute for the History of Medicine, 1984).

Mytum, H. C., 'Functionalist and Non-Functionalist Approaches in Monastic Archaeology', in *The Archaeology of Rural Monasteries*, ed. by R. Gilchrist and H. Mytum, British Archaeological Reports, British series 203 (Oxford: BAR, 1989), pp. 339–57.

Newman, M. G., *The Boundaries of Charity: Cistercian Culture and Ecclesiastical Reform 1098–1180* (Stanford: Stanford University Press, 1996).

The New Medievalism, ed. by M. S. Brownlee, K. Brownlee, and S. G. Nichols (Baltimore: Johns Hopkins University Press, 1991).

Niermeyer, J. F., *Mediae Latinitatis Lexicon Minus. Abbreviationes et Index Fontium* (Leiden and New York: E. J. Brill, 1993).

Noisette, P., 'Usage et représentation de l'espace dans la *Regula Benedicti*', *Regulae Benedicti Studia*, 14/15 (1985/86), 69–80.

Nolan, B., *The Gothic Visionary Perspective* (Princeton: Princeton University Press, 1977).

Nora, P., 'Between Memory and History: Les Lieux de Mémoire', *Representations*, Special Issue, 26 (1989), 7–25.

Orme, N., and M. Webster, *The English Hospital 1050–1570* (New Haven and London: Yale University Press, 1995).

Oxford, A., *The Ruins of Fountains Abbey* (Oxford: Oxford University Press, 1910).

The Oxford History of the Prison: The Practice of Punishment in Western Society, ed. by N. Morris and D. J. Rothman (New York and Oxford: Oxford University Press, 1995).

Panofsky, E., *Abbot Suger on the Abbey Church of Saint-Denis and its Art Treasures*, 2nd edn (Princeton: Princeton University Press, 1979).

———, *Gothic Architecture and Scholasticism* (New York: Meridian Books, 1957).

———, *Studies in Iconology: Humanistic Themes in the Art of the Renaissance* (New York and Oxford: Oxford University Press, 1939).

Park, C., *Sacred Worlds: An Introduction to Geography and Religion* (London and New York: Routledge and Kegan Paul, 1994).

Parkes, M. B., *The Medieval Manuscripts of Keble College Oxford: A Descriptive Catalogue with Summary Descriptions of the Greek and Oriental Manuscripts* (London: Scolar Press, 1979).

Parsons Lillich, M., 'Constructing Utopia' in *Studies in Cistercian Art and Architecture II*, ed. by M. Parsons Lillich (Kalamazoo: Cistercian Publications, 1984), pp. xi–xiv.

Paster, G. Kern, *The Body Embarrassed: Drama and the Discipline of Shame in Early Modern England* (Ithaca and London: Cornell University Press, 1993).

Patterson, L., *Negotiating the Past: The Historical Understanding of Medieval Literature* (Madison: University of Wisconsin Press, 1987).

———, 'On the Margin: Postmodernism, Ironic History and Medieval Studies', *Speculum*, 65 (1990), 87–108.

Paxton, F., 'Anointing the Sick and the Dying in Christian Antiquity and the Early Medieval West', in *Health, Disease and Healing in Medieval Culture*, ed. by S. Campbell, B. Hall, and D. Klausner (New York: St Martin's Press, 1992), pp. 93–102.

———, *Christianizing Death: The Creation of a Ritual Process in Early Medieval Europe* (Ithaca and London: Cornell University Press, 1990).

Peers, C., *Byland Abbey* (London: HMSO, 1952).

———, *Rievaulx Abbey* (London: HMSO, 1967).

Pelikan, J., *The Growth of Medieval Theology (600–1300)* (Chicago and London: University of Chicago Press, 1978).

Penco, G., 'Monasterium-Carcer', *Studia Monastica*, 8 (1966), 133–43.

Pevsner, N., *The Buildings of England, Yorkshire North Riding* (Harmondsworth: Penguin Books, 1966).

Platt, C., *The Monastic Grange in England: A Reassessment* (New York: Fordham University Press, 1969).

Porter, H. B., 'The Origin of the Medieval Rite for Anointing the Sick and the Dying', *Journal of Theological Studies*, 7 (1956), 211–25.

Porter, R., 'History of the Body', in *New Perspectives on Historical Writing*, ed. by P. Burke (Cambridge: Polity Press, 1991), pp. 206–32.

Postles, D., 'Monastic Burials of Non-Patronal Lay Benefactors', *Journal of Ecclesiastical History*, 47:4 (1996), 620–37.

Poucelle, M.-C., *The Body and Surgery in the Middle Ages*, trans. by R. Morris (Cambridge: Polity Press in association with Basil Blackwell, 1990).

The Life of Aelred of Rievaulx by Walter Daniel and the Letter to Maurice, trans. by F. M. Powicke and intro. by M. Dutton (Kalamazoo: Cistercian Publications, 1994).

Prescott, E., *The English Medieval Hospital 1050–1640* (London: Seaby, 1992).

Pressouyre, L., 'St. Bernard to St. Francis: Monastic Ideals and Iconographic Programmes in the Cloister', *Gesta*, 12 (1973), 71–92.

Progress and Problems in Medieval England: Essays in Honour of E. Miller, ed. by R. Britnell and J. Hatcher (Cambridge: Cambridge University Press, 1996).

Pugh, R., *Imprisonment in Medieval England* (London: Cambridge University Press, 1968).
Raciti, G., 'Une allocution familière de S. Aelred conservée dans les Mélanges de Matthieu de Rievaulx', *Collecteanea Cisterciensia,* 47 (1985), 267–80.
Ramsay, N., 'Artists, Craftsmen and Design in England 1200–1400', in *Age of Chivalry. Art in Plantagenet England 1200–1400,* ed. by J. Alexander and P. Binski (London: Royal Academy of Arts in association with Weidenfeld and Nicholson, 1987), pp. 49–54.
Rhodes, J. T., and C. Davidson, 'The Garden of Paradise', in *The Iconography of Heaven,* ed. by C. Davidson (Kalamazoo: Medieval Institute Publications, 1994), pp. 69–109.
Riddle, J. M., 'Theory and Practice in Medieval Medicine', *Viator,* 5 (1974), 157–84.
Rider, J., 'Other Voices: Historicism and the Interpretation of Medieval Texts', *Exemplaria,* 1:2 (1989), 293–312.
Riesner, A. J., *Apostates and Fugitives from Religious Houses,* Canon Law Studies 168 (Washington: Catholic University of America, 1942).
Rigg, A. G., *A History of Anglo-Latin Literature 1066–1422* (Cambridge and New York: Cambridge University Press, 1992).
Rigold, S. E., *Chapter House and Pyx Chamber, Westminster Abbey* (London, HMSO, 1988).
Robinson, D., *The Cistercian Abbeys of Britain: Far from the Concourse of Men* (London: B. T. Batsford, 1998).
Roehl, R., 'Plan and Reality in a Medieval Monastic Economy', *Studies in Medieval and Renaissance History,* 9 (1972), 83–113.
Roper, S. E., *Medieval English Benedictine Liturgy: Studies in the Formation, Structure and Content of the Monastic Votive Office, c.950–1540* (New York and London: Garland, 1993).
Rösener, W., 'Abbot Stephen Lexington and his Effort for Reform of the Cistercian Order in the Thirteenth Century', in *Goad and Nail: Studies in Medieval Cistercian History 10,* ed. by E. R. Elder (Kalamazoo: Cistercian Publications, 1985), pp. 46–55.
Rossi, M., and A. Rovetta, 'Indagini sull spazio ecclesiale immagine della Gerusalemme celeste', in *La Gerusalemme Celeste. Catalogo della mostra, Milano Universita Cattolica del S. Cuore, 20 maggio–5 giugno 1983,* ed. by M. L. Gatti Perer (Milan: Vita e Pensiero, 1983), pp. 77–118.
Rough Register of Acquisitions of the Department of Manuscripts, British Library, 1971–75 (London: Swift, 1977).
Rouse, R. H., and M. A. Rouse, 'Biblical Distinctions in the Thirteenth Century', *Archives d'Histoire Doctrinale et Littéraire du Moyen Age,* 41 (1974), 27–37.
Rubin, M., *Corpus Christi: The Eucharist in Late Medieval Culture* (Cambridge: Cambridge University Press, 1991).
Rudolph, C., 'Bernard of Clairvaux's *Apologia* as a Description of Cluny, and the Controversy over Monastic Art', *Gesta,* 27 (1988), 125–32.
——, *The Things of Greater Importance: Bernard of Clairvaux's Apologia and the Medieval Attitude Toward Art* (Philadelphia: University of Pennsylvania Press, 1990).
——, *Violence and Daily Life: Reading, Art, and Polemics in the Citeaux Moralia in Job* (Princeton: Princeton University Press, 1997).

Sayers, J., 'Violence in the Medieval Cloister', *Journal of Ecclesiastical History,* 41 (1990), 533–42.

Schaefer, T., *Die Fusswaschung in Monastischen Brauchtum und in der Lateinische Liturgie*, Texte und Arbeiten herausgegeben durch die Erzabtei Beuron, I, 47 (Beuron in Hohenzollern: Erzabtei Beuron, 1956).

Schama, S., *Landscape and Memory* (London: Harper Collins, 1995).

Schilla, C. E., 'Meaning and the Cluny Capitals: Music as Metaphor', *Gesta,* 27 (1988), 133–48.

Schmitt, J.-C., *Les revenants: les vivants et les morts dans la société médiévale* (Paris: Editions Gallimard, 1994).

Schneider, B., 'Cîteaux und die Benediktinische Tradition: Die Quellenfrage des *Liber Usuum* im Lichte der *Consuetudines Monasticae*', *ASOC,* 16 (1960), 169–254; *ASOC,* 17 (1961), 73–111.

Sedlmayr, H., *Die Entstehung der Kathedrale* (Graz: Akademische Druck- u. Verlagsanstalt, 1976).

Seidel, L., 'Medieval Cloister Carving and the Monastic Mentalité', in *The Medieval Monastery*, ed. by A. MacLeish (St Cloud, MN: North Star Press of St Cloud, 1988), pp. 1–16.

Seiler, R., 'Mittelalterliche Medizin und Probleme der Jensietsvorsorge', in *Himmel, Hölle und Fegefeuer. Das Jenseits in Mittelalter*, ed. by P. Jezler et al. (Zurich: Schweizerisches Landesmuseum, 1994), pp. 117–24.

Sharpe, J. A., 'The History of Violence in England: Some Observations', *Past and Present,* 108 (1985), 206–15.

Siegel, R. E., *Galen's System of Physiology and Medicine: An Analysis of his Doctrines and Observations on Bloodflow, Respiration, Humors and Internal Diseases* (Basel and New York: Karger, 1968).

Simson, O. von, 'The Cistercian Contribution', in *Monasticism and the Arts*, ed. by T. Verdon (New York: Syracuse University Press, 1984), pp. 115–37.

——, *The Gothic Cathedral: Origins of Gothic Architecture and the Medieval Concept of Order* (New York: Pantheon Books, 1956).

Skeat, T. C., *The Catalogues of the Manuscript Collections, The British Museum* (London: Trustees of the British Museum, 1962).

Smith, J. Z., *Map is Not Territory* (Chicago: University of Chicago Press, 1993).

Sommerfeldt, J. R., 'Epistemology, Education and Social Theory in the Thought of St. Bernard of Clairvaux', in *Saint Bernard of Clairvaux: Studies Commemorating the Eighth Centenary of his Canonization*, ed. by M. B. Pennington (Kalamazoo: Cistercian Publications, 1977), pp. 169–79.

——, *The Spiritual Teaching of Bernard of Clairvaux: An Intellectual History of the Early Cistercian Order* (Kalamazoo: Cistercian Publications, 1991).

Southern, R. W., 'Between Heaven and Hell', *Times Literary Supplement*, 18 June 1982, pp. 651–52.

——, *The Making of the Middle Ages* (London: The Cresset Library, 1987).

Stalley, R., *The Cistercian Monasteries of Ireland: An Account of the History, Art, and Architecture of the White Monks in Ireland from 1142–1540* (London and New Haven: Yale University Press, 1987).

Standaert, M., 'La doctrine de l'image chez S. Bernard' *Ephemerides Theologicae Lovainenses,* 23 (1947), 70–129.

Stegmuller, F., *Repertorium Biblicum Medii Aevi* (Madrid: Casimiro, 1940–50).

Stemmler, T., *Liturgische Feiern und geistliche Spiele: Studien zur Erscheinungsformen des Dramatisches im Mittelalter* (Tübingen: M. Niemeyer, 1970).

Stepsis, R., 'Fulfilment of the Self and Union with God in the Writings of Bernard of Clairvaux', *American Benedictine Review,* 24 (1973), 348–64.

Stephen of Sawley Treatises, ed. by B. K. Lackner and trans. by J. K. O'Sullivan, Cistercian Fathers Series 36 (Kalamazoo: Cistercian Publications, 1984).

Stewart, S., *The Enclosed Garden: The Tradition and Image in Seventeenth-Century Poetry* (Madison: University of Wisconsin Press, 1966).

Stiegman, E., 'Analogues of the Cistercian Abbey Church', in *The Medieval Monastery*, ed. by A. MacLeish (St Cloud, MN: North Star Press of St Cloud, 1988), pp. 17–33.

Stock, A., 'A Sounding Vase at Fountains Abbey?', *Cistercian Studies,* 23 (1988), 190–91.

Stock, B., *The Implications of Literacy: Written Language and Models of Interpretation in the Eleventh and Twelfth Centuries* (Princeton: Princeton University Press, 1983).

Stoddard, W., *Monastery and Cathedral in France: Art and Architecture in Medieval France* (Middletown, CT: Wesleyan University Press, 1972).

Stone, L., 'Interpersonal Violence and English Society, 1300–1980', *Past and Present,* 101 (1983), 22–33.

Stookey, L. H., 'The Gothic Cathedral as the Heavenly Jerusalem: Liturgical and Literary Sources', *Gesta,* 8 (1969), 35–41.

Talbot, C. H., 'A List of Cistercian Manuscripts in Great Britain', *Traditio,* 8 (1952), 402–16.

——, 'Monastic Infirmaries', *St. Mary's Hospital Gazette*, 67:1 (1961), 14–18.

——, and E. A. Hammond, *The Medical Practitioners of Medieval England: A Biographical Register* (London: Wellcome Historical Medical Library, 1965).

Taylor, J., *Medieval Historical Writing in Yorkshire* (York: St Anthony's Press, 1961).

Teasdale, W., 'A Glimpse of Paradise: Monastic Space and Inner Transformation', *Parabola,* 18 (1993), 59–62.

Temkin, O., *Galenism: Rise and Decline of a Medical Philosophy* (Ithaca: Cornell University Press, 1973).

Tentler, T. T., *Sin and Confession on the Eve of the Reformation* (Princeton: Princeton University Press, 1977).

Thompson, B., 'From Alms to Spiritual Services: The Function and Status of Monastic Property in Medieval England', in *Monastic Studies II*, ed. by J. Loades (Bangor: Headstart History, 1991), pp. 227–62.

Thompson, S., 'The Problem of Cistercian Nuns in the Twelfth and Early Thirteenth Centuries', in *Medieval Women*, ed. by D. Baker (Oxford: Basil Blackwell, 1978), pp. 227–52.

——, *Women Religious: The Founding of English Nunneries after the Norman Conquest* (Oxford: Oxford University Press, 1991).

Thorndike, L., and P. Kibre, *A Catalogue of Incipits of Mediaeval Scientific Writings in Latin* (London: Medieval Academy of America, 1963).

Tobin, S, *The Cistercians: Monks and Monasteries of Europe* (London: Herbert, 1995).

Turner, V., *Drama, Fields and Metaphors* (Ithaca: Cornell University Press, 1974).

——, *The Forest of Symbols: Aspects of Ndembu Ritual* (Ithaca: Cornell University Press, 1967).

———, *The Ritual Process* (Chicago: Aldine, 1969).
Van Damme, J. B., 'Les pouvoirs de l'abbé de Cîteaux aux XIIe et XIIIe siècles', *Analecta Cisterciensia,* 24 (1968), 47–85.
Van Engen, J., 'The "Crisis of Cenobitism" Reconsidered: Benedictine Monasticism in the Years 1050–1150', *Speculum,* 61 (1986), 269–304.
Veilleux, A., 'The Interpretation of A Monastic Rule', in *The Cistercian Spirit: A Symposium in Memory of Thomas Merton,* ed. by M. B. Pennington (Washington: Cistercian Publications, 1973).
Venturi, R., *Complexity and Contradiction in Architecture* (New York: Museum of Modern Art, 1977).
The Victoria History of the County of York, ed. by W. Page (London: A. Constable, 1907–1913).
Vogüé, A. de, 'Eucharistie et vie monastique,' *COCR,* 48 (1986), 120–30.
Waddell, C., 'Peter Abelard's Letter 10 and Cistercian Liturgical Reform', in *Studies in Medieval Cistercian History 2,* ed. by J. Sommerfeldt (Kalamazoo: Cistercian Publications, 1976), pp. 75–86.
———, 'The Cistercian Institutions and their Early Evolution: Granges, Economy, Lay Brothers', in *L'Espace cistercien,* ed. by L. Pressouyre (Paris: Comité des Travaux Historiques et Scientifiques, 1994), pp. 27–38.
———, 'The Glorified Christ, Presence and Future: the Eschatological Dynamic of the Spiritual Life', *Analecta Cisterciensia,* 46 (1990), 327–40.
———, 'The Origin and Early Evolution of the Cistercian Antiphonary: Reflections on Two Cistercian Chant Reforms', in *The Cistercian Spirit: A Symposium in Memory of Thomas Merton,* ed. by M. Basil Pennington (Washington: Cistercian Publications, 1973).
———, 'The Place and Meaning of Work in Twelfth Century Cistercian Life' *Cistercian Studies,* 23:1 (1988), 25–44.
———, 'The Pre-Cistercian Background of Cîteaux and the Cistercian Liturgy', in *Goad and Nail: Studies in Medieval Cistercian History 10,* ed. by E. R. Elder (Kalamazoo: Cistercian Publications, 1985), pp. 109–32.
———, 'Towards a New Provisional Edition of the Statutes of the Cistercian General Chapter, c.1119–1189', in *Studiosorum Speculum: Studies in Honour of Louis J. Lekai O. Cist.,* ed. by F. R. Swietek and J. Sommerfeldt, Cistercian Studies Series 141 (Kalamazoo: Cistercian Publications, 1993), pp. 384–419.
Wakefield, W. L., 'Burial of Heretics in the Middles Ages', *Heresis,* 5 (1985), 29–32.
Walbran, J., 'On the Recent Excavations at Sawley Abbey in Yorkshire', in *Memorials of the Abbey of St Mary's of Fountains,* ed. by J. Walbran, 3 vols (London: Publications of the Surtees Society, 1876), 67, pp. 159–77.
Wallis, F., 'Medicine in Medieval Calendar MSS', in *Manuscript Sources of Medieval Medicine,* ed. by M. R. Schleissner (New York and London: Garland, 1995), pp. 105–43.
Wander, S. H., 'The York Chapter House', *Gesta,* 17:2 (1978), 41–49.
Ward, H. D. L., and J. A. Herbert, *A Catalogue of Romances in the Department of Manuscripts in the British Museum,* 3 vols (London: Trustees of the British Museum, 1883–1910).
Wardrop, J., *Fountains Abbey and Its Benefactors 1132–1300,* Cistercian Studies Series 91 (Kalamazoo: Cistercian Publications, 1987).

Watson, A., *The Manuscripts of Sir Henry Savile of Banke* (Oxford: Bodleian Library, 1969).

Watson, A. G., *Catalogue of Dated and Datable Manuscripts, c. 700–1600 in the Department of Manuscripts, the British Library*, 2 vols (London: The British Library, 1979).

———, *Dated and Datable Manuscripts c. 435–1600 in Oxford Libraries*, 2 vols (Oxford: Clarendon Press, 1984).

Webb, G., *Architecture in Britain: The Middle Ages* (Harmondsworth: Penguin Books, 1956).

Werner, E., 'Bemerke zu einer neuen These über die Herkunft der Laienbrüder', *Zeitschrift fur Geschichtswissenschaft,* 6 (1958), 355–59.

Whittingham, S., *Salisbury Abbey Chapter House* (Salisbury: The Dean and Chapter, 1989).

Williams, D., 'Layfolk within Cistercian Precincts', in *Monastic Studies II*, ed. by J. Loades (Bangor: Headstart History, 1990), pp. 87–117.

Wilmart, A., *Auteurs spirituels et textes dévots du Moyen Age latin: études d'histoire littéraire* (Paris: Etudes Augustiniennes, 1932).

Wilson, C., 'Cistercians as "Missionaries of Gothic"', in *Cistercian Art and Architecture in the British Isles*, ed. by C. Norton and D. Park (Cambridge: Cambridge University Press, 1986), pp. 86–116.

Wirth, J., *L'Image médiévale. naissance et développements (VIe–XVe siècles)* (Paris: Méridiens Klincksieck, 1989).

Wood, S., *English Monasteries and their Patrons in the Thirteenth Century* (Oxford: Oxford University Press, 1955).

Wrathmell, S., *Kirkstall Abbey The Guest House* (Leeds: West Yorkshire Archaeology Service, 1987).

Yorkshire Monasticism: Archaeology, Art and Architecture, from the 7th to 16th Centuries, ed. by L. R. Hoey (London and Leeds: British Archaeological Association and W. S. Maney, 1995).

Zaleski, C., *Otherworld Journeys: Accounts of Near-Death Experience in Medieval and Modern Times* (New York and Oxford: Oxford University Press, 1987).

———, 'St. Patrick's Purgatory: Pilgrimage Motifs in a Medieval Otherworld Vision', *Journal of the History of Ideas*, 46:4 (1985), 467–85.

Zinn, G. A., Jr., 'Hugh of St. Victor and the Art of Memory', *Viator,* 5 (1974), 211–34.

Index

Abbey Dore 107, 113
Abelard 3, 58
Adam and Eve 200
Adam, forester of Clifford 186
Adam, Abbot of Fountains 113
   Medical Practioners of Medieval
      England (MPME) 146
Aelred of Rievaulx 18, 37, 95, 113,
   197, 212 n. 58
   and chapel 80
   death of 219-21, 222
   holiness of 68
   MPME 146
   on music 100, 101
   shrine of 240
   Speculum Caritatis 39
Aiguebelle
   extumulato of *conversus* 230
Aire, river
   Brother Roger of Rievaulx drowned
      in 222
Alan of Meaux
   poem on Susanna and the elders 67
Alexander, monk
   friendship with lay Brother Roger
      192
Alexander IV, pope 120
Alexander VIII, pope 179
Alice de Gant 236
altar 59, 61, 76, 99, 239 *see also*
   Byland abbey, chapter house,

   church, Durham cathedral,
      Fountains, Rievaulx
   in chapter house 105
   tombs near 218
ambition 199-200
Ampleforth 17
Anselm of Gembloux 147 n. 37
Apologia ad Guillelmum Abbatem 73
apostasy 21, 31, 37 n. 41, 198-99
   meanings 202-08
   responses to 209-16
apostate(s) 249, 252
   arrest of 185
   beating of 126
   body of 198
   fugitives 196-202
Aristotle
   De Physiognomia, Meteorum anon.
      commentary on 157
Aurelian on the Psalms 101 n. 80

Beaulieu, Cistercian house 31-32 n. 20,
   173
Bebenhausen, abbot of 232
Bell, David 11 n. 46, 16, 133
Benedict, monk from Fountains 224
Benedictine preachers 64
Bernard of Clairvaux (St Bernard) 3,
   13, 14, 39, 69, 101 n. 80, 213
   Apologia 94
   body 163-64

death of 221
importance of seeing 98
Bernard, prior of Newburgh 145 n. 28
Beverley Minster 65
　prior of 163
Biddick, Kathleen 8
body 4, 127, 199
　and bloodletting 147-60
　custodians of 144-47
　and death 225, 227-29, 233, 237, 251, 252, 240, 280
　diseased 144
　humiliation of 209
　humours 151
　and landscape 164-7
　in meditation 41 n. 58
　monastic, 98
　pain of 219
　punishment 125
　pleasures of 36
　social and corporate 129-32
　subjective, the 160-64
body of Christ 93, 98
Bonacumbae, Brother William of 183
　lay brother exiled from 123
Boniface VIII
　Liber Sextus 202
Bonneval, monk of 230
Boscanio, stealing of abbatial seal from 183
Boston (Lincs.), fair at 177 n. 33, 178
boundaries 4, 9, 21, 23, 164, 193, 195, 215, 216, 240, 249, 252
　between life and death 223
　imagined 169
　monastic 45, 58
　physical 25, 28-32
　of social systems 33
　violating the 196-98
Bourdieu, Pierre 5, 6-7
Brown, 'Capability' 17
burial 21, 217 n. 1, 218, chapter 8 *passim*
Burke, Peter 6 n. 22
Burton, Janet 12 n. 49
Byland Abbey 10, 12, 17, 95
　altar 76, 78 fig. 3.3
　chapter house 107, 108, 114
　church 76, 77 fig. 3.2, 79 fig. 3.4, 82 fig. 3.5
　cloister 49, 61

east end 90
gatehouse 28, 29 fig. 1.4, 30
infirmary 133 n. 1, 134, 137, 138 fig. 5.3
lay brothers' 'lane' 52, 55 fig. 2.6, 173, 174 fig. 6.3
manuscript 157
precinct 26 fig. 1.2, 28
tombs 252
west range 174 fig. 6.2

Caesarius of Heisterbach 68, 196, 197
　Dialogus Miraculorum 70, 168, 188, 203-05, 212, 225
　lay brother from 123
Cambridge, Jesus College Library MS QB17 44
Cantor, Peter 73
Carciloci, abbot of 142
Carta Caritatis 28, 111
Casanova, abbey of 141
Casey, Michael 129-31
Cayton village 235
cemetery 1, 2, 70, 217, 229, 230, 237
Certeau, Michel de 3-5, 200-01
chapter house 10, 21, 34, 49, 56 fig. 2.7, 54, Chapter 4 *passim*, 143, 144, 148, 171, 249, 252
　altar in 105
　tombs 218
charity 39
Christ 128, 181, 182 *see also* body of Christ
　burial 42
　flogging of 131
　journeying 60
　in Majesty, figure of 80
　meditation on Passion and life 41-42, 44, 96-98
　praise of 142
　resurrection 42
　source of light 81, 99, 160
　strength in 162
　trial 118
　vision of Gerard 222
　washing feet 63
Christian, the monk 47
Christiana de Arncliffe 233
church 14, 21, 34, 49, 61, Chapter 3 *passim*, 70, 71, 217, 226, 227, 240, 249

Index 285

altar 59, 61, 76, 99, 218, 239
architectural iconography 91-96
east end 74, 76-81, 89, 90, 102, 160, 173, 218, 249
as *patria* 60
pulpitum 80
visual importance of 73-76
Cistercian monastery precincts *see also* cemetery, chapter house, church, cloister, granges, infirmary
  court 33
  gatehouse 45
  landscape of 164-65
  precinct 33, 54, 57-58, 64, 196, 197, 249
  refectory 10, 49, 54, 59, 63, 124, 144, 148, 175, 209, 227
  west range 59, 171
Cistercian order
  'Anglicizing' of 185 n. 76
  monastic topos different from France 185 n. 76
Cîteaux 14, 19, 180, 185
  Cluny-Cîteaux dichotomy 94
  General Chapter meetings at 184, 237 n. 86
  heavenly vision 47-48, 71
  west range 173
Clairlieu 232 n. 61
Clairvaux abbey 32 n. 22, 70, 81, 164, 192
  death of St Bernard at 221
  grange 187, 190
  Regula conversorum 169
  west range 173
claustrum 54
cloister 9, Chapter 2 *passim*, 93, 106, 170, 171
  Cluniac plan 48
  infirmary cloisters 141
  inhabitants of infirmary separated from 142, 160
  as 'holy ground' 60
  like a prison 199
  processions in 58-61, 70
  rites in 60-64, 124
  as symbol of paradise 21, 48, 65-71, 75, 200
  of transcendence 45, 48 n. 3
  tombstones 217
  tranquillity of 175

Cockersand 107
Coleman, Janet 38
commemoration 113-16, 127, 132, 237-40, 249
community 21, 25, 31, 43, 52. 213, 214, 219, 220, 226-27, 237, 249, 251, 252
  boundaries 23-33
  in chapter house 106, 108, 111-13, 116, 117, 126
  within cloister 63, 71
  dead as members of 217 n. 1, 224
  discipline 129-34
  lay brother as part of 190
  novices 34-36
  processions 58-61
  ritual 61-63
  segregation of diseased 144
confession 21, 217-22
  advice to novices 35-37
  in chapter house 108, 116, 119-22, 124, 126, 127, 129, 130, 132
Conrad of Eberbach 47, 48
  Exordium Magnum 225
Conrad of Mainz
  decree of 1259 142 n. 15
conversi 54, 63, 117, Chapter 6 *passim*, 209, 214, 223, 251, 252 *see also* lay brothers
  barbati (nickname) 170
Coppack, Glyn 17, 176
cosmos 22, 134, 164, 165, 251
  Africa 159
  body and 151, 155 fig. 5.9, 156 fig. 5.10, 157-60
  Europa 159
  London, BL MS Sloane 793 159 n. 66
Crossley, Paul 91
curiositas 95
  'of the ears' 100

Daniel, Walter 67, 68, 196-97, 199, 212 n. 58, 219-21
  MPME 146
  novitiate 35, 37
  Vita Aelredi 30, 35, 37
Darel, Godfrey, monk of Rievaulx 195, 213, 215
De Apostata Capiendo 185-86, 210-12

death 1, 21, 204, 217 n. 1, Chapter 8
    passim
  of Aelred and other abbots 219-21
  burials 21
  infirmarer's duties 144
  of monks 222-23
  torture 130
decline 14, 14 n. 57, 15, 32, 95
De Pulsu de Philaretus
  commentary on 158
Dialogus Miraculorum (Caesarius of
    Heisterbach) 70, 193
Dijon (Conference 1951) 15
Dimier, Anselme 183
discipline 7, 8, 21, 54, 69 n. 68, 108,
    111, 113, 116-26, 189, 191, 193,
    252
  bodily 160
  breaches of 184
  institutional 130-31
  interior and exterior 150-51
Divine Office 34, 36, 43, 47, 70, 74,
    124, 142, 171, 231
  after bleeding 148
  Matins, Lauds and Prime
    method of psalmody 96
  meditation during 96-100
domesticity 54, 71
Dominicans 47
Donkin, R. A. 9
Donnelly, James 183
Draper, Peter 80 n. 12, 92
Dubois, Jacques 182
Durandus, William 64
Durham, cathedral of
  chapel of nine altars 81
  east end 90 n. 31

Ecclesiastica Officia 20, 30, 31, 63, 63,
    131, 237, 238
  beating 126
  bloodletting 148
  chapter house 105-06, 111, 113, 116,
    120, 122 n. 54
  infirmary, chanting of Hours 142
  instruction for the community on
    death of a monk 226-29
  novices 34-36
  Office of the Dead 237
  servitor infirmorum 144
Eden 66

Eleanor and Lambrinus de Studley
  burial of 233
enclosure 1, 7, 8, 22, 25, 54, 57-58, 59,
    69 n. 68, 165,195-97, 200, 245, 252
  cloister 54, 57-58, 59
  forcible 210
Esau, departure of 204
Eucharist 93
Europe 11
  Europa 159
  insurrection in monasteries 180
excommunication 124
exorcism 61
Exordium Magnum, Conrad of
    Eberbach 70-70, 15, 193, 225

Fergusson, Peter 10
Fitzhenry, Ralph
  effigy of 236
Fitzralph, Baldwin and wife
  son, Adam, given to Kirkstall 239
Foigny, monk from 123
Fontenay
  Cistercian church 102
  grange 183
Ford [abbey] church 231
Fossanova, Italy
  infirmary 141
Foucault, Michel 5, 7, 108, 111, 122 n.
    53, 150, 161, 251
  on discipline 130, 131
Fountains abbey 1, 12, 17, 19, 24 fig.
    1.1, 25, 92, 102, 125, 178, 186,
    191-93, 221, 233, 235, 244 fig. 9.1,
    245
  altars 239
  apostasy 211
  Chapel of Nine Altars 76, 80 n. 12,
    86 fig. 3, 9, 87 fig. 3.10, 88, 89,
    90, 98, 239, 252
  chapter house 56 fig. 2.7, 108, 113
  church 80-81, 85 fig. 3.8, 243
  cloister 25, 49, 51, fig. 2.2, 61, 70,
    243
  court 25, 28
  debts 179
  dispersal of monks 31
  establishment of 23, 67, 111, 161,
    205, 206
  gatehouse 25, 28, 243

Index 287

granges of Morcar and Beverley 176, 179 n. 45 (Morcar)
infirmary 134, 137, 140 fig. 5.5, 141
   lay brothers' infirmary 172
mutiny 167
precinct 25, 28, 243
property 178 n. 37
refectory, serving hatch 175
sheep 177 n. 33
west range 171,172 fig. 6.1
Fourth Lateran Council
Omnis utriusque sexus decree 119
France
   Cistercian houses
      distance between west range and cloister 173
   violence of lay brothers 169, 180
Franciscans 47
freedom 1, 22, 69 n. 68, 103, 131, 165, 181, 196, 245, 252
friendship 129
   after death 225
fugitives 196-202, 203, 209-16
Furness abbey 178 n. 37
   rebellious monks and lay brothers 181

Galen
   medical theories 151, 157, 159
   Tegni 158
Galenic physiology 165
gatekeeper (*portarius*; porter) 30, 31
   subportarius (sub-porter) 30
Geary, Patrick 6 n. 23, 217 n. 1, 224
General Chapter of Cistercians 117, 167-68, 182, 184, 189, 196, 202, 208 n. 46, 237 n. 86
   responses to apostates 209-16
   statutes (*statuta*) 12, 19, 31, 180-81, 184-86, 201, 239
      bloodletting 150
      commemoration 239-40
      on confession 119, 120, 122
      debts 180
      discipline 131
      lay burials 232, 235
      on light 99
      on punishment 124,126
      on singing 100
   wool trade 176, 177 n. 33, 178

Geoffrey of Monmouth
   Historia Regum Anglorum 158
Geoffrey, abbot of St Mary's, York 207, 208
Gerard, monk of Clairvaux 220, 222
Germany 68, 99 n. 67
Gervase of Canterbury 94 n. 47
Gervase and Radulf, monks
   apostasy of 198-99, 207-08
Gethsemane 97
Gilchrist, Roberta 8, 93, 107
   gender and material culture 92
Gospel of St John 221
Gothic 91, 94
   cathedral 92, 93
   church 10
   French Gothic 80
   vaulting 107
granges 141, 171, 179, 180 *see also* Cistercian monasteries, Clairvaux abbey, Fontenay, Fountains
   economy of 178
   as markets 175
   prohibition of drinking on 184
Gratian
   Decretum Gratiani 202
Gregory IX
   Decretals 202, 203 n. 26
   Ne Religiosi Vagandi 212
Gurevich, Aaron 182

Haget, Geoffrey 233
   Usus Conversorum 169, 170, 171
Harrison, Stuart 171
heaven 1, 60, 64, 66, 68, 89, 127, 205
   choirs of 101
   on earth 92, 164
   imagined space of 252
   monk, fool in life 118
heavenly space 47, 63, 69, 81, 91, 164, 249
Hélinand of Froidmont 105
hell 1, 89
   descent into 98
Hemmenrode 203
Henry III
   secret message via lay brother 178
   visit to Fountains abbey 106 n. 5
Henry de Elland 178
Herbert de Arches 236 n. 83

Herbert of Clairvaux 147 n. 37
  Liber Miraculorum 168, 189 n. 95,
    190
Herod 41
Hervey the carpenter 179
Hippocrates
  Prognostica, gloss on 158
Holm Cultram abbey 178 n. 37
Holy Land 81, 238 n. 88
Honorius Augustodunensis 64
Horn, Walter 25
Hovedoe, abbot of 231
Hugh of Kirkstall 19, 23, 31, 67, 205-
    07, 213
  Narratio 161, 168, 191-93, 235
Hugh de Matham, monk 214
humility 38-40, 63, 112, 193
  lay brother epitome of 168, 170
  of St Bernard 186 n. 82, 193
Hungary *see* Sanctae Crucis
iconography
  of Cistercian church 93-96
  of medieval buildings 91-93
  music 101
ideals 13-14, 32, 45
imprisonment 123
infirmary 21, 60, 70, chapter 5 *passim*,
    249, 252 *see also* medical practices
  and discipline 143
  minutor 148, 150
  servitor infirmorum 144
Innocent III, pope 119
intelligence 41 n. 58
intercession 88
Ireland 16
  Jerpoint and Mellifont, abbots of 124
Isabella de Roos 233, 236
Isaiah 41

Jacobus de Voragine 128 n. 88
Jarrow 48
Jeremiah of Ecclesfield 186
Jerusalem 64, 66
  heavenly 92-93
Jervaulx abbey 12, 17, 243, 244, 245
  apostasy 210
  chapter house 56 fig. 2.7, 107, 108,
    252
  church 80, 81, 82 fig. 3.5, 235, 236
  cloister 248 fig. 9.4
  east end 90

  markets and fairs 177-78
  medical MS 158
  Peter de Quincy, expert in medicine
    145
  poverty 179
  west range 247 fig. 9.3
Jocelin of Furness 188
  Life of Waldef 69-70, 117, 189, 193
  on St Patrick's Purgatory 127 n. 83
Johannitius
  and Galenic principles 159
  Isagoge, commentary on 158
John, apostle 70
  fleeing 97
John the Baptist, hymn to 101
John the Evangelist, hymn to 101
John of Ford
  Life of Wulfric of Haselbury 231
John of Kent, abbot 25, 134
John, king 31, 31-32 n. 20
John of Palfleteby, cellarer of Louth
    Park
  apostasy of 185
John, abbot of Rievaulx 115
John le Romeyn, Archbishop of York
    195, 215
Jordan de Normanton, monk 201
Jouy, monk of 123
Judas, betrayal 97

Kinder, Terryl 11, 14
Kirkstall Abbey 12, 17, 89, 121, 233,
    238-39, 250 fig. 9.5, 251 fig. 9.6
  Barsnoldswick grange, case of lay
    brother Peter 186
  chapter house 56 fig. 2.7, 107
  church 245
  cloister 49, 245
  gift of land and villeins 178, 179
  infirmary 134, 135 fig. 5.1, 137
  lay brothers' 'lane' 52
  Medulla Philosophorum manuscript
    157
  precinct 27 fig. 1.3, 25, 28
  presbytery 80, 81 fig. 3.7
  wool trade with Italian merchants 177
    n. 32
Kirkstead abbey 178 n. 37
Klosterkamp, lay brother at 188

labour 54, 71, 251

Index

Lateran Council *see* Fourth Lateran Council
lay brothers (*conversi*) 21, 30, 37 n. 41, 63, 63, 69, 90, 120, 123, Chapter 6 *passim*, 215, 251, 252
  bleeding 148, 150
  death of 222-25
  infirmarer 137, 141
  moral tales 186-91
  punishment 126
  in processions 58, 227, 229
  quarters 49, 52, 55 fig. 2.6
  role in wool trade 176-78
  segregation from choir 102
  status of 169-71, 186
Lazarus 42
Leclercq, Jean 15, 130-31, 183, 219
lectio divina 49
  not for lay brothers 171
Leeds 243
Lefebvre, Henri 2-4, 8
Lekai, Louis 184
Leo, novice 203
Lescher, Bruno 182
Liber Translationibus S. Cuthberti 158
Life of Godric of Finchdale 70
light 6, 7, 10, 93, 107, 214, 227, 231, 249
  at Byland church 76
  of Cistercian church 74
  and death 220
  east as source of 81, 89
  General Chapter, concern of 99
  as Gothic feature 91
  at Rievaulx church 80
  and visualizing 98
liturgy 10, 11, 40, 54, 63, 71, 240, .250
  in chapter house 106, 108, 131
  commemoration 113
  of lay brothers 173
Locedio, abbot of
  relocation of bones of 230
London, abbatical agents to 175
Lotharingia, Duke of 232 n. 61
Louth Park abbey 178 n. 37
Lysa, Norway
  Cistercian house at 140

Malachy, archbishop of Armagh 220, 223 (Gerard and Malachy of Clairvaux)

Manley, Sir Peter and wife, Isabella 238
Margam, Wales 107, 113
markets in Yorkshire and North of England 175-78
  East Witton manor179
  Guisborough manor 177
  Kingston-on-Hull 177
  Pocklington manor 177
Maris Stellae 211
Martha and Mary (of Bethany) 169
Mary Magdalen 42
Mary, Virgin 41, 60, 70, 74, 81, 93, 127, 164, 224
  as agent of intercessionary prayer 88-90
  central role of 89
  divine intervention of 188, 190, 196
  mass for 238
  miracles of 147
Mass 105, 142, 189, 209, 218 *see also* Mary, Virgin
  private 161
Matilda, countess of Warwick 233
Matthew of Rievaulx, precentor 18, 231 n. 59, 250
  ambition 199-200
  bodily pain 161-61
  letters and poetry of 65-66, 71, 74, 116, 128, 161, 213
  music 101
Maurice of Askern 186
Maurice of Rievaulx, abbot, resignation of 213
Meaux abbey 12, 17 n. 65, 65, 234
  cemetery 230, 231
  chronicle 238
  cloister 49
  infirmary 134
  markets 177
  property in Boston 178 n. 37
medical manuscripts 158-99
medical practices:
  bloodletting (*Minutione*) 21, 134, 147-51, 152 fig. 5.6, 157-60, 165, 250
  practitioners: *medicus, physicus* 145, 146
    Adam of Fountains, Benedict of York, Peter de Quincy of Rievaulx, Richard of York,

Stephen 145
Simon 147
medicine 132, 144, 161, 164
  theory and practice 158
meditation (*meditatio*) 40-45, 65, 95 n. 52, 205, 249
  on Christ's life and Passion 96-98
  during Divine Office 96-100
  in reading of *Regula S. Benedicti* 112
Meditations on the Virgin (Stephen of Sawley) 88
Melrose, abbot of 69
  Walter, lay brother of 188
memory (*memoria*) 35-40, 42 n. 62, 126, 191, 249
  of the dead 218
  memories as intrusions 35-37
  mnemonic strategies 43-45
  reformation of 40-43
  spaces of Chapter 8 *passim*
Molesmes, monks from 207
Monkwearmouth 48
Montacute [abbey], monks of 231
Mont-Dieu 44
Montpeyroux, reinterment of monk of 230
Moses 101
Mowbray family 176

Newman, Martha 32, 45
Newminster Abbey 178 n. 37
Nicholas of York, prior 201
Nora, Pierre 5-6, 40
Northgrange 231
Norton, Christopher 16
Norway *see* Lysa 140
novice 21, 25, 69, 95, 96, 118, 197, 198, 199, 203, 212, 213, 249, 252
  acquisition of humility 38-40
  admission 34
  advice on apostasy 204
  bodily discipline 160-61
  confession 121
  mandatum rite 61-63
  meditation 40-45
  memory 36-38
  in processions 58
  reflecting on death 222-23
  transition to monastic world 34-35
novice master 34

Odo of Cluny 207
Odo, servant 238 n. 91

Pachomius 25
Panofsky, Erwin 10
  on light 99
paradisal spaces 75
paradise 187, 205, 225
  cloister as symbol of 21, 48, 64-71, 131
  'compromised' 66
  earthly 243
Paris, Matthew 177
Park, David 16
Paulinus of Leeds 145
pauperes 61
Paul, apostle 95
  hymn to 101
penance 127
Percy family 176
  Henry, Lord of Alnwick 235
  Lord William de 233, 236
Peter of Meaux, John son of 238 n. 91
Peter, abbot of Rievaulx 114
Peter, apostle 70
  denial of Jesus 97, 131
  hymn to 101
Philip, abbot of Jervaulx, murder of 201, 210
Pilate [Pontius] 41, 118
Pilis monastery, abbot of 124
  monks and lay brothers 183
Pincerna, Amandus, body of 230
Pontigny, abbot of 143
  west range 173
portarius (porter) *see* gatekeeper 30, 31
prayer 71, 169, 170
  intercessionary 74, 88-91
pride 36, 38, 39
prison 7, 123, 130, 205
  of illness 219
processions 48, 57, 58-61, 249
  in cloister 143
  funeral 227, 229
  lights in 99
  Palm Sunday 58-60, 112
Pseudo-Dionysius, 'metaphysics of light' 99
punishment 21, 106, 108, 123-26, 129, 131, 222, 249 *see also* discipline
  of apostates 209-10, 211, 213, 215

Index

of the dead 217 n. 1
purgatory 1, 89-90, 102, 127-29, 225
  purgatorial spaces 75
  representations of 88
  vision of 187

Ralph, abbot of Fountains 145 n. 28,
    191-93
  death of 221
reality 13-14, 32
reason (*intelligentia*) 40
Regula S. Benedicti 19, 34, 106, 111-
    13, 169, 195, 204, 207
  beating 125
  body 150
  discipline 131
  mandatum rite 61
Revesby, daughter house of Rievaulx
    146, 178 n. 37
Ribble, river, brother of Meaux
    drowned in 222
Richard, first abbot of Fountains 120 n.
    46, 167, 207
  death of 221
Richard, second abbot of Fountains
    213-14, 221
Richard, third abbot of Fountains 221
Richard, abbot of Meaux 180 n. 49
Richard, prior 111, 161
Rievaulx abbey 1, 10, 12, 17, 37, 62
    fig. 2.9, 65, 66, 68, 88, 95, 102,
    196-97, 203
  altar 233, 239
  chapter house 105, 107, 108, 109 fig.
    4.1, 110 figs. 4.2, 4.3
    remodelling of 113
    tombs 114
  church 80, 81, 83 fig. 3.6, 84 fig. 3.7,
    233
  cloister 49, 52 fig. 2.3, 53 fig. 2.4, 54
    fig. 2.5, 63
  east end 90
  infirmary 133 n. 1, 134, 135 fig. 5.2
  lay brothers' quarters 137 n. 10
  'magic' walls 213
  manuscripts from 44, 147
  and poverty 179
  property in Boston 178 n. 37
Rigniaci, abbot of 143
Ripalta, murder by *conversi* 183
Ripon, leper house 201 n. 21

ritual(s) 1, 6-7
  Benedictus Aquae 61
  in Chapter House 108, 112, 131
  cloister 49, 58, 61-63
  confession and punishment 124, 126
  mandatum 30-31, 58, 61-64
  of novices 34-35
  surrounding death and burial 226,
    228
Robert Ayling of Meaux, abbot 230
Robert Colcerullus of Coldona, body to
    be buried near Meaux 235
Robert, son of Helewise de Bourdon
    238
Robert of Hesding
  gift to Fountains 238
Robert of Pipewell, abbot of Fountains
    114 n. 24
  death of 221
Robert and Raganilda de Sartis, burial
    of 235
Robert de Sturtonis 211 n. 55
Roche abbey 12, 17
  chapter house 117
  gatehouse 28, 30
  infirmaries 134, 137, 139 fig. 5.4
  natural boundary 25
Roger of Byland 219
Roger, lay brother 192-93
Roger de Mowbray 235
Roger of Rievaulx, monk 222
Roger and Alexander, monks of
    Stratford Langthorne 224-25
Romanesque puritanism 94
Rome, seat of Church 2
Rüffer, Jens 11 n. 48
Rye river, Rievaulx situated on 1
  drowning of brother William Wulsy
    in 222
  valley 68

St-Denis, church at 92
  windows and light 99
St Gall cloister plan 48, 50 fig. 2.1
St Riquier, triangular cloister 49 n. 4
saints *see also* apostles: John, Paul,
    Peter
  Andrew, suffering of 128
  Augustine 38, 60, 102
    Confessions 220

Benedict 66, 89, 90, 161, 200, 204, 207 *see also Regula S. Benedicti*
Bernard of Clairvaux 38, 39, 58 n. 19, 73, 74, 88, 102, 130, 190, 191, 221
  importance of seeing 98
  letters 207-08
Cuthbert 90 n. 31
Godric of Finchdale 70-71
Jerome 38
Lawrence, suffering of 128
Patrick's Purgatory 127-28
relics of 221
Stephen, hymn to 101
suffering of 128-29
Saltrey, Huntingdonshire, monk of 127
Sampson, Henry, lay brother of Fountains abbey 185, 211, 212
Sanctae Crucis, Hungary, abbot of 126
Sawley abbey 12, 17
  church 236
  cloister 49
  fair at Guisborough manor 179
  property in Boston 178 n. 37
Schönau, conspiracy at 189 n. 95
Serlo, Cistercian abbot 196
Serlo of Fountains 23, 31, 67, 68, 167, 191, 198-200, 205, 214, 221, 235
  on granges 176 n. 26
Serlo of Pembroke 235
Serlo of York, canon 235
Shadwell 236 n. 83
Shechem and Dinah 199
shrines 108, 218, 240
Sicard of Cremona 64
Sicily, Cistercian transactions with 177
Silvan of Byland, abbot 68
Simon, Count of Ponthieu 178
Simon (Peter) 97
Sinnulph, lay brother of Fountains 191-93
Skell, river, site of Fountains infirmary 134
sodomy 123
Solomon, portico of 64
sorcery 117
sound 76
  singing 100-02
Soutra, Midlothian, Scotland
  site of Augustinian hospital 149 n. 45
space, introduction *passim*

abstract 1, 2, 4, 8, 21, 216, 245, 249, 252
blurring of spatial divisions 252
church 102
of/in death 234, 240
eschatological 69, 112
heavenly 47, 63, 69, 81, 164, 249
imagined 1, 2, 45, 134, 193, 216
of infirmary 142
intercessionary spaces 88-91
manipulation of to segregate monks and lay brothers 173
material 1, 103, 134
of memory 237
mental 100
metaphorical 48
of the other-world/afterlife 21, 223
regulation of 1, 7, 20, 21, 245
sacred 90
spiritual 180, 186
transitional 32-35
Speculum Caritatis, Aelred of Fountains 39, 95
Speculum Novitii, attrib. Stephen of Sawley 35-44, 96, 112, 127, 128, 129, 131, 222-23
  advice to novices 37-38, 39 n. 49, 43-44, 96-98, 119
  apostasy 204-05, 212
  bodily purity in death 219 n. 4
  confession 121
  light and vision 100
  Meditations on the Virgin 88
  on responses to accusation 117-18
  tomb 114
spirituality, Cistercian 182
Stalley, Roger 16
Stapylton, Martin 17
Stephen de Catton 233 n. 65
Stephen of Hamerton 233
Stephen of Sawley, abbot of Newminster and Fountains 18, 35, 68, 74, 121, 204-05, 219 n. 4 *see also Speculum Novitii*
Stephen of Tournai 202
Stratford Langthorne monastery 187, 192, 193, 224
Suger, abbot of St-Denis 3, 92, 99
Swine convent 230

Tamar, daughter of David 204

Index 293

Tertullian 130, 200 n. 14
Theophilus, commentary on 158
Thomas, abbot of Jervaulx 201
Thomas Scot 179
Thorpe village 233
Thurstan, Archbishop of York 111
   letter 161, 206, 207
tombs 108, 113, 114, 218, 229, 240, 249, 252
Tosti of York, canon 235
'total institution' 129, 130, 182
transcendence 39-40, 45, 71, 102-03, 193, 245, 249, 252
transgression(s) 31, 36
Turner, Victor 33

Vallis Sanctae Mariae, abbot of 232
Vaudey abbey
   conversi selling wool 177 n. 33
   tomb of Stephen of Sawley 114
Venturi, Robert 57
Vieuville, abbot of 232
Villard de Honnecourt, architect 73
Villarium, Brabantia 123
violence 117
   of lay bothers 169, 181 n. 51, 182, 183, 186
   of medieval society 184
violent punishment 126
Virgin Mary *see* Mary
vision, Cistercian understandings of 74, 249
visual, the
   importance of, in Cistercian monasteries 74
visualization 96, 249

Waddell, Chrysogonous 14 n. 53
Waldef 189, 192 *see also* Jocelin of Furness, *Life of Waldef*
Wales, abbey Dore and Margam 107
washing of the feet 61–63 *see also* rituals: *mandatum*
Wardrop, Joan 232
Warner, lay brother of Maris Stellae punishment of 211
Waverley abbey 12
Western Michigan University 15, 16
Westminster 107
Werner, Ernst 181–82
Whitby 48

will (*voluntas*) 38, 41 n. 58, 196, 199
   training of 37
William de Acton (William of Akerton)
   lay brother at Rievaulx
   stabbing of monk 185, 201
William le Brun of Kirkstall, death of 222
William, prior of Byland 161, 163
William of Corbeil 206
William de Modwither, monk of Jervaulx 201, 210
William of Rievaulx, [first] abbot
   death of 221
   holiness of 68, 69
   shrine of 108, 110 fig. 4.3, 116, 240, 252
   veneration of 113
William Rufus, monk 225
William of St Thierry 14, 44
William de Stuteville, tomb of 233
William Wulsy of Rievaulx, monk 222
Woburn abbey, death of abbot Robert 114 n. 24, 221
wool trade, role of lay brother in 176–78
Worcester 107
Wulfric of Haselbury, body of 231

York
   St Mary's abbey 111, 161, 198, 206
Yorkshire
   abbeys *see* Byland, Fountains, Jervaulx, Kirkstall, Meaux, Rievaulx, Roche, Sawley
   Assizes 186, 222
   Cistercian monasteries of 12, 15, 17–18, 65, 67–68, 81, 88, 89, 102, 133, 145, 146, 171, 211, 218 *see also* monastic houses of Northern England
   demographic change 177 n. 33
   flight from 200
   granges 176
   grants of land 238
   income from wool 176, 177
   lay brothers 193
   lay burial 232
   poverty of houses 180
   society
      women as benefactors 236
   West Riding 145